Atom-probe field ion microscopy

Field ion emission and surfaces and interfaces at atomic resolution

Atom-probe field ion microscopy

Field ion emission and surfaces and interfaces at atomic resolution

TIEN T. TSONG
Distinguished Professor of Physics   The Pennsylvania State University

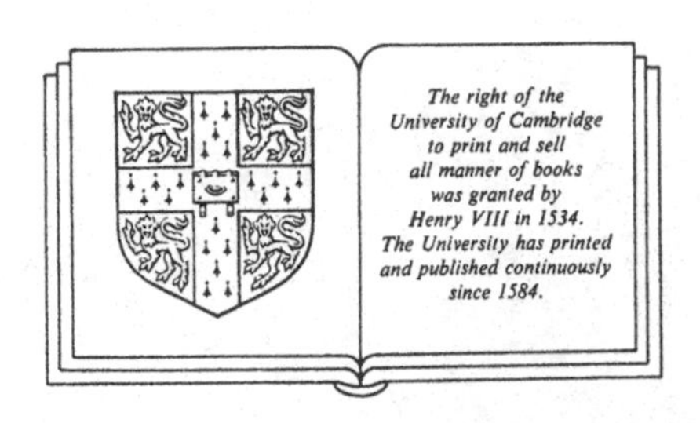

CAMBRIDGE UNIVERSITY PRESS
Cambridge
New York   Port Chester
Melbourne   Sydney

CAMBRIDGE UNIVERSITY PRESS
Cambridge, New York, Melbourne, Madrid, Cape Town, Singapore, São Paulo

Cambridge University Press
The Edinburgh Building, Cambridge CB2 2RU, UK

Published in the United States of America by Cambridge University Press, New York

www.cambridge.org
Information on this title: www.cambridge.org/9780521363792

© Cambridge University Press 1990

This publication is in copyright. Subject to statutory exception
and to the provisions of relevant collective licensing agreements,
no reproduction of any part may take place without
the written permission of Cambridge University Press.

First published 1990
This digitally printed first paperback version 2005

A catalogue record for this publication is available from the British Library

ISBN-13 978-0-521-36379-2 hardback
ISBN-10 0-521-36379-9 hardback

ISBN-13 978-0-521-01993-4 paperback
ISBN-10 0-521-01993-1 paperback

Contents

Preface

The field ion microscope (FIM) was invented by Erwin W. Müller in 1951. By 1957 he was able to show atomically resolved images of surfaces of tungsten and other refractory metals. The FIM is very simple in design and it is hard to imagine that one can image atomic structures of solid surfaces routinely with this instrument. Of course, the simplicity is somewhat deceptive. While atomic images can be routinely obtained for refractory metals with the FIM, to widen its application to other materials has not been easy. It is no wonder that even now, after the FIM has been in existence and in active research for nearly forty years, I am still often asked by scientists outside the field, some of them are very prominent and are also very knowledgeable in science, whether or not FIM can now be used to image materials besides tungsten. The truth of the matter is that the field ion microscope has long been successfully used to study most metals and many alloys, and recently good field ion images of some semiconductors and even ceramic materials, such as high temperature superconductors and graphite, etc., have also been successfully obtained.

The atomic resolution aspect is no longer unique to field ion microscopy. Other microscopies have now also achieved atomic resolution. Some of these microscopes promise to have a great versatility and are currently very actively pursued. On the other hand, there are still some experiments unique to field ion microscopy. An example is the study of the behavior of single atoms and single atomic clusters on well characterized surfaces where quantitative data on various atomic processes can be derived routinely with the field ion microscope. In addition, there is the atom-probe field ion microscope with which single atoms and atomic layers of one's choice, chosen from field ion images, can be chemically analyzed. Also, defects buried inside the bulk can be brought up to the surface by low temperature field evaporation. Thus both the structure and the chemistry of these defects can be studied with a spatial resolution of a few angstroms. The material applicability of this instrument is much wider than that of the FIM, and there is as yet no other microscope capable of doing chemical analysis with the same single atom detection sensitivity, which is the ultimate sensitivity in chemical analysis.

There already exist a few books written on field ion microscopy. Most of these either were published before 1970 when most works were concerned with techniques and methods, or are later ones which emphasize applications to materials science. While some of the basic principles of field ion microscopy remain unchanged from those twenty years ago, when Müller and I wrote a book on the subject, there have been many important new theoretical and technical developments and applications, and also many more detailed studies of a variety of problems in surface science and materials science. In the book just referred to, the subject of atom-probe field ion microscopy was only barely touched. This is of course where most of the new developments are made, and is also the instrument now most actively employed by investigators in the field. In the present volume I try to emphasize basic principles of atom-probe field ion microscopy, field ion emission and applications to surface science. As books emphasizing applications to materials science already exist, only selected topics in this area are presented here. They are used to illustrate the various capabilities of atom-probe field ion microscopy in materials science applications.

This book is intended for general scientists who are interested either in having some knowledge of studies of solid surfaces and materials on atomic scale or in working in related fields, for graduate students who either are using this technique for their scientific studies or are entering this field, and for scientists working in this field who would like to have a convenient book of reference. A book of this size cannot cover all the important subjects in the field. The choice of subjects reflects very much my personal interest, and even so it is impossible to cover all these subjects with equal depth. Some of the views expressed in this book are somewhat personal, and a few errors are unavoidable. Nevertheless, I hope that it will be useful to those who read it. Scientists are now more than ever interested in atomistic understanding of physical phenomena. Field ion microscopy should be able to contribute further to this endeavor. If this book can be of some help in such an effort, all the labor of preparing it will be amply compensated. Finally, I would like to thank many of my colleagues for their fruitful discussions and contributions over the years; especially to those who have kindly supplied some of the materials used in this book. Without their support and help this book could not have been completed.

TIEN T. TSONG
University Park, Pennsylvania

1

Introduction

1.1 Early developments

The field ion microscope (FIM)[1] is the first microscope to have achieved atomic resolution. It is a surprisingly simple instrument (though the simplicity is somewhat deceptive), consisting of a sample in the form of a tip and a phosphorus screen some 10 cm away. To obtain an atomic image, one has only to introduce about 10^{-4} Torr of an image gas such as He into the system, cool down the sample tip below liquid nitrogen temperature and apply a positive voltage to the tip and raise the tip voltage gradually. When the field at the tip surface reaches about 4 V/Å, an atomic image will start to appear. This simple instrument is an outgrowth of field emission microscope, invented by Müller in 1936.[2] In a field emission microscope,[3] the sample is also a sharp tip. In ultra-high vacuum, a negative voltage is applied to the tip. When the field at the nearly hemispherical tip surface reaches a value above ~0.3 V/Å, electrons are emitted out of the surface by the quantum mechanical tunneling effect. These electrons are projected onto a phosphorus screen some 10 cm away to form an image of the surface. As the electron current density depends very sensitively on the work function of the surface, the greatly magnified, radial projection image of the field emitted electrons represents a map of the work function variation of the surface. These field emitted electrons originate mostly from the vicinity of the Fermi level. They have a relatively large kinetic energy and therefore a relatively large velocity component in the lateral direction of the surface. The lateral motion of field emitted electrons combined with the additional large diffraction effect of the electron de Broglie wave limits the lateral resolution of this microscope to no better than 20–25 Å. The lateral velocity component of these electrons cannot be reduced by cooling of the tip since electrons are fermions; cooling will do little to lower the electron energy.

To improve the lateral resolution of the microscope, Müller tried in 1951 to use desorbed ions from a surface to form an image by reversing the polarity of the tip voltage, and also by introducing hydrogen into the system. Field desorption of barium by a high positive field was found by Müller around 1941.[4] The idea was to have gas molecules adsorbed on the surface and then to desorb them by a high positive field. As adsorbed atoms and molecules should have a very small lateral velocity component, the lateral resolution of the microscope should be improved. This was the first field ion microscope.[1] Further investigation demonstrated that ions were formed by field ionization above a critical distance of hopping gas molecules.[5,6,7] The realization of the need to lower the image gas temperature led to the introduction of a cold finger to cool the tip,[8] and therefore also the hopping gas atoms by thermal accommodation. Atomic resolution was then achieved for the first time.[8,9] Since that year there has been no major improvement in the atomic resolution of the FIM even though many technical developments on image intensification, vacuum setups and tip refrigeration have been introduced over the years to perfect the FIM and to facilitate its operation and to improve the versatility of the instrument.

When the FIM achieved atomic resolution for the first time in 1956, it was then the only microscope capable of seeing individual atoms. The situation has gradually changed over the years. Around 1970, a scanning electron microscope with a field emission source also succeeded in imaging heavy atoms such as uranium atoms which were either deposited on a thin carbon film or stained on biological molecules which were supported on such a film.[10] Around 1983 the scanning tunneling microscope also achieved atomic resolution,[11] and since then atomic structures of silicon surfaces and graphite surfaces have been successfully obtained. Atomic resolution is no longer unique to field ion microscopy. In fact, owing to some intrinsic limitations of the field ion microscope, such as the need of a very high electric field to form an image, so that the sample has to be limited to a tip shape, etc., the instrument is considered by many to be somewhat limited in scope. However, there are still many capabilities and quantitative atomic resolution experiments unique to field ion microscopy. These studies will be described in later chapters.

Seeing atoms and atomic structures of a sample is only one aspect of high-resolution microscopy. One would also want to be able to tell what kind of atoms one is seeing, or there is the chemical analysis aspect. For this purpose, Müller and co-workers in 1967 introduced the atom-probe field ion microscope.[12] It is a combination of a field ion microscope and a single-ion detection sensitivity mass spectrometer, most conveniently the time-of-flight (ToF) spectrometer. With this instrument, atoms selected by an observer from a field ion image can be chemically identified one by one by subjecting them to pulse field evaporation and measuring their flight times to reach an ion detector. During the last 20 years most technical developments and applications in field ion microscopy have focused on atom-probe field ion microscopy. The single-atom chemical analysis aspect is still unique to field ion microscopy, and in time scientists will discover more and more about the power of this instrument.

There is another important aspect of field ion microscopy which unfortunately has been considered by some field ion microscopists as a hindrance rather than a new frontier of knowledge and research. It is the investigation of high electric field phenomena occurring at and near the surface. It is true that the high electric field needed for producing a field ion image will often interfere with the effect one intends to study, and thus either experiments must be carefully designed to avoid the effect of the applied field or the experimental data must be properly interpreted by including the effect of the applied field. One must also recognize that the field strength achievable in field ion microscopy, up to about 5.5 V/Å, cannot be produced by other man-made techniques. This field strength is comparable to that encountered inside ionic crystals and much higher than that encountered at the surfaces of electrodes or achievable in a gas discharge tube. Under such high electric fields, many interesting phenomena can occur. Some of the better known high electric field phenomena are field ionization, field desorption and evaporation, field adsorption, field dissociation, field gradient induced surface migration and field induced polymerization, etc. It is also known that chemical reactions can be altered by a high electric field. High electric field effects are a new frontier of research and knowledge, and they are not of purely academic interest. A good example is the development of the liquid metal ion source, which may find technological applications. One should also recognize that dust particles in interstellar space, under the constant bombardment of charged energetic particles and high-energy photons, must be highly charged. A spherical atomic cluster containing 20 to 100 atoms needs only a charge of a few e to produce a field of ~2 V/Å at the cluster surface, enough to produce field adsorption, field desorption, and field induced chemical reactions. Thus new molecules or ions, such as H_3^+, NH_2^+ and HeW^{3+}, etc., can be formed, as has been found in field ion emission experiments. H_3^+ and NH_2^+ are known to be two of the most abundant ionic species in interstellar space. Field ion microscopy, or rather field ion emission, is a technique where the field strength and the physical environment can be controlled on atomic level, and one should make good use of such capabilities for studying high field phenomena in well-controlled environments.

There have been many monographs and review articles written on field ion microscopy.[13] The earlier ones by Müller & Tsong either have little discussion of atom-probe field ion microscopy, or the materials are somewhat out of date. Many review articles focus on narrow aspects of field ion microscopy, especially its applications. It is the intention of this author to describe here the fundamentals of field ion emission phenomena, the basic principles of the field ion microscope and the atom-probe, instrumentations, applications to surface science, and selected topics in applications to materials science. This book, it is hoped, will be instructive for someone new to the field or who has only a related interest, while at the same time serving also as a monograph. Therefore while some of the more specialized materials covered in earlier reviews will be omitted, many of the fundamentals will be repeated here. There is no doubt that field ion microscopy will continue to find many applications and also continue to advance in instrumentation. The basic aspects, however, will not change very much and it is on these that the author will try to focus in this book.

1.2 Basic principle of the field ion microscope

In the field ion microscope, shown schematically in Fig. 1.1, the sample is a sharp tip with a radius in the 100 to 2000 Å range. It is attached to a cold finger so that the tip can be cooled to below the liquid nitrogen temperature. A phosphorus screen is placed opposite the tip at a distance of about 10 cm. The microscope, after being evacuated to ultra-high vacuum, is back-filled with an image gas such as helium or neon or argon. A high positive voltage of several to over 20 kV is applied to the tip. The sample tip is usually prepared by electrochemical polishing of a thin wire. The surface so prepared is often very irregular with many asperities. However, when the tip voltage, and therefore the electric field on the tip surface, is gradually raised, the electric field above these asperities will be much higher than those above the smooth parts of the tip surface. When the field reaches a few V/Å, surface atoms start to field evaporate, i.e. evaporation is by high electric field rather than by thermal effect. Field evaporation is self-regulating, since any part of the surface with a higher local field will be field evaporated first. Eventually the field ion emitter surface becomes 'atomically smooth', as shown schematically in Fig. 1.2.

In the field of a few V/Å, image gas atoms will be attracted to the tip by a dipole force. Above each of the more protruding surface atoms will be field adsorbed with an image gas atom. In general, image gas atoms will hop around the tip until they either escape along the tip shank to the free space again or are field ionized. Field ionization is achieved by tunneling of an atomic electron into the metal, and it can occur only if the gas atoms are about 4 Å away from the surface and within a zone of width of ~0.2 Å at the best image field condition, as will be discussed in Chapter 2. Also, field ionization occurs most readily above a protruding surface atom, as indicated by the field ionization disk in Fig. 1.2. An image gas atom once field ionized will be accelerated toward the screen in the radial direction

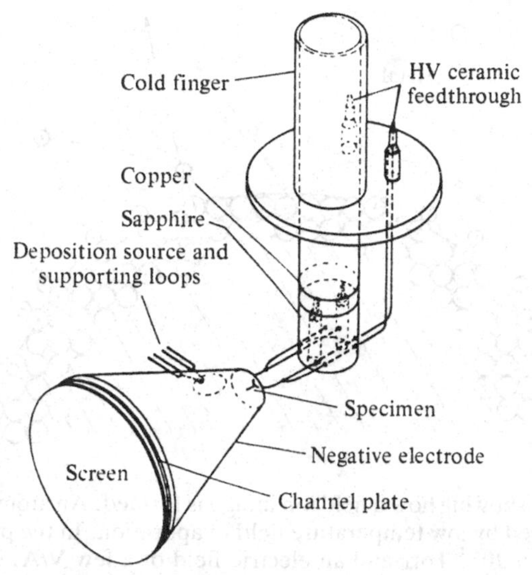

Fig. 1.1 Schematic diagram of a field ion microscope. A sample tip is mounted on a tip heating loop, a Pt or Mo wire of about 0.2 mm diameter, by spot welding. The loop is also spot welded with two Pt potential leads of about 0.025 mm. These wires are tightly plugged into a sapphire which is attached to a cold finger. A high positive voltage can be applied to the tip with the screen kept at ground potential, or the polarity can be reversed. The cone-shaped electric shield is needed only if the tip is kept at the ground potential for easy monitoring of the tip temperature by a resistivity measurement. Tip temperature can be controlled by passing a heating current through the tip mounting loop, and monitored by measuring the resistance with the two potential leads. An adatom deposition coil source is included in this FIM.

of the tip hemisphere. Each second, about 1000 to 10 000 ions originate from the same ionization disk above a surface atom. Thus an image spot is formed at the screen which represents the image of the surface atom.

In the field ion microscope, not all surface atoms are imaged, but rather only those atoms in the more protruding positions. These include plane edge atoms of low index planes and all atoms in the loosely packed high index plane. In Fig. 1.3 is shown a hard ball model of a nearly hemispherical tip surface of fcc crystal with a tip radius of about 100 Å. Atoms painted in dark color are the imaged ones. A neon field ion image of a gold tip of about twice the radius is shown at the left hand side. The one-to-one correspondence of the FIM image and the ball model is quite clear. It is important to recognize that surfaces prepared by low temperature field evaporation are not thermally equilibrated surfaces. For metals, the surfaces so prepared usually have the (1×1) structure, or the structure a solid will have when the solid is truncated. This is why the FIM of gold resembles the ball model so well. For semiconductors such as

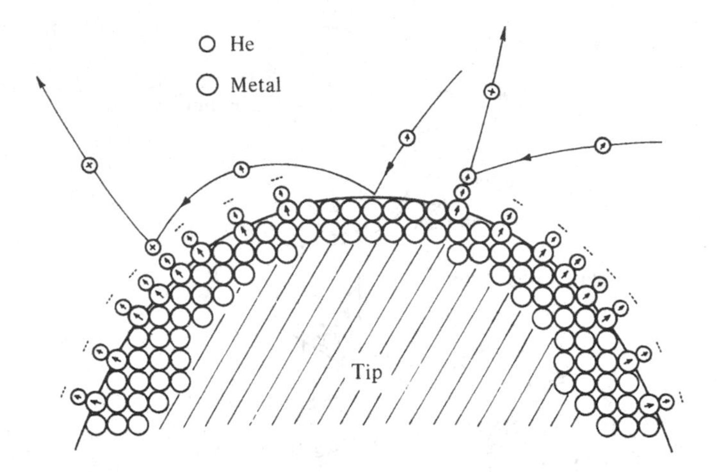

Fig. 1.2 Diagram showing how a field ion image is formed. An atomically perfect surface is developed by low temperature field evaporation. In the presence of an image gas of about 10^{-4} Torr and an electric field of a few V/Å, each of those atoms in the more protruding positions will have an image gas atom field adsorbed on it. The hopping gas atoms, during their hopping motion, may be field ionized by having one of their electrons tunneled into the tip. Once an ion is formed, it is accelerated by the applied field toward the screen some 10 cm away to produce an image of the surface atom. Every second about 10^3 to 10^4 image gas ions originate from the same atomic site of the surface.

silicon, a field evaporated surface is usually slightly disordered because of the very large field penetration depth: thus atoms inside a plane have nearly the same probability of being field evaporated as plane edge atoms. Details of field evaporated surfaces and thermally equilibrated surfaces will be discussed in Chapter 4.

1.3 Basic principle of the atom-probe field ion microscope

The atom-probe FIM combines a field ion microscope with a probe-hole and a single ion detection sensitivity mass spectrometer, in general a time-of-flight (ToF) spectrometer. As shown in Fig. 1.4, the tip is now mounted on a gimbal system (either an external or an internal gimbal) so that its orientation can be finely adjusted. At the screen there is a small probe-hole covering a few atomic diameters of the field ion image. The tip orientation is adjusted in such a way that atoms chosen by an operator for chemical analysis will have their images falling into the probe-hole. The entire field ion emitter surface is then slowly field evaporated using ns-width high voltage pulses or laser pulses. However only those atoms having their images falling into the probe-hole can go through the probe-

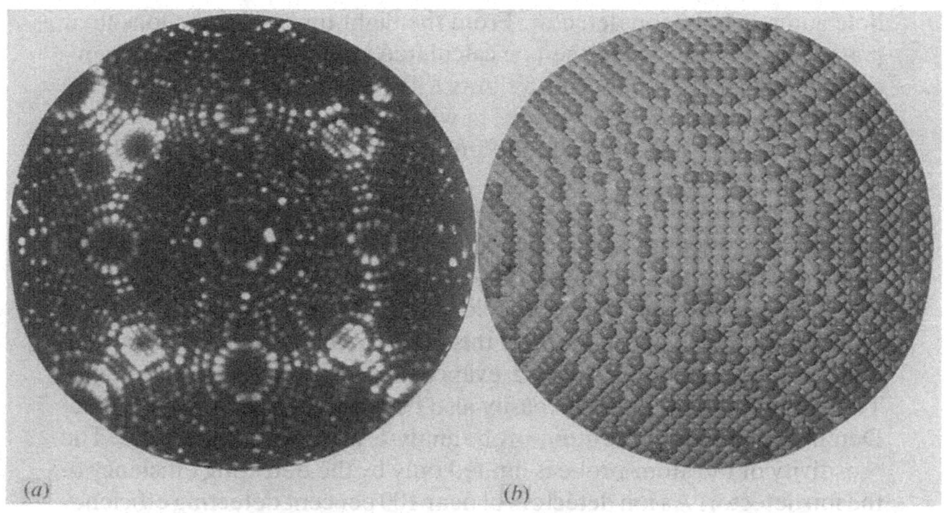

Fig. 1.3 (*a*) Neon field ion image of a gold tip taken at about 6 kV with 40 K cooling of the tip. (*b*) A hard ball model of a nearly hemispherical (001) oriented fcc crystal. Along different crystallographic poles are small facets of surfaces of different Miller indices. In the field ion microscope, not all atoms are imaged. The electric field above nearly closely packed smooth planes is too uniform, thus field adsorption of image gas atoms will not occur. Field adsorption occurs at apex sites of edge atoms of these planes and at surface atoms of high index surfaces. These atoms are painted in dark color; they are also the imaged atoms. The one-to-one correspondence of the field ion image and the ball model is obvious.

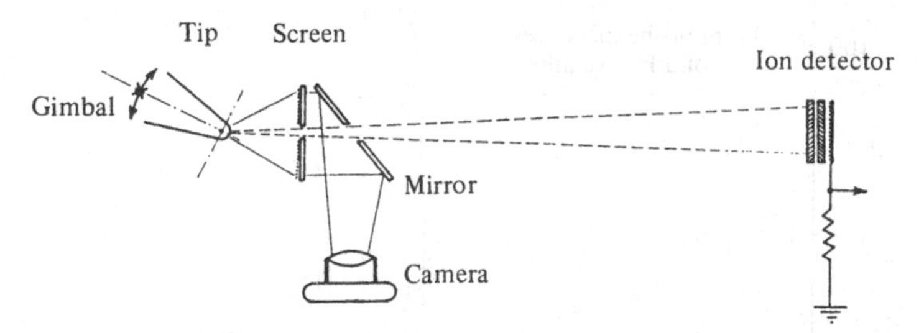

Fig. 1.4 Diagram showing the principle of operation of a time-of-flight atom-probe. The tip is mounted on either an internal or an external gimbal system. The tip orientation is adjusted so that atoms of one's choice for chemical analysis will have their images falling into the small probe-hole at the screen assembly. By pulse field evaporating surface atoms, these atoms, in the form of ions, will pass through the probe hole into the flight tube, and be detected by the ion detector. From their times of flight, their mass-to-charge ratios are calculated, and thus their chemical species identified.

hole and reach the ion detector. From the flight times of these ions their mass-to-charge ratios can then be calculated, and therefore their chemical species identified. Figure 1.5 shows a ToF atom-probe mass spectrum obtained from a Pt–5at.% Au tip, in which Pt and Au atoms are clearly separated. There are, of course, different types of atom-probe, which will be described in Chapter 3.

Using the atom-probe with sufficient care, atoms on a surface can be analyzed one by one, and atomic layers one by one also. Field evaporation of metals and alloys almost always starts from the edges of a surface layer. Thus by aiming the probe-hole at the edge of the top surface layer only atoms field evaporated from this layer are collected. For adsorbed atoms, they almost always field evaporate before the substrate layer. Thus adatoms on a plane can easily also be mass analyzed one at a time. Details of methods of the atom-probe analysis will be described later. The sensitivity of the atom-probe is limited only by the detecting efficiency of the ion detector. As ion detectors of near 100 per cent detecting efficiency are available, the atom-probe is capable of doing chemical analysis with single atoms. This represents the ultimate sensitivity in chemical analysis. In contrast, a well known ultrasensitive tool for chemical analysis, the electron microprobe, still needs hundreds of atoms to achieve sufficient signal intensity. Although atom-probe analysis can be greatly facilitated by the availability of the field ion image, so that atoms of one's choice can be analyzed one by one, this visual effect is not a prerequisite. Many

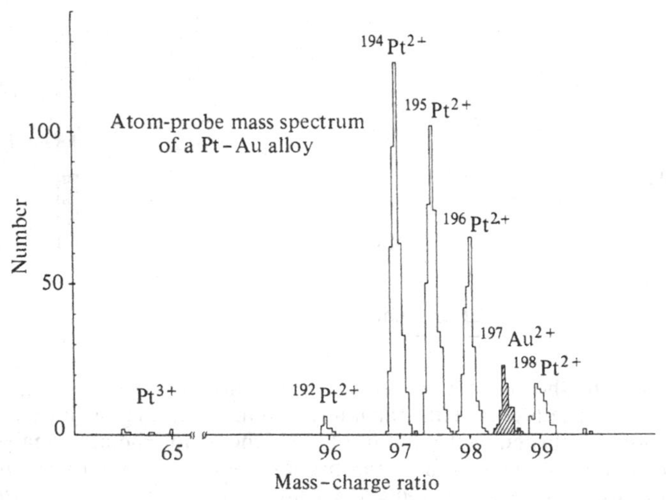

Fig. 1.5 Mass spectrum of a Pt–5% Au sample taken with a flight-time-focused ToF atom-probe. Note the complete separation of all the isotopes, and the validity of the composition of the sample derived.

chemical species such as hydrogen and oxygen do not appear in the field ion image, but can still be detected. The advantage of the time-of-flight atom-probe is that all the chemical species on the surface can be detected at once with nearly equal detection efficiency. With the availability of fast electronic timers good resolution can also be achieved. Detailed designs of the atom-probe will be discussed in Chapter 3, and applications of this unique technique are described in Chapters 4 and 5.

References

1 E. W. Müller, *Z. Physik*, **131**, 136 (1951).
2 E. W. Müller, *Z. Physik*, **106**, 541 (937).
3 R. Gomer, *Field Emission and Field Ionization*, Harvard University Press, 1961.
4 E. W. Müller, *Naturwiss.*, **29**, 533 (1941).
5 M. G. Inghram & R. Gomer, *Z. Naturforschung*, **109**, 863 (1955).
6 E. W. Müller & K. Bahadur, *Phys. Rev.*, **102**, 624 (1956).
7 T. T. Tsong & E. W. Müller, *J. Chem. Phys.*, **41**, 3279 (1964).
8 E. W. Müller, *J. Appl. Phys.*, **27**, 474 (1956).
9 E. W. Müller, *J. Appl. Phys.*, **28**, 1 (1957).
10 A. V. Crewe, J. Wall & J. Langmore, *Science*, **168**, 1338 (1970).
11 G. Binnig, H. Rhorer, Ch. Gerber & E. Weibel, *Phys. Rev. Lett.*, **50**, 120 (1983).
12 E. W. Müller, J. A. Panitz & S. B. McLane, *Rev. Sci. Instrum.*, **39**, 83 (1968).
13 (a) E. W. Müller & T. T. Tsong, *Field Ion Microscopy, Principles and Applications*, Elsevier, New York, 1969.
(b) E. W.Müller & T. T. Tsong, *Field Ion Microscopy, Field Ionization and Field Evaporation*, Pergamon Press, Oxford, 1973.
(c) K. M. Bowkett & D. A. Smith, *Field Ion Microscopy*, North-Holland, Amsterdam, 1970.
(d) R. Wagner, *Field Ion Microscopy*, Springer, Berlin, 1985.
(e) M. K. Miller & G. D. W. Smith, *Atom-Probe Microanalysis: Principles and Applications to Materials Problems*, to be published by Mat. Res. Soc.
(f) J. J. Hren (ed.), *Field Ion, Field Emission Microscopy and Related Topics*, Surface Science, Vol. 23, 1970.
(g) P. H. Cuter & T. T. Tsong (eds.), *Field Emission and Related Topics*, Surface Science, Vol. 70, 1978.
(h) J. A. Panitz, in *Methods of Experimental Physics*, Vol. 22, Solid State Phys.: Surfaces, p. 349, ed. R. Celortta & J. Levine, Academic Press, New York, 1985.
(i) T. Sakurai, S. Sakai & H. W. Pickering, *Atom-Probe Field Ion Microscopy and its Applications*, Academic Press, New York, 1989.
(j) T. T. Tsong, *Surface Sci. Rept.*, **8**, 127 (1988); *Rept. Prog. Phys.*, **51**, 759 (1988).

2

Field ion emission

2.1 Field ionization

2.1.1 Basic mechanisms and field ionization rate

Field ion images are formed by field ionization. Without this process there would simply be no field ion microscope. Field ionization is therefore the most important physical process in field ion microscopy. Theories of field ionization actually preceded field ion microscopy by over twenty years. In 1928, when quantum mechanics was still in its infancy, Oppenheimer[1] found by detailed quantum mechanical calculations that under a field of about 2 V/Å the electron in a hydrogen atom could tunnel out of the atom into the vacuum. This is one of the earliest theories of quantum mechanical tunneling phenomena. In the same year Fowler & Nordheim[2] used the quantum mechanical tunneling effect to explain field emission phenomena, and Gamow, Gurney and Condon also explained α-particle decay by tunneling of helium nucleus out of a heavy nucleus.[3] These theories represented early triumphs of quantum mechanics outside of atomic systems. Tunneling from excited states of hydrogen atoms was treated by Lanczos[4] to explain field induced quenching of spectral lines in Stark effect at a field near 10^6 V/cm, the highest attainable in gaseous discharge tubes. Gurney[5] considered inverse tunneling in the much higher fields that were encountered in the cathodic neutralization of H^+ ion in an electrolyte, and was already discussing how the atomic level would line up with the electronic levels, including the Fermi level, of the electrode at different distances. Although these studies were probably known to some of the early pioneers in field ion microscopy, there is no evidence that they directly influenced the early development of the field ion microscope. It seems to this author that connections were found later during the time when simple theories were developed to explain field ion image formation, after the field ion microscope was already developed. Of course, field ionization, which requires a field above 2 V/Å, was not experimentally observed until the advent of the field ion microscope. Using tip geometry, the high electric field needed for field ionization was reached in 1951.

The well accepted theory of field ionization near a metal surface, based on the WKB method, was initiated by Inghram & Gomer[6] in 1954 after experimental evidence of field ionization near a metal surface and also in free space was established by Müller in 1951 when the first field ion microscope was introduced. Further work was reported by Müller & Bahadur[7] in 1956. In the absence of an applied field, an electron in a free atom finds itself in a potential trough of infinite width, as shown in Fig. 2.1(*a*). An energy I, corresponding to the ionization energy of the atom, must be supplied to excite the electron and ionize the atom. When a high electric field of a few V/Å is applied, the potential barrier is reduced on one side, and the electron can now tunnel out of the atom even without the excitation by an external energetic particle, as shown in Fig. 2.1(*b*). Near a metal surface, owing to the work function of the surface and interactions of the electron with the surface, the potential barrier is further reduced, as shown in Fig. 2.1(*c*). As is clear from this diagram, the potential barrier depends on the distance from the atom to the surface. There is a critical distance, x_c, below which the atomic level will fall below the Fermi level of the substrate crystal, and tunneling of the atomic electron into the metal cannot occur since all the electronic levels are almost completely filled at low tip temperature. The WKB barrier penetration is given by

$$D(E, V(x)) = \exp\left\{-\left(\frac{8m}{\hbar^2}\right)^{1/2}\int_{x_1}^{x_2}(V(x)-E)^{1/2}\,dx\right\}, \qquad (2.1)$$

where $V(x)$ and E represent respectively the potential and the total energies of the electron, m is its mass, $\hbar$ is Planck's constant divided by 2π, and x_1 and x_2 are the classical turning points for the electron of energy E. Numerically it is given by

$$D(E, V(x)) = \exp\left\{-1.024\int_{x_1}^{x_2}(V(x)-E)^{1/2}\,dx\right\}, \qquad (2.2)$$

if the energies are expressed in eV and the barrier width is measured in Å. When a gas atom is near the surface of a conductor, the electron potential is reasonably well represented by[7]

$$V(x) = -\frac{e^2}{|x_i - x|} + eFx - \frac{e^2}{4x} + \frac{e^2}{x_i + x}, \qquad (2.3)$$

where the first term represents the Coulomb potential due to a positive ion of charge e located at a distance x_i from the plane of the conductor, the potential of the electron in the applied field is given by the second term, the third term is the image potential of the electron and the fourth term is

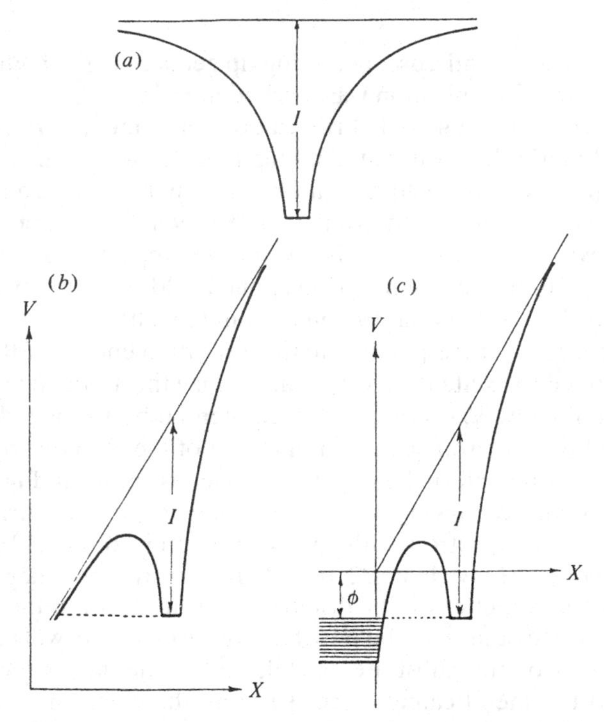

Fig. 2.1 (*a*) Potential energy of an electron of an atom. I is the ionization energy, and the potential profile is Coulombic if it is a hydrogen atom.

(*b*) Potential energy of an electron of an atom in an applied electric field in free space. If the field is strong enough, the width of the potential barrier on one side will be reduced to become comparable to the de Broglie wavelength of the electron. The probability of the electron tunneling out of the potential barrier will become appreciable.

(*c*) Potential energy of an electron of an atom in an applied electric field near a metal surface. The atomic electron can tunnel into the metal only if the atomic level lines up with an energy level above the Fermi level of the metal. This condition imposes a critical distance of field evaporation, as discussed in the text. If the atom is closer to the metal than the critical distance, the atomic level will fall below the Fermi level, and tunneling cannot occur since the metallic electronic states are already occupied. On the other hand, if the atom is very far away from the surface, the potential barrier will be that corresponding to free space field ionization. It is wider and the barrier penetration probability is smaller.

the interaction of the electron and the image charge of the ion. While the barrier penetration probability can easily be calculated with a numerical method, for the purpose of visualizing the basic mechanisms, it is more convenient to have an analytical form even if it is only an approximation. If one assumes an equilateral triangular potential well of width $(I - 2\sqrt{e^3F})$, where $2\sqrt{e^3F}$ represents the Schottky barrier reduction by the superposition of the Coulomb field and the applied field, and a base of width $(I-\phi)/eF$, then the maximum barrier penetration probability, which occurs at the critical distance of field ionization, is given by

$$D(x_c, F) = \exp\left\{-\left(\frac{8m}{\hbar^2}\right)^{1/2} \tfrac{2}{3}(I - 2\sqrt{e^3F})^{1/2}\frac{(I-\phi)}{F}\right\}$$

$$= \exp\left\{-0.683\,(I - 7.59F^{1/2})^{1/2}\frac{(I-\phi)}{F}\right\}, \qquad (2.4)$$

where I is the ionization energy in eV, ϕ is the work function in eV, and F is the field strength in V/Å. In this calculation we have omitted the effect of field penetration.

In field ion microscopy there is a field strength at which the FIM gives the sharpest and clearest image of the emitter surface. This field strength is called the best image field (BIF) and the imaging voltage corresponding to this field is called the best image voltage (BIV). The BIF depends on the image gas used. Values of the BIF have been obtained from many years of observations by various investigators, and are in general agreed to be 4.4 V/Å for He, 3.75 V/Å for Ne and 2.2 V/Å for Ar and H_2, as will be discussed in Chapter 3. The BIF obviously depends not only on the barrier penetration probability but also on many other factors such as the hopping motion of the image gas atoms above the emitter surface, thermal accommodation of these atoms to the tip temperature, image contrast effects from field adsorption of image gas atoms and from a two-dimensional variation in the field ionization rate along the lateral directions of the surface, etc. A rough estimate of the best image field for different image gases, however, can be made by assuming a constant value of the argument of the exponential function of eq. (2.4), or a constant value of $(I - 7.59F^{1/2})^{1/2}\cdot(I-\phi)/F$, and by taking the BIF of He to be 4.4 V/Å and ϕ of the emitter surface to be 4.5 eV. The best image fields of Ne, Ar and H_2, so calculated from their values of ionization energies, are 3.5 V/Å, 1.9 V/Å and 1.9 V/Å, respectively. These values agree with experimental values to within 15%.

The critical distance for field ionization is related to I and ϕ by[16]

$$eFx_c \simeq I' - \phi - e^2/4x_c + \tfrac{1}{2}(\alpha_a - \alpha_i)F^2 \simeq I - \phi, \qquad (2.5)$$

where α_a and α_i are respectively the polarizabilities of the atom and the ensuing ion and I' is the effective ionization energy of the atom at x_c. In this one-dimensional model, the ionization rate $\varkappa$ and the life time before ionization τ are related to the barrier penetration probability $D(x, F)$ and the frequency, ν, the electron strikes the barrier according to

$$\varkappa = 1/\tau = \nu D(x, F). \qquad (2.6)$$

The probability of ionization during a time interval from 0 to t is given by

$$P(t) = 1 - \exp(-t/\tau). \tag{2.7}$$

If this probability is to be calculated for a given hopping path between x_a and x_b, then

$$P(x_a, x_b) = 1 - \exp\left\{-\int_{x_a}^{x_b} \frac{dx}{v(x)\tau(x)}\right\}, \tag{2.8}$$

where $v(x)$ is the velocity of the atom at x. More detailed calculations with transfer Hamiltonian methods have also been developed for field ionization near a metal surface. We will mention them again in connection with the measurement of field ion energy distributions.

2.1.2 Gas supply function

In the strong electric field of non-uniform strength near the tip surface, gas atoms are polarized, and are attracted to the tip surface by a polarization force $\alpha F \cdot dF/dx$, thus the gas flux impinging on the tip surface is not the simple kinetic gas flux $p/(2\pi MkT)^{1/2}$, where M is the mass of the atoms and p is the gas pressure and T is the gas temperature, but rather greatly enhanced by the applied field. The enhancement factor is very difficult to evaluate exactly because of the complicated tip geometry. For a spherical emitter, it was shown by Southon[8] that the total gas supply function is given by

$$Z_s = 4\pi r_t^2 \frac{p}{\sqrt{2\pi MkT}} [\exp(-\phi) + (\pi\phi)^{1/2} \operatorname{erf}(\sqrt{\phi})], \tag{2.9}$$

where r_t is the tip radius and ϕ the ratio of polarization energy and kT, or $\phi = \alpha F^2/2kT$, and erf(x) is the error function given by

$$\operatorname{erf}(x) = \frac{2}{\sqrt{\pi}} \int_0^x e^{-t^2} dt \tag{2.10}$$

and p is the gas pressure far removed from the field ion emitter surface; at the emitter surface the pressure as defined by the rate of momentum transfer per unit area is much larger because of the enhanced gas supply and the much larger kinetic energy of the image gas atoms. Under normal operational conditions of the FIM, ϕ is much greater than one, and the total gas supply reduces to

$$Z_s \approx 4\pi r_t^2 \frac{p}{\sqrt{2\pi MkT}} (\pi\phi)^{1/2}. \tag{2.11}$$

Similarly, the gas supply to the unit length of a cylindrical emitter of radius r_t is given by

$$Z_c \approx 2\pi r_t \frac{p}{\sqrt{2\pi MkT}} \left(\frac{4}{\pi}\phi\right)^{1/2}. \tag{2.12}$$

For the tip geometry, van Eekelen[9] approximates the emitter by a hyperloid of revolution with a radius of curvature at the apex given by the tip radius. The field along the axis of revolution is shown to be

$$F(r) = \frac{F_0}{2\left(\frac{r}{r_t}\right) - 1} = \frac{F_0}{1 + \frac{2x}{r_t}}, \qquad x = r - r_t. \tag{2.13}$$

A spherical emitter with a hypothetical r dependence of field given by the above equation gives a total gas supply of

$$Z_h = 4\pi r_t^2 \frac{p}{\sqrt{2\pi MkT}}[\tfrac{1}{4}(\phi + 2.7\phi^{2/3} + 2.7\phi^{1/3} + 1)]. \tag{2.14}$$

In reality, none of these gas supply functions are satisfactory. Experiments show that dominant gas flow is along the tip shank,[8,10] which is most difficult to treat theoretically.

2.1.3 The hopping motion and thermal accommodation

To achieve good lateral resolution of the field ion microscope, the tip has to be cooled to low temperature. As will be discussed in Section 2.5.2, one of the dominant terms limiting the resolution of the microscope is the lateral component of the velocity of the image gas atoms just before they are field ionized. The incoming image gas atoms, usually with the thermal energy of the microscope chamber, which is normally kept at the room temperature, are attracted to the tip surface. At the tip surface their kinetic energy is the sum of the thermal energy and the polarization energy, $\frac{1}{2}\alpha F_0^2$. The polarization energy is of the order of 0.15 eV, larger than the thermal energy by a factor greater than 5. The fact that the resolution of the field ion microscope can be improved drastically by cooling down the tip to around 20 K clearly indicates that this large kinetic energy of the image gas atoms can be greatly reduced by thermal accommodation to the tip temperature. This accommodation process takes a large number of hopping motions of the image gas atoms. In fact, theories of the resolution of field ion microscope implicitly assume that the image gas atoms are fully accommodated to the tip temperature. This is a good approximation only at or below the best image field where the resolution of the FIM is at its optimum.

The energy exchange between an impinging gas atom or molecule and a surface can be represented by a coefficient, the accommodation coefficient, as defined by Knudsen,[11]

$$a = \frac{E_i - E_r}{E_i - E_s}, \tag{2.15}$$

where E_i and E_r are the average energies brought to and carried away by the impinging gas atoms from the surface, and E_s is the energy the rebounding atoms will carry away if they are already in thermodynamic equilibrium with the surface. Instead of energies, the corresponding temperatures can also be used, more correctly by the definition

$$a = \lim_{\Delta T \to 0} \frac{T_i - T_r}{T_i - T_s}. \tag{2.16}$$

For a very high impact energy, where the impact energy is much larger than the binding energy of the surface atoms, a classical energy transfer factor averaged over all incident angles, $a = \frac{1}{2} \cdot 4mM/(m + M)^2$, can be used.[12] But for low impact energies such as those encountered in field ion microscopy, use of experimental values is more appropriate. In Table 2.1, a few values appropriate for FIM conditions are listed.[13] As can be seen, the accommodation coefficients for He on clean surfaces are usually only a few per cent. They increase as the gas atom mass and surface atom mass ratio decreases. The accommodation coefficient also increases for surfaces chemisorbed with light gases. The accommodation coefficient usually depends on the surface temperature, but the relatively small dependence is omitted from the table.

As mentioned earlier, the fact that the resolution of the FIM can be

Table 2.1 *Accommodation coefficients*

Gas	Metal	Surface condition or adsorbant	Accommodation coefficients
He	W	Clean	0.020
	W	H_2	0.041
	W	D_2	0.046
	W	O_2	0.185
	W	N_2	0.040–0.064
	Mo	Clean	0.026
	Be	Evaporated	0.145
Ne	W	Clean	0.055
	W	H_2	0.112
	W	D_2	0.117
	W	O_2	0.406
	W	N_2	0.117
	Mo	Clean	0.055
	Be	Evaporated	0.315
	Fe	Clean	0.056
Ar	Mo	Clean	0.315
H_2	W	H_2	0.165

drastically improved by cooling down the tip to around 20 K indicates that most of the field ions are produced from those almost fully thermally accommodated image gas atoms. The average temperature of gas atoms after n hoppings T_n is given by

$$T_n = (T_0 - T_s)(1 - a)^n + T_s, \tag{2.17}$$

where T_n is the temperature equivalent of the impinging gas atoms, and T_0 is the initial kinetic energy of the gas atoms. From this equation it can easily be shown that it takes over 50 hoppings to cool down the gas temperature below the room temperature, and over 200 hoppings to cool down the gas to near the surface temperature when the tip is kept at about 20 K. There is an experimental observation showing clearly that the image atoms are not really fully accommodated to the tip temperature when they are field ionized, at least under the best image field of the microscope. Nishikawa and Müller[14] found that the resolution of the field ion images can be noticeably improved if a small amount of hydrogen is added to the image gas. As can be seen from Table 2.1, the accommodation coefficient is improved by a factor of almost 2 when the surface is chemisorbed with the very light hydrogen. Thus with the same number of hoppings the image gas atoms will achieve a much lower temperature; this in turn will improve the resolution of the FIM.

The hopping motion of image gas atoms has been discussed in detail by Gomer[15] and by Müller & Tsong.[16]. Here we will describe only some of the basic aspects and derive some of the equations without presenting details of the derivation. We will consider two extreme regimes of the field strength. First, when the field is below the best image field the ionization rate is so low that the removal of hopping gas atoms by field ionization is negligible and does not appreciably change the equilibrium gas density n_t near the tip. Thus one may assume that a themodynamic equilibrium distribution of image gas atoms is maintained. Under such conditions, from statistical mechanics, this density is larger than the gas density n_g far away from the tip by

$$n_t = n_g \left(\frac{T_g}{T_t}\right)^{1/2} \exp\left(\frac{\alpha F_0^2}{2kT_g}\right), \tag{2.18}$$

where T_g and T_t are respectively the gas temperature far away from the tip and at the tip surface. The ionization current formed within a volume element of the hemispherical tip cap $2\pi r^2\,dr$ is then given by

$$di = 2\pi r^2 e n_t \nu D(r)\,dr. \tag{2.19}$$

The total current is obtained by integrating this expression over the entire distance range, i.e. from x_c to ∞,

$$i = 2\pi n_g e \left(\frac{T_g}{T_t}\right)^{1/2} \int_{r_t + x_c}^{\infty} r^2 \nu D(r) \exp\left(\frac{\alpha F^2(r)}{2kT_g}\right) dr. \tag{2.20}$$

As will be discussed in Section 2.1.5, below the best image field, field ionization occurs only in a very narrow spatial zone of width 0.2 to 0.3 Å. For most practical purposes, $D(r)$ may be approximated by a step function

$$\begin{aligned} D(r) &= D(x_c) && \text{for } r_t + x_c < r < r_t + x_c + \Delta x, \\ &= 0 && \text{for } r > r_t + x_c + \Delta x. \end{aligned} \tag{2.21}$$

With $F \simeq F_0$, we have

$$i \approx 2\pi r_t^2 n_g e \left(\frac{T_g}{T_t}\right)^{1/2} \nu D(x_c) \exp\left(\frac{\alpha F_0^2}{2kT_g}\right) \Delta x. \tag{2.22}$$

Therefore in this field regime the ion current should increase exponentially with F_0^2. In I–V characteristics, when the field is very low, the total ion current is found to increase with about the thirtieth power of the field, which in fact may better be represented by an exponentially dependent function of the field strength. Thus this calculation is in fairly good agreement with experimental I–V characteristics at very low field strength.

In the high field regime when the ionization rate is very high, the image gas density near the tip surface is no longer in thermodynamic equilibrium with its surroundings since the image gas atoms are continuously depleted by the field ionization process. The total ion current is gas supply limited, and the field dependence should be well described by the gas supply function. Nevertheless, it is worthwhile to consider the hopping motion of the image gas atoms near the tip surface and also to try to visualize the hopping heights of the gas atoms. When an image gas atom reaches the tip surface from infinity, the kinetic energy of the atom is $\sim(3kT/2 + \frac{1}{2}\alpha F_0^2)$. By impinging on the surface, it will lose a fraction of its kinetic energy and will either bounce off the surface or be bound to the surface and be field adsorbed on the surface in a field adsorption site. Field adsorption energy for He is of the order of 0.2 to 0.3 eV as will be discussed in Section 2.3. The mechanism with which an incoming image gas atom is trapped into a field adsorption site has not been discussed so far, and is a possible subject of future research. For the present discussion, we will assume that all the field adsorption sites on the field ion emitter surface are already being occupied, so that the incoming image gas atom can only hop around the tip surface and eventually lose most of its kinetic energy. Let us consider the hopping height of an image gas atom with kinetic energy E at the surface.

$$\tfrac{1}{2}Mv^2 = E - \tfrac{1}{2}\alpha\{F_0^2 - F^2(x)\}, \tag{2.23}$$

where $F(x)$ is the field strength at height x. Using eq. (2.13), we can easily show that the hopping height of the image gas atom is given approximately by

$$x \approx Er_t/2\alpha F_0^2. \qquad (2.24)$$

Obviously if the hopping height is less than the critical distance of field ionization then the image gas atom cannot be field ionized, and it may eventually be lost by escaping along the tip shank. If the surface temperature is low enough then the hopping height of well-accommodated atoms will become too low to be field ionized, and field induced condensation of adsorbed layers will occur.

2.1.4 Current–voltage characteristics

There are basically two kinds of experiments which can be used for studying the mechanisms of the field ionization process near a field ion emitter surface and also the field ion image formation process. They are the measurement of field ion current as functions of tip voltage, tip temperature, and other experimental parameters, and the measurement of the ion energy distribution.

Before the advent of the low-temperature field ion microscope, Müller and Bahadur[7] measured the ion current as a function of the tip voltage and gas pressure for various gases at room temperature. They realized the current collected at the screen contained not only field ion current but also secondary electrons released from the screen by the bombardment of the ions. Thus they later suppressed this secondary electron current by a negatively biased secondary electrode.[17] Southon & Brandon[8,18] first measured the current–voltage characteristics of the field ion microscope under the normal operating conditions of the FIM. They also suppressed the secondary electron current by biasing the accelerating electrode with respect to the screen. Their log *I*–log *V* plot displays two distinctive linear branches, as shown in Fig. 2.2. In the low field region the current increases very rapidly with the field, amounting to about the thirtieth power of the field. As the field is further increased, a transition region is reached. Beyond the transition region, the current increases with the voltage at a much slower rate, only about a third power of the voltage. The ion current increases linearly with the gas pressure, and the current at 63 K is found by them to be only about twice that at 273 K.

Some of the essentials of these findings were later confirmed with an image photometry by Tsong & Müller,[10] although the temperature dependence data were found to be inaccurate because of the poor tip temperature control in the experiment, and even the two distinctive branches of the *I*–*V* characteristics were found to exist only at low tip temperature. The slope of the high field regime was found to depend on the tip cone angle, and ranged from about 3 for a large tip cone angle of about 40° to about 5 for a very small tip cone angle of about 10°. The ion current decreases rapidly with increasing tip temperature. The ratio

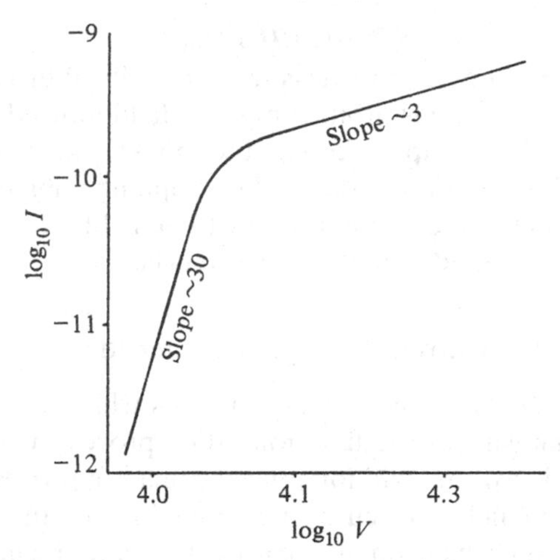

Fig. 2.2 Current–voltage characteristic for helium at 78 K. Small deviations from straight-line behavior are usually present in the high field regime.

$I(T_1)/I(T_2)$ at two temperatures depends very sensitively on the field strength. For $T_1 < T_2$, this ratio becomes very large at low field, and approaches a value close to T_2/T_1 at high field. At the high field limit where the ion current should be gas supply limited, both eqs (2.11) and (2.12) predict the total ion current to depend on the T^{-1} power. Experimental data indeed agree with this dependence at the high field regime where the total ion current is gas supply limited. On the other hand, these equations do not give an F^3 dependence of the total ion current. Since the experimental data of Tsong & Müller show that the slope of the I–V characteristics is dependent on the tip cone angle, and there is also a time delay of the ion current following a tip temperature change, the total ion current is really determined by a dynamical balance between the gas supply coming from the tip shank and a gas flux escaping also along the tip shank. A study of the angular distribution of the ion current above the emitter surface by Beckey *et al.*[19] also concludes that the most important part of the gas supply comes from a flow of low hopping gas flux from the tip shank. An attempt has been made by Southon to account for the gas supply of a realistic tip shape, but no gas supply by field gradient induced hopping motion and surface diffusion of the image gas atoms along the tip shank was made. The mathematics involved seems formidably complicated. It is fair to say that although a quantitative agreement between experimental data on I–V characteristics and theoretical calculations has not been reached, the basic mechanisms of gas supply process in field ion microscopy is now well understood.

2.1.5 Field ion energy distributions

An equally important and perhaps a more direct method for experimentally checking theories of field ionization and theoretical models of field ion image formation is the measurement of the field ion energy distribution. An ion formed right at the field ion emitter surface, after being accelerated by an applied voltage of V, should have a kinetic energy equal to the full energy of the accelerating voltage, or eV. On the other hand, an ion formed at a distance x away from the surface will have a kinetic energy smaller than eV by an amount $\int_0^x eF(x)\,dx$. A measurement of the ion kinetic energy distribution will therefore locate the spatial origin where the ions are formed, and hence provide information, particularly as regards the energetics, on the field ionization mechanism and field ion image formation.

The first measurement of ion kinetic energy distribution was reported by Inghram & Gomer using a magnetic sector mass spectrometer.[6] Although the energy resolution of the system was only about 20 eV, and therefore could not confirm the existence of the critical distance of field ionization, they were able to find that around the best image field, the ion energy distribution width was very narrow. At a sufficiently high field, the ion energy distribution broadened because the ionization rate was high enough to ionize all the incoming image gas atoms before they could reach the tip surface, or field ionization could occur in space far above the emitter surface. Müller and Bahadur,[7] using a spherical electrostatic retarding potential analyzer, confirmed the existence of the critical distance of field ionization, although the resolution of the analyzer was not yet good enough to determine this distance accurately. Tsong & Müller,[20] using an improved spherical retarding potential analyzer, established quantitatively the critical distance of field ionization above tungsten surfaces of several image gases to within ~0.5 eV. In addition, they found the distribution width below the best image field to be very narrow at only about 1 eV, which corresponds to about a 0.2 Å width for He. They also found that the distribution width widens rapidly as the field is raised above the best image field. These features were confirmed in a study of Hanson & Inghram.[21] Jason *et al.*[22] and Jason,[23] using a magnetic sector mass spectrometer combined with a retarding potential grid analyzer, found additional resonance secondary structures in the ion energy distributions. These secondary peaks are separated from each other by several to about 10 eV. Existence of these secondary structures was later confirmed by the work of Müller & Krishnaswamy,[24] although the field calibration of Jason was found to be off by about 30%. Utsumi & Smith[25] found fine structures in the main peak of the He ion energy distribution from W (110) and (001) surfaces. These fine structures are separated by about 0.5 eV from each other. They attribute these fine structures to originate from the structure in the density of unoccupied states of the substrate crystal. Additional structures are found for surfaces exposed to nitrogen at 300 K. However, none of these observations by Utsumi & Smith have been confirmed by others and few details of the experiment are given in the publication. As the pulsed-laser time-of-flight atom-probe has now a sufficient energy resolution and more importantly, reproducibility to study these fine structures, it will be most interesting to look into this problem further.

Ion energy distributions can now be obtained with great accuracy for field ionization and field desorption either using a system similar to Jason *et al.*'s i.e. combined with a magnetic sector mass spectrometer and a retarding potential energy filter as actively pursued by Ernst *et al.*[26] and others,[27,28] or with a pulsed-laser time-of-flight atom-probe as pursued by Tsong and co-workers.[29,30] Figure 2.3 shows the system used by Ernst *et al.*, Fig. 2.4 is a potential energy diagram of the ion in a retarding potential energy analyzer and in Fig. 2.5 an ion energy distribution of He^+ ions in pulsed-laser stimulated field desorption, obtained from a high-resolution pulsed-laser time-of-flight atom-probe, is shown. In such desorption, He atoms are thermally desorbed from their field adsorption

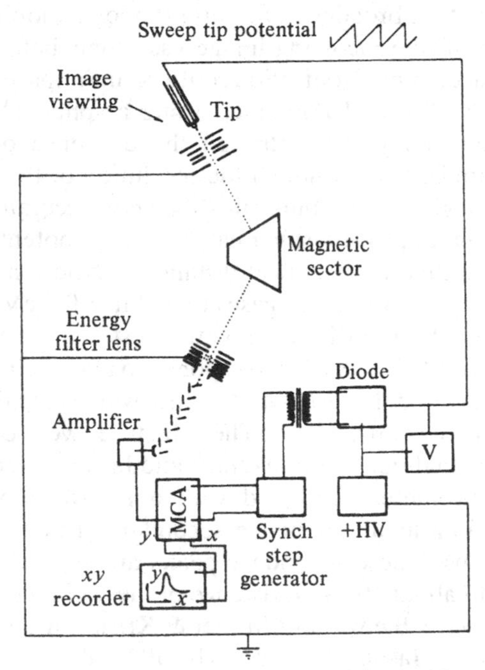

Fig. 2.3 Magnetic sector mass spectrometer combined with a retarding potential ion energy filter used by Ernst *et al.*[26] for measuring ion energy distributions in field ionization and field evaporation.

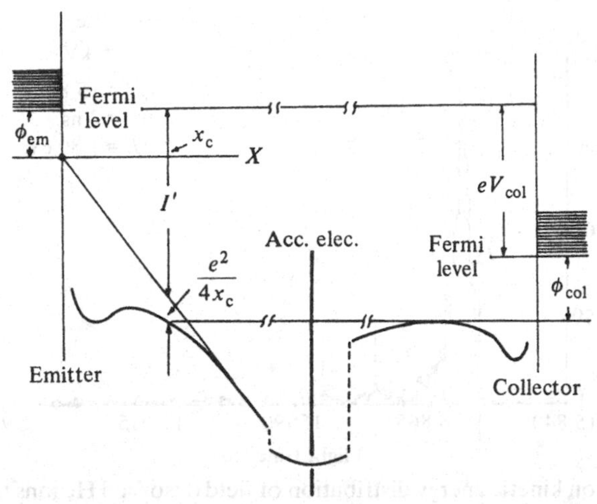

Fig. 2.4 Potential energy diagram of an ion in the retarding potential energy analyzer when the ion is formed right at the critical distance of field ionization, x_c. Note that the retarding potential, or equivalently the collector voltage V_{col}, is related to the work function of the collector even though the critical distance of field ionization is related to the emitter work function. In a time-of-flight ion kinetic energy analyzer, the collector work function should be replaced with the average work function of the flight tube. In the diagram, I' is the ionization potential of the atom near the metal surface, which is presumably the free space ionization energy I reduced by the image potential energy.

states, and are field ionized when they pass through the field ionization zone. Thus their ion energy distributions are identical in almost all aspects to those in field ionization.

The potential energy diagram for an ion formed right at the critical distance of field ionization, x_c (i.e. the most energetic ion), in a retarding potential analyzer is shown in Fig. 2.4. From this diagram, one obtains

$$eV^0_{col} = I' - \phi_{col} + e^2/4x_c,$$
$$= I - \phi_{col}, \quad (2.25)$$

where I' is the ionization energy of the image gas atom in the presence of atom–surface interaction in the applied field, ϕ_{col} is the work function of the collector or the retarding grid, and the last term of the first equation is the potential energy of the ion interacting with its own image. For other ions formed further away from the surface,

$$eV_{col} \simeq I - \phi_{col} + e \int_{x_c}^{x} F(x)\, dx, \quad (2.26)$$

or

$$V_{col} - V^0_{col} \simeq (x - x_c)F_0. \quad (2.27)$$

Thus the onset voltage, or the retarding potential, of the ion energy

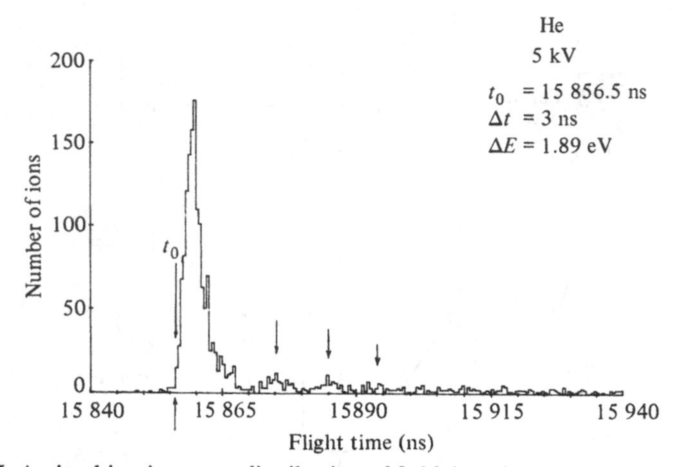

Fig. 2.5 An ion kinetic energy distribution of field desorbed He ions taken with a pulsed-laser time-of-flight atom-probe. In pulsed-laser stimulated field desorption of field adsorbed atoms, atoms are thermally desorbed from the surface by pulsed-laser heating. When they pass through the field ionization zone, they are field ionized. Therefore the ion energy distribution is in every respect the same as those in ordinary field ionization. Beside the sharp onset, there are also secondary peaks due to a resonance tunneling effect as discussed in the text. The onset flight time is indicated by t_0, and resonance peak positions are indicated by arrows. Resonance peaks are pronounced only if ions are collected from a flat area of the surface.

distribution in a retarding potential energy analyzer is related to the collector work function, not the emitter work function. In a time-of-flight system, an ion spends most of its flight time in the environment of the flight tube, thus the kinetic energy of the ion as determined from the flight time is related to the average work function of the flight tube ϕ, and not to the work function of the emitter surface. As is evident from Fig. 2.5, an ion energy distribution usually consists of a sharp onset, a distribution width of about 1 to several eV, depending on the field strength where the data are taken, and possibly a few secondary resonance peaks a few eV apart. The onset energy of the ion kinetic energy distribution is now well established, to within 0.2 eV, to be given by

$$E_{\mathrm{on}} = I - \phi. \tag{2.28}$$

The width of the ion kinetic energy distribution is very narrow, at only about 1.0 eV, when the field strength is below the best image field, and widens rapidly as the field strength is increased.[20] Thus at the best image field most of the field ions are formed within a spatial disk of thickness 0.2 to 0.3 Å and with a diameter of about 2 Å.

The narrowness of the field ion energy distribution, though it was initially quite unexpected, has been explained in the theoretical calcu-

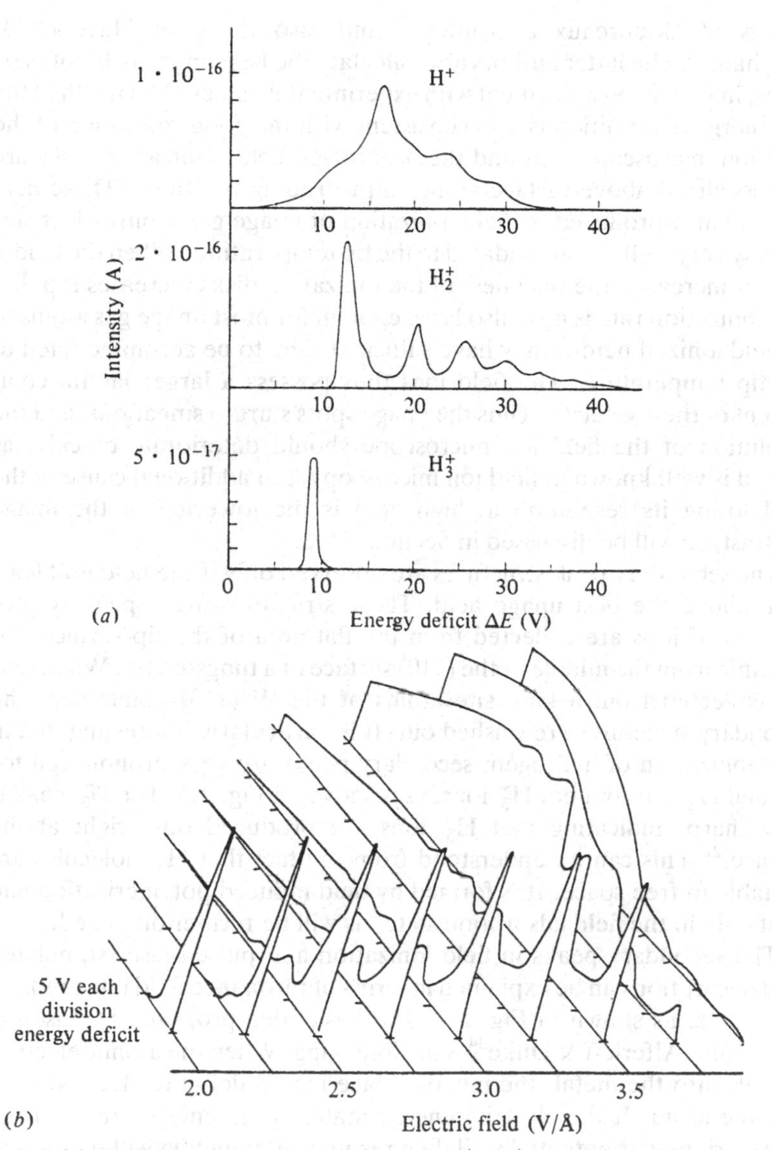

Fig. 2.6 (*a*) Field ion energy distributions of H^+, H_2^+ and H_3^+ ions obtained by Jason *et al.*[22] Secondary peaks due to resonance field ionization are most pronounced for H_2^+. H_3^+ are formed right near the surface, and no low energy tail is found.

(*b*) Ion energy distributions of H_2^+ obtained by Jason,[23] but with the field strength slightly readjusted to agree better with the result of Müller & Krishnaswamy.[24] At a very low field the secondary peaks are very small. At a very high field, free space field ionization becomes the dominant process, and the low energy tail dominates.

lations of Boudreaux & Cutler,[31] and also those of Haydock & Kingham.[32] The latter authors also calculate the best image fields of inert gases, in very good agreement with experimental values. The width of the ion energy distribution is also consistent with the good resolution of the field ion microscope. Around the best image field, ionization disks are well localized above surface atoms in protruding positions. These field ions are also produced by field ionization of image gas atoms which are already very well accommodated to the tip temperature. When the field is further increased the thickness of the ionization disks increases rapidly. The ionization rate is now also large enough for most image gas atoms to be field ionized before they have sufficient time to be accommodated to the tip temperature. The field ions thus possess a larger lateral component of their velocity. Thus the image spots start to smear out, and the resolution of the field ion microscope should deteriorate quickly, as indeed is well known in field ion microscopy. An additional cause of the FIM losing its resolution at high field is the lowering of the image contrast, as will be discussed in Section 2.5.2.

The secondary peak structures are observed only if the field is at least 10% above the best image field. These structures are especially pronounced if ions are collected from the flat area of the tip surface, for example from the middle of the (110) surface of a tungsten tip. When ions are collected from a kink site atoms of the W (110) plane step, the secondary structures are washed out. It is particularly interesting that in field ionization of hydrogen, secondary peaks are very pronounced for H^+ and H_2^+ ions but not H_3^+ ions, as is shown in Fig. 2.6. The H_3^+ peak is very sharp, indicating that H_3^+ ions are produced only right at the surface.[22] This can be understood from the fact that H_3 molecules are unstable in free space. It is formed by field induced polymerization and exist only in the field adsorption state, as will be further discussed.[33]

The secondary peaks in field ionization and pulsed-laser stimulated field desorption can be explained in terms of resonance field ionization of gas atoms, as shown in Fig. 2.7. The basic idea proposed by Jason *et al.*[22,23] and Alferieff & Duke[34] is the following. When the atomic electron tunnels into the metal, the metal surface may reflect the electron back into the atom. If the electron energy matches the energy states of the nearly triangular potential well then resonance tunneling will occur, and the field ionization rate will be greatly enhanced. A one-dimensional calculation by Müller & Krishnaswamy[24] agrees quite well with their experimental data on the field dependence of the peak separations. Appelbaum & McRea[35] use experimental data of the reflectivity of low-energy electrons for the tungsten (110) surface from a LEED study[36] to evaluate the heights and positions of the secondary peaks, and find their result to agree well with the experimental data of Müller &

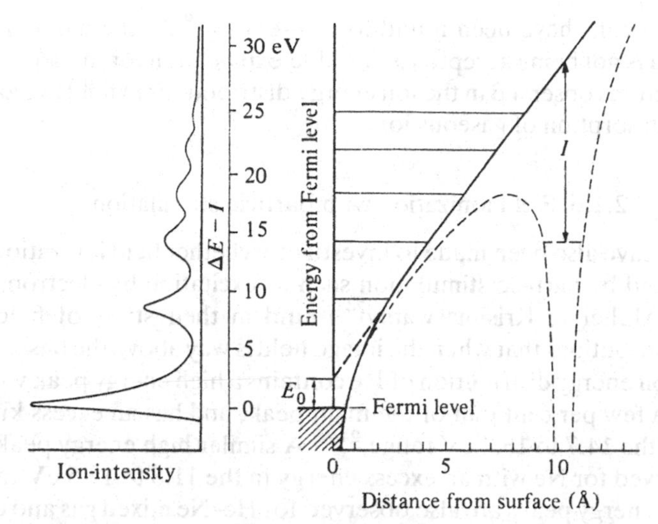

Fig. 2.7 Diagram showing how resonance field ionization occurs. When an image gas atom is field ionized, the tunneling electron may be reflected right back to the atom. Field ionization is enhanced if the atomic level lines up with an energy level formed between the metal surface and the potential barrier of the applied field, as shown in the figure. The potential barrier is approximately triangular in shape.

Krishnaswamy. They emphasize the importance of a phase factor in the complex reflection coefficient of electrons from the surface, and propose that field ion energy distributions may be used to measure the phase of electron reflection at the crystal surface. However, neither is the potential barrier known accurately enough nor are the available experimental data good enough, and to this date no such information has been obtained.

An alternative explanation of the secondary peak structures has been given by Lucas.[37,38] He proposes that these structures are the result of a loss of energy of field ions in exciting surface plasmons, either at the instant the ion is created or during the time it is accelerated away from the surface. The slight variations in secondary peak separations can be explained in terms of energy dispersion of the plasmons. Although this is a very simple and elegant explanation, it is not substantiated by all available experimental data. First, plasmon excitation cross-section should be greater if ions are created closer to the surface. Yet in field ionization and field desorption of hydrogen, while both H^+ and H_2^+ ions show pronounced secondary peaks, H_3^+ ions which are produced right at the surface as indicated by the lack of low energy tail in the ion energy distribution do not show any secondary peak structure. Also, in field evaporation of metals and semiconductors, multiply charged ions are formed right at the surface and are then accelerated away, yet no secondary peaks have been found for these ions.[39] At the moment this clever idea is not being accepted as a viable explanation for the secondary peak structures observed in the ion energy distributions in field ionization and field desorption of gaseous ions.

2.1.6 Field ionization with particle stimulation

Attempts have also been made to investigate whether field ionization can be enhanced by particle stimulation such as excitation by electrons and photons. Müller & Krishnaswamy[24] found in their study of field ion energy distributions that when the image field is way above the best image field the ion energy distribution of He contains a high energy peak with its intensity a few per cent that of the main peak, and has an excess kinetic energy in the 14.7 to 16.4 eV range.[24,40] A similar high energy peak was also observed for Ne with an excess energy in the 11.8 to 13.5 eV range. Such high energy peaks are also observed for He–Ne mixed gas and other gases such as Ar. As ions in the high energy peak have to originate closer to the surface than the critical distance in ordinary field ionization, or where field ionization of image gas atoms in their ground states is forbidden, these high energy ions are called 'ions from the forbidden zone'. The high energy peak is much more pronounced if the electric field is above the best image field of the image gas which is now mixed with a low ionization energy gas such as hydrogen.[41] Ions in the high energy peak originate from image gas atoms in their field adsorption states. In the high applied field, the low ionization energy hydrogen molecules can be field ionized far above the surface. The electrons released in the tunneling process of the hydrogen molecules will gain sufficient kinetic energies when they reach the surface and will be able to excite the field adsorbed image gas atoms to their excited states. These electronically excited atoms in effect have a lower ionization energy, and can thus be field ionized much closer to the surface. In an ion energy distribution they will show up as the high energy peak. The term 'field ions from the forbidden zone' is therefore not an accurate description of the phenomenon. The phenomenon is really field ionization of electronically excited image gas atoms, and such ions are still formed beyond the critical distance for field ionization of these 'low ionization energy' gas atoms. The excitation is produced by electron shower from the low ionization energy gas atoms which are field ionized far above the emitter surface. When the tunneling electrons reach the surface they will have gained enough energy to excite field adsorbed image gas atoms. This electron stimulated field ionization, or rather electron stimulated field desorption, has been used to study the location of the image plane and field adsorption of inert gas atoms, as will be discussed in Section 2.3.[42]

Field ionization can also be enhanced by photons. Tsong *et al.*[43] searched for this effect and found that Ar field ion current near the threshold field of field ionization from an aluminum oxide tip could be enhanced by a factor of about 5 when the tip was irradiated with 4.16 eV photons from laser pulses of 2 μs width at a pulse repetition rate of 25 Hz, or a duty cycle of only 0.005%. As the photon energy is much too small to

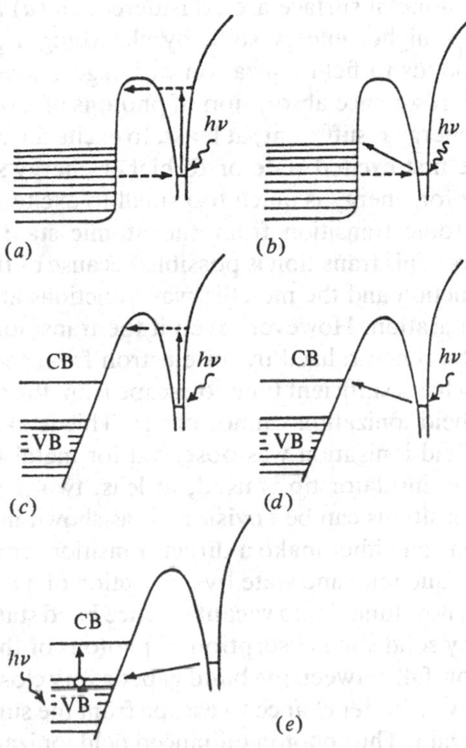

Fig. 2.8 (*a*) Photon-stimulated field ionization above a metal surface can occur by electronic excitation of the atom to an excited state. This process is essentially field ionization of atoms in their excited states. But the ion formed in this process will be quickly reneutralized by an electron from the metal before it can escape from the surface.

(*b*) As the metal wavefunctions overlap with the atomic wavefunction at such a close distance, photon stimulated field ionization can also occur by a direct electronic transition from the atomic state to a vacant metallic state above the Fermi level. However, the ion will be quickly reneutralized before it can leave the surface. Thus photon stimulated field ionization cannot be observed from a metal tip. However, if the photon intensity is excessively high then a high energy tail with an energy in excess of a few hundreds of eV can be observed.

(*c*) to (*e*) Similar processes above a semiconductor or an insulator surface. With the existence of a band gap between the conduction band and the valance band, an atom, once ionized by photon stimulation, may be able to escape away from the surface before it is reneutralized again.

excite Ar atoms to their first excited state, the photon enhanced field ionization was thought to involve with direct electronic transition from the atomic state to vacant states in the conduction band. Tsong *et al.* did not observe any detectable enhanced field ion current from metal tip surfaces. Several possible mechanisms for the enhancement were considered as shown schematically in Fig. 2.8. In (*a*) and (*b*), possible mechanisms above a metal surface are considered. In (*a*) an image gas atom is excited to a higher energy state by absorbing a photon. This mechanism corresponds to field ionization of image gas atoms in their excited states. Such resonance absorption of photons of atoms can occur only if the photon energy is sufficient, at least, to excite an atom from the ground state to the first excited state or to higher energy states. In the experiment the photon energy is much too small to excite Ar atoms. In (*b*), a direct electronic transition from the atomic state to a vacant metallic state occurs. This transition is possible because of the overlap of the atomic wavefunction and the metallic wavefunctions at a very small atom-to-surface separation. However, even if the transition occurs, the ionized atom will be reneutralized by an electron from the conduction band before the ion has a sufficient time to escape from the surface. Thus photon enhanced field ionization cannot occur. This is exactly why no photon enhanced field ionization was observed for metal tips. When a semiconductor or an insulator tip is used, at least two possibilities for direct electronic transitions can be envisioned, as shown in (*d*) and (*e*). The atomic electron can either make a direct transition from the atomic state to a vacant conduction band state by absorption of a photon, or the atomic electron can now tunnel into vacant valence band state; these hole states are created by solid state absorption of photons of the tip. As the atomic state may now fall between the band gap, or very close to the band gap, the ion may have a better chance to escape from the surface without being neutralized again. Thus photon enhanced field ionization could be observed from an aluminum oxide tip surface.

Field ionization can also be treated as a direct electronic transition from the atomic state to a vacant state at the surface. Following the Fermi golden rule, the rate of field ionization is given by[31]

$$\varkappa = \frac{2\pi}{\hbar}\int_{E_F}^{\infty} \rho(E_k)|\langle\psi_k|eV|\psi_a\rangle^2\delta(E_k - E_a)\,\mathrm{d}E_k, \qquad (2.29)$$

where $\rho(E_k)$ is the electronic density of states at the surface, ψ_k is the surface wave function, ψ_a is the atomic wave function, and eV is the perturbation of the applied field. In photon-stimulated field ionization, E_a in the δ-function is replaced by $(h\nu + E_a)$; otherwise the equation and mechanisms are identical to those in field ionization. $h\nu$ is the photon energy.

Photon enhanced field ionization was later also reported by Niu *et al.*[44]

from tungsten emitter surfaces for benzene, naphthalene, anthracene and O-nitroaniline using a 200 W tungsten lamp combined with a chopper. Since neither did the experiment of Tsong *et al.*[43] find any enhanced field ionization from a tungsten tip surface, and nor could theoretical considerations predict such an enhancement, Niu *et al.*'s result was questioned by Viswanathan *et al.*[45] A convincing method of establishing the occurrence of photon enhanced field ionization is by measuring the ion kinetic energy distribution. Ions formed by photon stimulation should have less energy deficit than $I - \phi$ as they can be formed closer to the surface. No such studies have so far been reported.

2.2 Field evaporation and field desorption

2.2.1 Introduction

When the applied electric field reaches a few volts per angstrom range, atoms on a surface, irrespective of whether they are lattice atoms or adsorbed atoms and of whether the surface temperature is high or low, may start to emit out of the surface in the form of ions. This high electric field produced evaporation phenomenon is usually called field evaporation if the surface atoms are lattice atoms, and is called field desorption if they are adsorbed atoms. From a theoretical point of view there are no fundamental differences. We will use the term 'field desorption' for general purposes, especially for theoretical discussions, since desorption is the term used in many other adsorption–desorption phenomena. When we specifically mean removal of lattice atoms by electric field the term 'field evaporation' will be used. Sometimes field evaporation is used where it may mean both field evaporation and field desorption.

Field evaporation is important in field ion microscopy for many reasons. First, the field ion microscope uses radial projection of field ions to produce a greatly magnified and atomically resolved image of the emitter surface. The image magnification can be reasonably uniform on the atomic scale over the entire surface only if the emitter surface is smooth on the atomic scale. The surface of a field ion tip, usually prepared from electrochemical polishing of a piece of thin wire, is by no means smooth but contains many asperities, sharp edges and protrusions. Fortunately, when the applied voltage is gradually raised, the local electric fields above these protrusions are higher than those above the smooth parts of the emitter surface. Surface atoms at these protrusions will be field evaporated before atoms at the smooth parts of the surface. Field evaporation is thus a self-regulating process with which the emitter surface can be processed to become the smoothest surface possible, and the emitter surface so prepared can thus serve as a good projecting surface for the field ions to produce a greatly magnified field ion image of very small image distortion. Low temperature field evaporation, when the temperature is low enough so that contaminants adsorbed on the tip shank cannot diffuse from the shank to the tip apex region, is also a good method for removing surface contamination to produce a clean surface at the tip apex. Because of the capability of field evaporating surface atoms and surface layers in a very regular fashion, it is possible to reach into the bulk of the sample by gradual field evaporation, and lattice defects or impurities inside the bulk can be revealed; in fact this capability is one of the most valuable assets of field ion microscopy which has not been fully appreciated by scientists outside the field. Field evaporation is also essential for doing chemical analysis of surface atoms and surface layers in the atom-probe. In this instrument, surface atoms are slowly pulse field evaporated and only ions of atoms of one's choice, chosen from the field ion image, are allowed to pass through a small probe-hole to reach an ion detector through a flight tube. From the flight times of these ions, their mass-to-charge ratios can be calculated and thus their chemical species identified. In liquid metal ion sources, it is now generally agreed that the dominant ion formation mechanism is field evaporation, although some of the ions may be produced by electron shower bombardment induced ionization. Field evaporation is therefore in every respect as important a process in field ion microscopy as field ionization.

2.2.2 Basic mechanisms

(a) *Image-hump model*

In his early study of field desorption, Müller[46] viewed the process as the thermal activation of a metal ion, of charge n, over a potential barrier known as the Schottky saddle[47]. This potential is the superposition of the potential of the applied field, $-neFx$, and the image potential, $-(ne)^2/4x$, of the ensuing ion. This superposition reduces the maximum in the potential energy of the ion, called the Schottky saddle or hump, by an amount $(ne)^{3/2}F^{1/2}$ below the zero field value. In the absence of an applied field, the energy needed to create an $n+$ ion in free space from a lattice atom can be derived by considering a Born–Haber energy cycle. First, an energy corresponding to the binding energy of the atom from its site, Λ, is needed to remove it to free space. An energy of $\Sigma_i I_i$ is needed to ionize the atom into an $n+$ ion. The n electrons are then returned to the metal whereby an energy of $n\phi_{em}$ is regained, where ϕ_{em} is the emitter work function. Thus the net energy needed, Q_0, is

$$Q_0 = \Lambda + \Sigma_i I_i - n\phi_{em}. \tag{2.30a}$$

The activation energy for forming an $n+$ charged ion in an applied field F from the site where the atom sits is therefore

$$Q_n(F) = Q_0 - (ne)^{3/2}F^{1/2}. \tag{2.30b}$$

As field evaporation is a thermally activated process, the field evaporation rate is given by

$$\varkappa = \nu \exp\{-Q_n(F)/kT\}. \tag{2.31}$$

The frequency factor, ν, should be on the order of the lattice vibrational frequency, or $\sim 10^{13}\ \mathrm{s}^{-1}$. The extreme simplicity of the model makes it very convenient for many applications. This theoretical model is now generally referred to as the image-hump model, or the Schottky-hump model, of field desorption. The potential energy curve of this model is not defined at all distances because of the crude nature of the argument; it is nevertheless shown schematically in Fig. 2.9(a).[48]

(b) *Charge-exchange model*

There is another theoretical model of field desorption called the charge-exchange model. This model was originally proposed by Gomer[50] for the field desorption of electronegative gases. In the paper written by Gomer and Swanson,[51] they described different theoretical models valid for field desorption of different chemical species. For field evaporation of metals, the image-hump model with a gradual draining of electronic charges during the desorption process was thought to be the valid one. Tsong[52] showed later that an estimate of the desorption fields for ions of different charge states based on this model was as good for predicting the charge states of ions in low temperature field evaporation of all metals as that given by Brandon[53] based on the image-hump model. He also used the charge-exchange model to analyze data on field evaporation rates of many metals.[48] Since then the charge-exchange model has become a generally accepted model for treating field desorption of all atomic species.

In this model, field evaporation is treated as a transition from the $A + M$ state to the $A^{n+} + M^{n-}$ state, where the former state is called the atomic state and the latter one the ionic state. As shown in Fig. 2.9(b), in the absence of an applied field the potential energy curve of the atomic state is represented by $U_a(x)$ and that of the ionic state is represented by $U_i(x)$. At a very large distance away from the surface, the ionic state should have an energy higher than the atomic state by $\Sigma_i I_i - n\phi$, as has been explained. In an applied field, the atomic curve, represented by $U_a(x, F)$, is lowered only slightly by the polarization binding, whereas the ionic curve, represented by $U_i(x, F)$, is now greatly reduced by an amount $-neFx$. If the applied field is strong enough, the ionic curve will intersect the atomic curve at a distance close to the equilibrium distance of the surface atom, which is denoted by x_c. At x_c, the atomic state and the ionic state have identical energy. Beyond x_c, the atomic state now has a higher energy than the ionic state. Thus when the amplitude of the thermal vibration of a surface atom exceeds x_c, the atom may make an

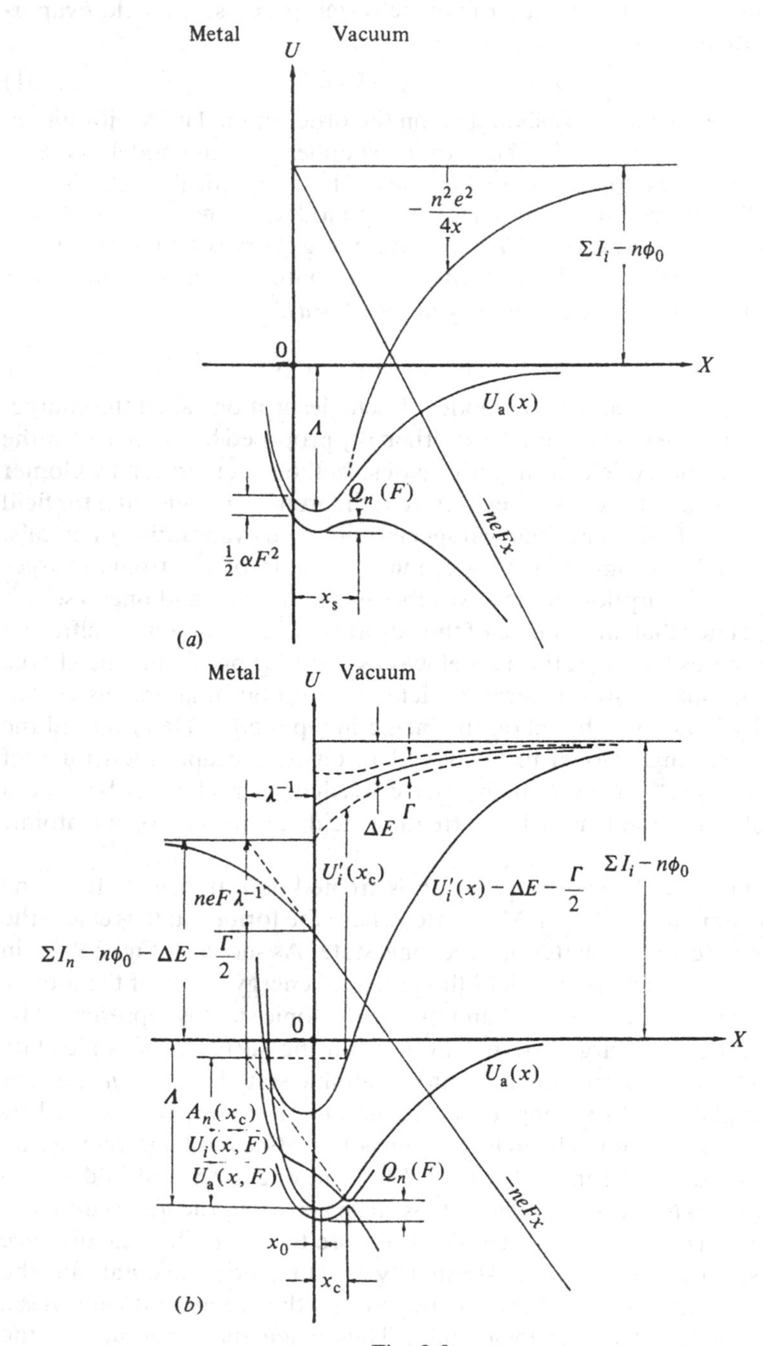
Metal
Vacuum
U
$-\frac{n^2e^2}{4x}$
$\Sigma I_i - n\phi_0$
0
X
$U_a(x)$
Λ
$Q_n(F)$
$-neFx$
$\frac{1}{2}\alpha F^2$
x_s
(a)
Metal
U
Vacuum
λ^{-1}
Γ
ΔE
$U_i'(x_c)$
$\Sigma I_i - n\phi_0$
$neF\lambda^{-1}$
$U_i'(x) - \Delta E - \frac{\Gamma}{2}$
$\Sigma I_n - n\phi_0 - \Delta E - \frac{\Gamma}{2}$
0
X
$U_a(x)$
Λ
$A_n(x_c)$
$U_i(x, F)$
$U_a(x, F)$
$Q_n(F)$
$-neFx$
x_0
x_c
(b)

Fig. 2.9

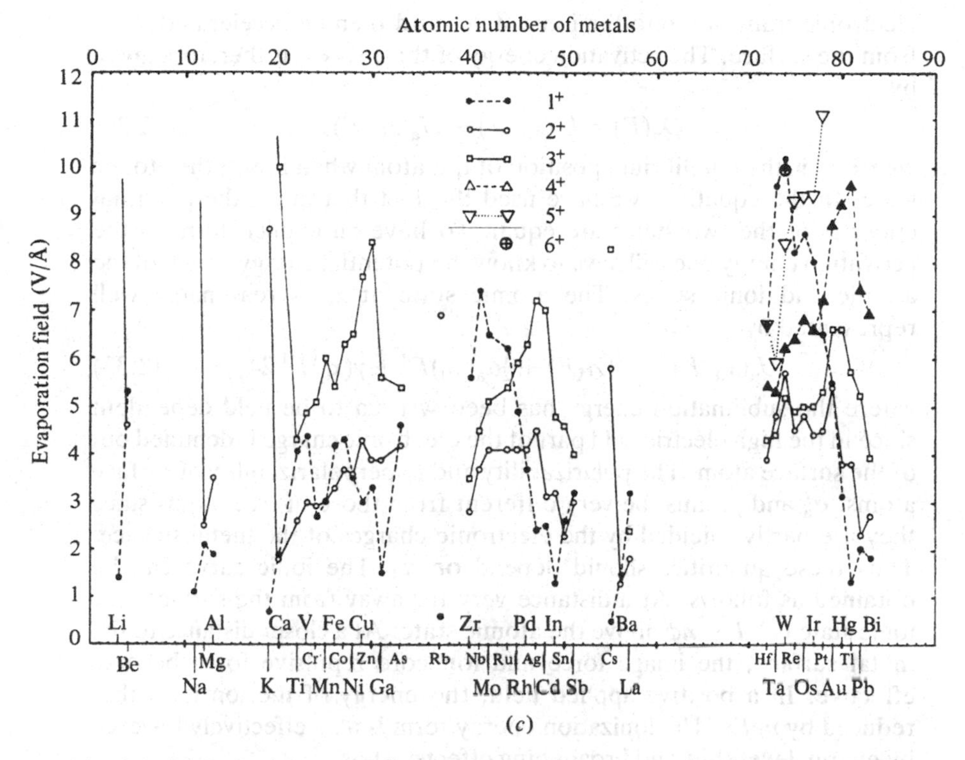

Fig. 2.9 (*a*) Potential energy diagram of a surface atom and an ion near a metal surface for the image-hump model. The ionic potential is represented by $(\Sigma I_i - n\phi - n^2e^2/4x)$. In this model, the potential in the intermediate distance is not clearly defined. It is simply assumed that in the applied field the surface atom comes off the surface with its charges gradually being drained by the applied field, and eventually approaches the ionic curve of the ion of the final charge state of the field evaporated ion, as represented by the ionic curve further reduced by $neFx$, or the thick curve.

(*b*) In the charge exchange model, field evaporation occurs by electronic transition beyond the critical distance where the atomic potential energy becomes larger than the ionic potential energy. In this potential energy diagram we also take into consideration the effect of field penetration, the depth of which is represented by λ^{-1}. U_i refers to ionic potential and U_a refers to atomic potential. For details see the text and reference 48.

(*c*) Low temperature evaporation fields for different charge states of metals according to the image-hump model. For low melting point metals, the evaporation fields are lowest for field evaporating as 1+ ions. Evaporation fields for the majority of metals are lowest for 2+ ions. For refractory metals such as W and Ta, evaporation fields are lowest for 3+ ions. If field evaporation is carried out at a very high field using high voltage pulses, higher charge state ions can also be produced.

electronic transition into the ionic state, and then be accelerated away from the surface. The activation energy of this process is therefore given by

$$Q_n(F) = U_i(x_c, F) - U_a(x_0, F), \tag{2.32}$$

where x_0 is the equilibrium position of the atom when it is in the atomic state. In the equation, we have used the fact that at x_c, the potential energies of the two states are equal. To have an explicit form of the activation energy one will have to know the potential energy curves of the atomic and ionic states. The atomic state at x_0 is reasonably well represented by

$$U_a(x_0, F) \simeq -\{\Lambda(F) + \tfrac{1}{2}\alpha_a(x_0)F^2 + \gamma(x_0)F^4/24\}, \tag{2.33}$$

where the sublimation energy has been written to be field dependent since in the high electric field part of the electronic charge is denuded out of the surface atom. The polarizability and hyperpolarizability of surface atoms, α_a and γ, must be very different from those of free atoms since they are partly shielded by the electronic charges of the metal surface. Thus these quantities should depend on x_0. The ionic curve can be obtained as follows. At a distance very far away from the surface, the ionic state is $\Sigma_i I_i - n\phi$ above the atomic state. At a closer distance to the metal surface, the image force and ion core repulsive force become effective. In a positive applied field, the energy of the ion is further reduced by $neFx$. The ionization energy term is also effectively lowered by energy level shift and broadening effects. Thus

$$\begin{aligned} U_i(F, x_c) = \Lambda + \sum_i^n I_i - \frac{\Gamma}{2} - \Delta E - n\phi(F) - neFx_c \\ - \tfrac{1}{2}\alpha_i F^2 - \tfrac{1}{24}\gamma_i F^4 - \frac{n^2e^2}{4(x_c + \delta)} + \sum_j \frac{K}{r_j^p(F)}, \end{aligned} \tag{2.34}$$

where ΔE and Γ represent respectively the energy level shift and broadening,[54] $\phi(F)$ represents the surface potential energy, or the work function of the surface but with the surface potential curve bent by the field penetration effect, α_i and γ_i are the ordinary and hyper polarizabilities of the n-fold charged ion, and δ is the field penetration depth.[55,56] The image potential has been modified because of field penetration.[57] The last term represents the repulsive interaction term between the ensuing ion and the ion cores of the metal. In the case of semiconductors and dielectric materials, the image potential term has to be multiplied by a factor $(\varepsilon - 1)/(\varepsilon + 1)$, where ε is the dielectric constant.[58] In eq. (2.34) the ion core repulsion has been represented by a Lennard–Jones type repulsive potential energy. Obviously, both the forms of the atomic potential and the ionic potential are very complicated and are not really known. For data analysis, even these equations are often too complicated to use. In that case, appropriate approximations have to be made.

2.2.3 Charge states, desorption fields and post field ionization

(*a*) *Charge states and desorption fields*

The field evaporation rate, as can be seen from eq. (2.31), increases several orders of magnitude by a small change in the activation energy of about $15kT$, or about 26 meV if the tip is kept at 20 K. The activation energy in most field evaporation must be quite small, of less than a few tenths of an eV. For a rough estimate of the evaporation field of materials, one may define an 'evaporation field F_e' where the activation energy is zero, or $Q_n(F_e) = 0$. Müller,[59] in fact, made such a calculation of evaporation fields for various metals for field evaporating as singly charged ions. All the calculated values were considerably higher, 40 to 100%, than what had been experimentally observed. This discrepancy was resolved when Brandon,[53] using the image-hump model, made a careful comparison of the evaporation fields for forming ions of different charge states and found that for most metals the evaporation fields for doubly charged ions were considerably lower than those for singly charged ones. In other words, the evaporation field as n-fold charged ions for a metal as calculated from

$$F_{en} \simeq (\Lambda + \Sigma_i I_i - n\phi)^2/n^3e^3 \tag{2.35}$$

is a minimum for a certain integer number n. This number is 2 for most metals and for some low melting point metals. Since for most metals F_{e2} is smaller than F_{e1}, when the tip voltage is gradually raised atoms will start to field evaporate as doubly charged ions before they can even be field evaporated as singly charged ions. In fact, for some highly refractory metals n can be as large as 3 or 4,[60] and the charge states of low temperature field evaporated ions should be 3 or 4. Table 2.2 lists values given by Tsong[60] for most metals. The existence of a Schottky hump in the presence of a predominant repulsive term very close to the surface is questionable.[47,48] Nevertheless, the evaporation fields and the most abundant ion species derived from this simple model calculation do agree surprisingly well with experimental data in low temperature field evaporation where the activation energy has to be very low. The experimental data are also listed in the table. A similar calculation by Tsong, based on the charge-exchange model, gives a similar result.[52] With this model, the evaporation field as n-fold charged ions is given by

$$F_{en} \approx \frac{1}{nr_0}\left(\Lambda + \sum_i^n I_i - n\phi - \frac{3.6n^2}{r_0}\right) \text{V/Å}, \tag{2.36}$$

Table 2.2 *Field evaporation parameters for various elements*

Atomic No.	Metal	Λ (eV)	I_1 (eV)	I_2 (eV)	I_3 (eV)	I_4 (eV)	ϕ (eV)	F_1 V/Å	F_2 V/Å	F_3 V/Å	F_4 V/Å	Ion species observed	F_{ob} V/Å
3	Li	1.65	5.392	75.638	122.451		2.5	*1.4*	52	100			
4	Be	3.33	9.322	18.211	153.893	217.713	3.9	5.3	*4.6*	77	162	Be^{2+}, Be^{+}	3.4
11	Na	1.13	5.139	47.286	71.64	98.91	2.3	*1.1*	21	36	50	Na^{+}	?
12	Mg	1.53	7.646	15.035	80.143	109.24	3.7	*2.1*	2.5	22	43		
13	Al	3.34	5.986	18.828	28.447	119.99	4.1	*1.9*	3.5	5.0	28	Al^{+}, Al^{2+}, Al^{3+}	<3.3
19	K	0.941	4.34	31.63	45.72	60.92	2.2	*0.7*	8.7	15	20		
20	Ca	1.825	6.113	11.871	50.91	67.15	2.7	*1.9*	*1.8*	10	18		
22	Ti	4.855	6.82	13.58	27.49	43.26	4.0	4.1	*2.6*	4.3	6.9	Ti^{2+}, Ti^{+}	2.5
23	V	5.30	6.74	14.65	29.31	46.71	4.1	4.4	*3.0*	4.9	8.1	V^{2+}, V^{3+}	?
24	Cr	4.10	6.766	16.50	30.96	49.1	4.6	*2.7*	*2.9*	5.1	8.6	Cr^{2+}, Cr^{+}	?
25	Mn	2.98	7.435	15.640	33.67	51.4	3.8	*3.0*	*3.0*	6.0	10	Mn^{2+}	
26	Fe	4.29	7.90	16.16	30.651	54.8	4.4	4.2	*3.3*	5.4	10	Fe^{2+}, Fe^{+}	3.5
27	Co	4.387	7.86	17.06	33.50	51.3	4.4	4.3	*3.7*	6.3	10	Co^{2+}, Co^{+}	3.6
28	Ni	4.435	7.635	18.168	35.17	54.9	5.0	*3.5*	*3.6*	6.5	11	Ni^{2+}, Ni^{+}	3.5
29	Cu	3.50	7.726	20.292	36.83	55.2	4.6	*3.0*	4.3	7.7	12	Cu^{+}, Cu^{2+}	3.0
30	Zn	1.35	9.394	17.964	39.72	59.4	3.8	*3.3*	3.9	8.4	14	Zn^{2+}, Zn^{+}	
31	Ga	2.78	5.999	20.51	30.71	64	4.1	*1.5*	3.9	5.6	13	Ga^{+}	
33	As	3.0	9.81	18.633	28.351	50.13	4.7	4.6	*4.2*	5.4	9.0	As^{+}	?
37	Rb	0.858	4.177	27.28	40	52.6	2.1	*0.6*	6.9	11	15		?
40	Zr	6.316	6.84	13.13	22.99	34.34	4.2	5.6	*2.8*	3.5	4.8	Zr^{2+}, Zr^{3+}, Zr^{+}	2.5
41	Nb	7.47	6.88	14.32	25.04	38.3	4.0	7.4	*3.7*	4.5	6.3	Nb^{2+}, Nb^{3+}	3.5
42	Mo	6.810	7.10	16.15	27.16	46.4	4.2	6.5	*4.1*	5.1	8.2	Mo^{2+}, Mo^{3+}	4.6
44	Ru	6.615	7.37	16.76	28.47	50	4.5	6.2	*4.1*	5.4	9.0	Ru^{2+}	4.3
45	Rh	5.752	7.46	18.08	31.06	48	4.8	4.9	*4.1*	5.9	9.0	Rh^{2+}, Rh^{3+}, Rh^{+}	4.8
46	Pd	3.936	8.34	19.43	32.92	53	5.0	*3.7*	4.1	6.3	10	Pd^{+}, Pd^{2+}	3.2

47	Ag	2.96	7.576	21.49	34.83	56	4.6	*2.4*	4.5	7.2	12	*Ag^{+}*, Ag^{2+}	2.5
48	Cd	1.160	8.993	16.908	37.48	59	4.1	*2.5*	3.1	7.0	12		
49	In	2.6	5.786	18.869	28.03	54.4	(4)	*1.3*	3.2	4.8	9.5		
50	Sn	3.12	7.344	14.632	30.502	40.734	4.4	2.6	*2.3*	4.6	6.7		
51	Sb	2.7	8.641	16.53	25.3	44.2	4.6	3.2	*3.0*	4.0	6.8		
55	Cs	0.827	3.894	25.1	35	46	(2)	*0.5*	5.8	8.3	11		
56	Ba	1.86	5.212	10.004		49	2.5	1.5	*1.3*				
57	La	4.491	5.577	11.06	19.175	52	3.3	3.2	*1.8*	2.4	6.8		
72	Hf	6.35	7.0	14.9	23.3	33.3	3.5	6.7	*3.9*	4.3	5.4	*Hf^{3+}*, Hf^{2+}	4.2
73	Ta	8.089	7.89	16	22	33	4.2	9.6	4.8	*4.4*	5.3	*Ta^{3+}*, Ta^{4+}, Ta^{2+}	4.5
74	W	8.66	7.98	18	24	35	4.5	10.2	5.7	*5.2*	6.2	*W^{3+}*, W^{4+}, W^{2+}, W^{5+}	5.7
75	Re	8.10	7.88	17	26	38	5.1	8.2	*4.5*	4.9	6.4	*Re^{3+}*, Re^{2+}	4.8
76	Os	(7)	8.7	17	25	40	4.6	8.6	*4.8*	5.0	6.8		
77	Ir	6.93	9.1	17	27	39	5.3	8.0	*4.4*	5.0	6.6	*Ir^{2+}*, Ir^{3+}	5.3
78	Pt	5.852	9.0	18.56	28	41	5.3	6.3	*4.5*	5.3	7.2	*Pt^{2+}*, Pt^{3+}	4.8
79	Au	3.78	9.225	20.5	30	44	4.3	*5.3*	*5.4*	6.6	8.8	*Au^{+}*, Au^{2+}	3.5
80	Hg	0.694	10.437	18.756	34.2	46	4.5	*3.1*	3.8	6.6	9.2		
81	Tl	1.87	6.108	20.428	29.83	50.7	3.7	*1.3*	3.8	5.7	9.6		
82	Pb	2.04	7.416	15.032	31.937	42.32	4.1	*2.0*	2.3	5.2	7.4		
83	Bi	2.15	7.289	16.69	25.56	45.3	4.3	*1.8*	2.7	3.9	6.9		
92	U	4	6				3.3	3.1					
6	C	7.37	11.256	24.376	47.871	64.476	$\chi \approx 4$	*11.0*	*10.3*	15.5		*C^{+}* (graphite)	?
12	Si	4.63	8.149	16.34	33.46	45.13	$\chi \approx 4.2$	4.7	*3.6*	6.4		*Si^{2+}*	3.8
32	Ge	3.85	7.88	15.93	34.21	45.7	$\chi \approx 4.2$	3.7	*3.1*	6.2			

if Λ, I_i, and ϕ are expressed in eV, and the atomic radius r_0 is in Å. A similar table calculated with this equation has already been given in the book by Müller & Tsong,[16] and will not be repeated here. As mentioned in the book, the agreement with experimental data is as good as calculations based on the image-hump model. For some metals, in fact, values based on eq. (2.36) are in better agreement with the experimental values. For other materials such as semiconductors or insulators, it is much more appropriate to use the charge exchange model. At the end of Table 2.2, calculations made for C (diamond), Si and Ge based on eq. (2.36) are listed. $(\varepsilon - 1)/(\varepsilon + 1)$ is taken to be one. Values of r_0 used are ~1, 1.173, and 1.242 Å respectively for C, Si and Ge. In field evaporation of semiconductors and insulator surfaces, electrons are returned to the conduction band edge and thus the emitter work function term is replaced with the electron affinity χ.

(*b*) *Post field ionization in field desorption*

As discussed above, we have shown that simple model calculations by Brandon, later extended by Tsong on the basis of the image-hump model, and also by Tsong on the basis of the charge-exchange model, give the charge states and 'evaporation fields' of elements in very good agreement with experimental observations. Very soft metals such as Al, Cu, Ag and Au are expected to field evaporate as singly charged ions, and most other metals should field evaporate as doubly charged ions. For a few highly refractory metals such as W and Ta, triply charged ions can be expected, as can be seen from Table 2.2. It is remarkable that such simple calculations agree with experimental results for almost all elements over the entire periodic table. The validity of these calculations has since then been severely questioned for the reason that in nanosecond high voltage pulse field evaporation at a very high evaporation rate highly charged ions such as Mo^{4+}, W^{5+} and W^{6+} have been observed by Müller & Krishnaswamy.[61] This observation is in general considered to be incompatible with the calculations of Brandon and Tsong. Ernst[62] and Haylock & Kingham[63] propose that these highly charged ions in high voltage pulse field evaporation are formed by post field ionization of field evaporated ions. In their original paper[63] Haylock & Kingham only considered further field ionization in space of field evaporated ions which were already multiply charged. However Kingham[64] later stated, as does Ernst in his first paper on post field ionization, that all ions in field evaporation come off as singly charged ions, and multiply charged ions are all produced by subsequent post field ionization of these ions. The possibility of post field ionization was in fact suggested before[16] but was later abandoned because of the calculation of Chamber *et al.*[65] and of Taylor.[66] These calculations showed that the probability of post field ionization was much too small to be important. Since the revival by Ernst and by

Haylock & Kingham of post field ionization, a series of experiments by Ernst & Jentsch,[67] Kellogg,[68] Konishi *et al.* and others[69] have been carried out, and all these results have been interpreted to support post field ionization of field evaporated singly charged ions. Tsong,[70] on the basis of a measurement of the ion energy distribution of field evaporated ions, argued in favor of the validity of the evaporation field calculations, but he further showed clear evidence of the occurrence of post field ionization at high field where ions of charge states higher than those expected from the calculations are also formed. Details of these studies of post field ionization will be further discussed in the next section.

2.2.4 Field evaporation rates and ion energy distributions

(*a*) *Field evaporation rate measurements*

As in the case of field ionization, experimental tests of theoretical models of field desorption consist mainly of two types, namely measurements of the *I–V* characteristics and the ion energy distributions. However, in field desorption the number of ions which can be desorbed from the surface is very small. Therefore the *I–V* characteristics take a different form of field desorption rates as functions of experimental parameters, such as the field strength and the tip temperature. In field evaporation of tip materials, surface atoms are removed atomic layer by atomic layer. Thus the tip radius continues to change, and the field strength cannot be kept constant during the entire length of the measurement. The field strength seen by kink site atoms and plane edge atoms at a given tip voltage varies even when only the size of the plane is changed. From the variation of the field in image intensity across the tip surface we can also conclude that there is a considerable local field variation on the emitter surface. Thus the measurement of field evaporation rate under well-specified physical conditions is at best a very difficult task.

Experimental measurements of field evaporation rates were reported by Brandon[71] and by Tsong & Müller[72] using a step-up d.c. method and a high voltage pulse method. In the d.c. method, the tip voltage is raised in steps of about 100 V each. The time to field evaporate a surface layer is then recorded for each of the successive layers. The experimental conditions, such as the tip temperature and gas pressure, can also be varied. A set of data is shown in Fig. 2.10(*a*). When the tip voltage is stepped up for say 100 V, field evaporation proceeds fairly rapidly and a W (110) layer can be field evaporated very quickly. As surface layers are gradually field evaporated, the tip radius starts to increase and the field strength becomes lower. When the field gets too low, the field evaporation rate becomes very slow and the voltage is stepped up by another 100 V. In this method the change in the field strength as a result of the gradual field evaporation of surface layers can be properly accounted for and can

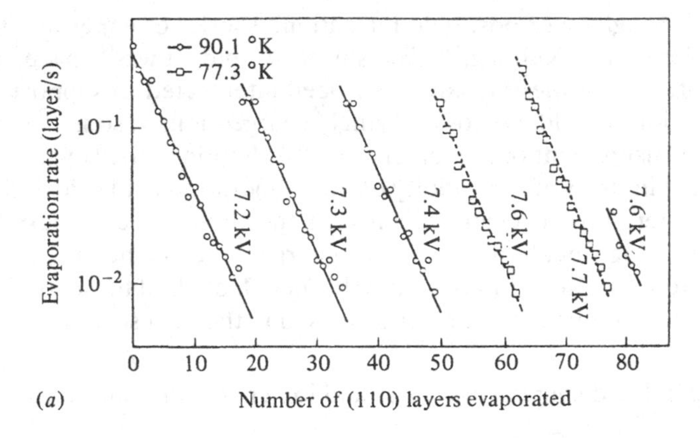

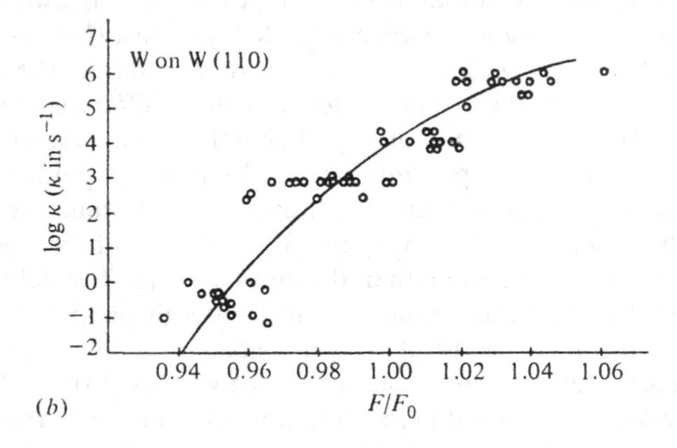

Fig. 2.10 (*a*) Field evaporation rates of W (110) layers at 90.1 K and 77.3 K, measured with a d.c. voltage step-up method. Note that when the temperature is changed by 12.8 K, the evaporation voltage changes by about 1.3 kV out of a total of about 7.6 kV.

(*b*) Field desorption rate of single W adatoms on a W (110) surface measured with a high voltage pulse method. Even though there are less than 60 adatoms in the measurement, the rate covered is over 8 orders of magnitude. However, statistical fluctuation of the data is large, as in most single adatom experiments.

(*c*) Field ion micrographs showing how adatoms are field evaporated one at a time using high voltage pulses. An adatom located near the edge of the W (110) plane sees a higher electric field and thus is field evaporated before other adatoms. In these figures, the adatoms dotted with black ink are the ones which are field desorbed next. From the field strength at the location of each adatom and the number of pulses needed to field desorb that atom, evaporation rates vs. field data are derived which are shown in the previous figure.

(*d*) Temperature dependence of field strength at a constant field evaporation rate, ~0.5 layer/s, for the W (110) layers.

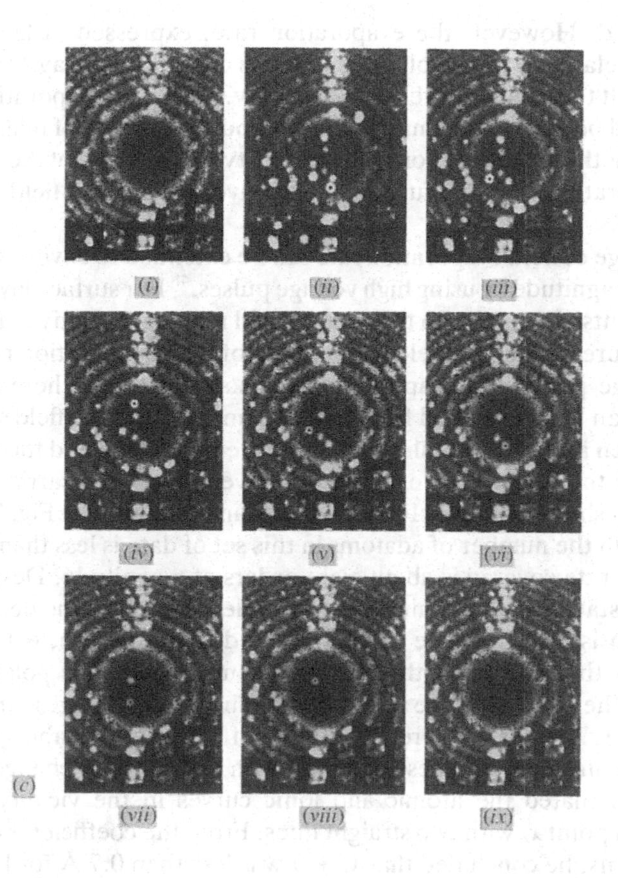
(i)
(ii)
(iii)
(iv)
(v)
(vi)
(c)
(vii)
(viii)
(ix)

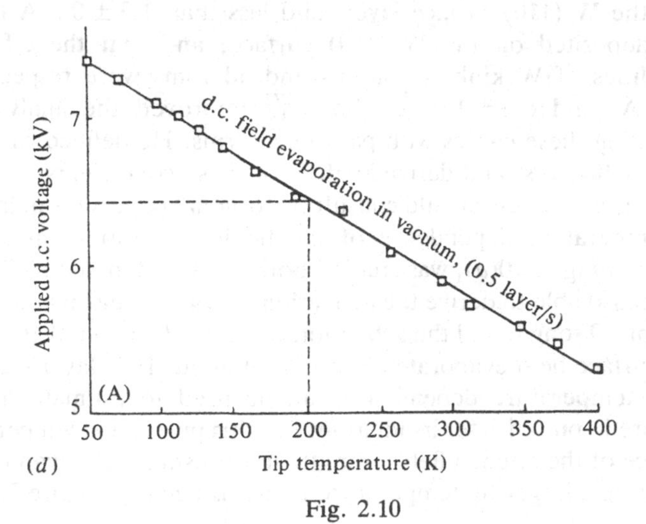
Applied d.c. voltage (kV)
7
6
5
d.c. field evaporation in vacuum, (0.5 layer/s)
(A)
50
100
150
200
250
300
350
400
(d)
Tip temperature (K)

Fig. 2.10

be corrected. However, the evaporation rate, expressed in layers per second, is related to the absolute rate, $\varkappa$, in a complicated way,[72] and it is very difficult to specify this relation explicitly. Also, the evaporation rate which could be covered is limited to only about two orders of magnitude. With this method, only data on field sensitivity, or the slope of the relative field evaporation rate as a function of the average applied field, can be obtained.

The range of field evaporation rate can be extended to cover over nine orders of magnitude by using high voltage pulses.[48] For surface layers, the rate constants obtained with this method still represent relative rates. An ideal measurement of the field dependence of field evaporation rate uses high voltage pulse field evaporation of adsorbed atoms where adatom positions can be determined from field ion images, and the field strength seen by each atom can be calibrated from the tip voltage and the adatom position on the plane. Such experiments have in fact been carried out by Tsong,[48] as shown in Fig. 2.10*b* and *c*. As can be seen from Fig. 2.10(*c*), even though the number of adatoms in this set of data is less than 60, the desorption rate covered is about eight orders of magnitude. Despite the very large statistical uncertainty of this single-atom data, the desorption rate, which is now truly the absolute field desorption rate, is found to depend on the field strength. This effect arises from the polarization binding of the adatoms in the applied field. The rate of change in the slope of the rate *vs*. field curve is related to the 'effective polarizability' of the adatom. Tsong analyzed these data based on the charge exchange model and approximated the atomic and ionic curves in the vicinity of the interaction point x_c with two straight lines. From the coefficients of the F and F^2 terms, he concluded that $x_c + \delta$ was less than 0.7 Å for kink site atoms of the W (110) surface layer, and less than 1.3 ± 0.2 Å for W adatoms deposited on the W (110) surface, and that the effective polarizabilities of W kink site atoms and adatoms were respectively 4.6 ± 0.6 Å^3 and 6.8 ± 1.0 Å^3. Forbes[73] improved the analysis by approximating these curves with parabolic forms. He defined carefully different coefficients, and derived values of these coefficients from the data. Interested readers should consult the original papers for details.

The temperature dependence of the field evaporation rate, as is obvious from Fig. 2.10(*a*), was studied both by Brandon and by Tsong. They were also able to derive the activation energy of field evaporation. For example, Tsong found the activation energy to be about 0.16 eV for W (110) surface field evaporated at a rate of about 10^{-2} layer/s at 77.3 K.[72] This temperature dependence can be used to estimate the tip temperature in pulsed-laser irradiation. For this purpose, a temperature dependence of the applied field strength at a constant field evaporation rate, covering a larger tip temperature range, is needed. Figure 2.10(*d*) gives a set of data obtained with an imaging atom-probe, obtained by Kellogg & Tsong, as will be discussed in Chapter 3.

In the above experiments, field evaporation was done in the presence of helium image gas. It is well known, particularly from the work of Nishikawa & Müller,[14] that 'field evaporation fields' of metals can be greatly reduced by the presence of gases. The reductions in the 'field evaporation fields' for Mo, W and Pt are found to be independent of the gas pressure, and are about 1 to 1.5% for He, about 6% for Ne with Mo and Pt, and about 10.5% for Ne with W. Tsong's field evaporating rate measurement with an ultra-high vacuum FIM, on the other hand, finds that the effect of high purity helium prepared by diffusion through a Vyco glass bulb, on the evaporation voltage is negligibly small while high purity Ne reduces the evaporation voltage by only 3 to 4% for W. Nishikawa & Müller interpret such a reduction of 'evaporation field' to arise from the impingement of surface atoms by image gas atoms. However, in view of the non-gas pressure dependence of this effect, and the subsequent finding of field adsorption effect as will be discussed in Section 2.3, this reduction has to be the result of field adsorption of the image gas atoms. With chemically reactive gases such as hydrogen, the 'evaporation voltage' can be reduced by over 40% for some materials. This effect is known as 'field induced chemical etching'. The best example is the reduction of 'evaporation' field of many metals and silicon by the presence of hydrogen where the field evaporation products are usually hydride ions.[74] There is little doubt that this reduction of the 'evaporation field' by a chemically reactive gas is an effect of chemisorption. As a result surface atoms are field evaporated at a much lower field in the form of compound ions. This method has been cleverly used to reduce the fields needed for surface processing of some low melting metals and to improve the quality of the image structure of field evaporated emitter surfaces of materials such as Fe, Ni, Nb, and Si etc. The field evaporation rate measurements of Tsong, fortunately, were carried out by using Vyco glass diffused helium as the image gas. Helium does not change the 'evaporation field' appreciably; thus the discrepancy is explained.

Visual aid with field ion images is not needed if one uses a single ion imaging capability Chevron channel plate for the screen assembly of the field ion microscope. With this setup, field evaporation of surface layers can be visualized from the field desorption image of the desorbed species even under ultra-high vacuum conditions as demonstrated by Walko and Müller.[75] In addition, Kellogg[68] found that the number of ions detected is proportional to the signal intensity of the Chevron channel plate, thus signal intensities in field desorption can be converted to the numbers of atoms field evaporated. There is still of course the uncertainty of where these atoms are being field evaporated from. As is clear from our discussions, the field seen by an atom depends sensitively on the position of the atom on the surface layer. For adsorbed atoms in the largest (110) plane, producible by low temperature slow d.c. field evaporation of tungsten emitter, the field strength seen by an adatom at the center of the plane at a given applied voltage is about 15% lower than that seen by an adatom near the edge of the plane.[48] Without the correction of this field variation and the continuous blunting of the tip by the field evaporation, the rate measured is at best only semi-quantitative. Nevertheless, using pulsed-laser imaging atom-probe, Kellogg was able to obtain linear Arrhenius plots of the field evaporation rates at different field strengths, covering a rate of about three orders of magnitude. From these plots, the activation energy of field evaporation for tungsten is found to decrease from 0.9 eV at 4.93 V/Å to 0.12 eV at 5.92 V/Å. These Arrhenius plots and activation energies as functions of field strength are shown in Fig. 2.11(*a*) and (*b*). Field evaporation rates of Rh as functions of temperature and field strength have earlier been investigated by Ernst[62] using a magnetic sector mass spectrometer combined with a retarding potential ion energy filter. The temperature range covered was from 100 to 600 K, and the field strength changed from 4.1 to 1.7 V/Å. In this experiment, field evaporation was done at such a high temperature that surface structure changes may have occurred. A nearly constant field evaporation current is probably maintained by a constant flow of atoms from the tip shank to the tip apex by a field gradient induced surface migration. The field and temperature dependence of the activation energy derived were similar to those obtained later by Kellogg. Again, the field strength of each field evaporating atom was neither known nor well defined. The field strength given was only an average value. The rates derived in these measurements are really relative rates since these atoms are field evaporated from surface layers even though these authors either did not try to distinguish the two rates or else considered them as absolute rates with the justification that these rates are expressed in terms of the number of ions detected by the ion detector per unit time. An ideal field desorption rate measurement should use single adatoms and the measurement should be carried out in vacuum using an atom-probe. As long as the ion detection efficiency of the Chevron channel plane is not 100%, and there are too many noise signals, such an experiment cannot be done. An additional difficulty of field evporation rate measurements is the possible influence of a very small amount of hydrogen in the system. Unless a system is carefully baked, a hydrogen partial pressure in the 10^{-10} Torr range cannot be avoided. Kellogg found that in a very good vacuum of $\sim 5 \times 10^{-10}$ Torr, adsorption of residual hydrogen occurs below 200 K, which greatly increases field evaporation rate and reduces dramatically the activation energy of field evaporation from 0.26 eV to 0.06 eV at a field of 5.56 V/Å.

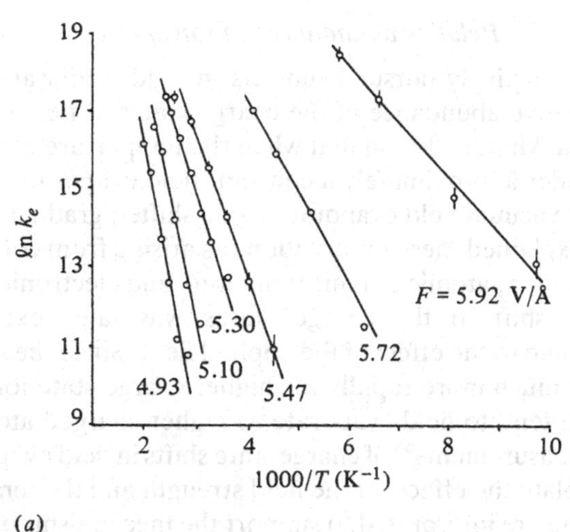

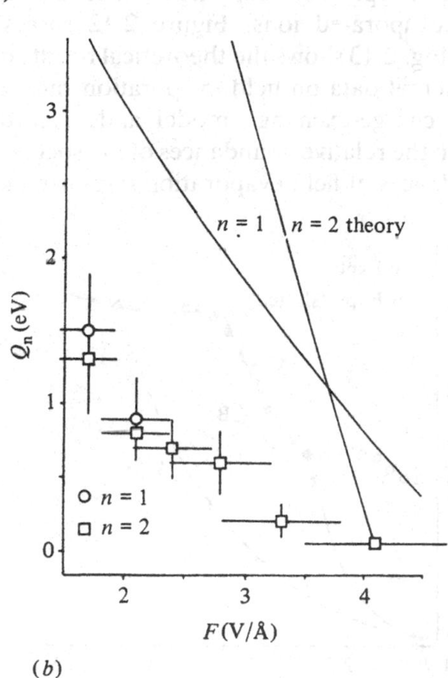

Fig. 2.11 (*a*) Arrhenius plots of field evaporation rates for tungsten obtained by Kellogg. The activation energy is found to decrease from 0.9 eV to 0.12 eV when the field is increased from 4.93 V/Å to 5.92 V/Å.

(*b*) Activation energy as a function of field for Rh, obtained by Ernst. According to Ernst, these data cannot be explained with the image-hump model of field evaporation. Calculated results based on this theoretical model are represented by the two solid lines.

(*b*) *Relative abundances of ion species*

One of the most actively pursued subjects in field evaporation experiments is the relative abundance of the charge states of field evaporated ions. Barofsky & Müller[76] found that when the temperture of a Be or Cu tip is raised, under approximately a constant field evaporation rate, the charge states of vacuum field evaporated ions shifted gradually from 2+ to +. Tsong[52] explained these observations as arising from combinations of thermal activation, atomic or ionic tunneling, and electronic transition processes. The shift in the charge states was later explained by McKinstry[77] as due to the effect of the applied field. Since the ionic curve will be lowered much more rapidly for higher charge state ions, at high field, atoms will tend to field evaporate as higher charged atoms. Later experimental measurements[68] of charge state shifts in field evaporation of metals try to isolate the effects of the field strength and the temperature, and all these data are interpreted to support the mechanisms of post field ionization of field evaporated ions. Figure 2.12 shows a few such measurements, and Fig. 2.13 shows the theoretical results of Kingham.[64] In general, experimental data on field evaporation rates can be better explained with the charge-exchange model and, according to these authors, data on both the relative abundances of ion species and field and temperature dependences of field evaporation rates can be explained in

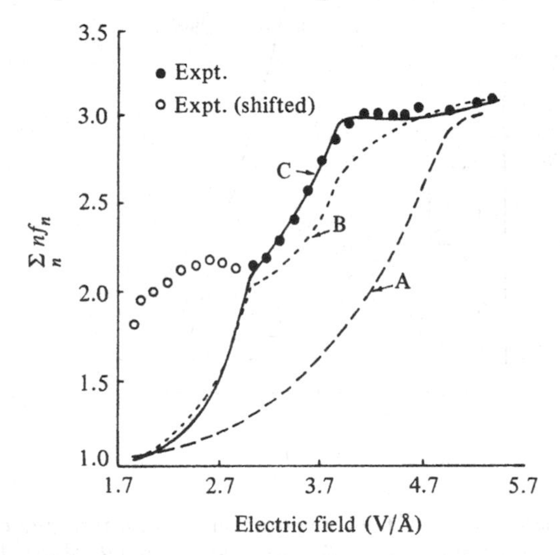

Fig. 2.12 Average charge state in field evaporation of Mo as a function of the applied field, obtained by Kellogg.[68] The field evaporation rate is kept nearly constant by adjusting the temperature of the pulsed-laser heating. Curves are theoretical results: *A* from the image-hump model, *B* from the same calculation with post field ionization correction. Calculations are based on those of Kingham.[64]

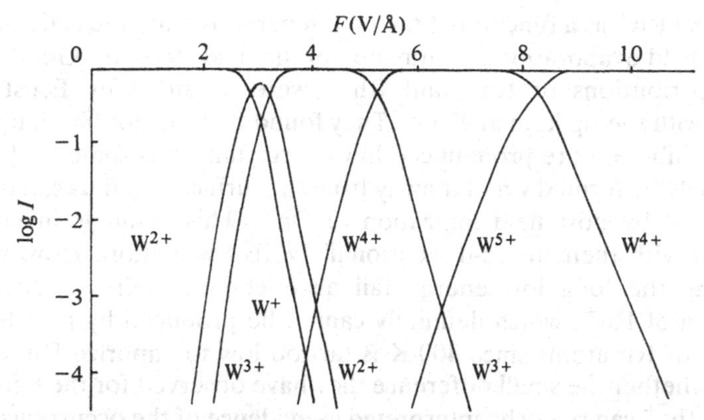

Fig. 2.13 Theoretical post field ionization probability for tungsten as a function of field obtained by Kingham.[64] He has presented similar curves for most of the metals in the periodic table.

terms of post field ionization, although for a few metals such as Mo significant discrepancies can be found.[63,67]

(c) *Ion energy distributions*

In field evaporation rate measurements, the greatest uncertainty lies in the difficulty of defining the field strength seen by each atom. Because of the very limited number of atoms available for a measurement, accurate rates are difficult to derive over a large range, and the effects of field strength and tip temperature are even more difficult to isolate. Also, in most field evaporation rate measurements the tip temperature is very high either for the purpose of sustaining a nearly constant flow of atoms for a measurable period of time, or for increasing the evaporation rate by pulsed-laser heating. Since even without post field ionization one can expect formation of singly charged ions for all metals at a sufficiently high temperature, where the activation energy can be very high, the changes of relative abundances of ionic species observed really do not support convincingly the occurrence of post field ionization. A more direct method for establishing the occurrence of post field ionization in field evaporation is by a measurement of the ion energy distribution. As has already been explained in connection with field ionization, an ion energy distribution locates the spatial zone in which ions are formed.

A measurement of the ion energy distribution of low temperature field evaporated ions was attempted by Waugh & Southon in 1977,[78] but the energy resolution of the instrument was not good enough to test field evaporation models. Ernst[62] then reported a measurement of ion energy distribution in high temperature, 100 to 600 K, field evaporation of Rh, obtaining good data on the appearance potentials, the relative abundance of Rh^{2+} and Rh^+ as a function of the tip temperature, and the activation energy of field evaporation as a function of the field strength. Good ion energy distributions of Rh^+ and Rh^{2+} were obtained by Ernst & Jentsch[67] with the tip kept at 400 K. They found that the doubly charged Rh ions exhibit a more pronounced low energy tail; thus some of these ions can only be formed very far away from the surface and thus can only be produced by post field ionization of Rh^+. This result is in good agreement with their 'three-dimensional' WKB calculation. However, considering the long low energy tail also seen in their ion energy distribution of Rh^+, which definitely cannot be produced by post field ionization of Rh atoms since 400 K is far too low to vaporize Rh, one wonders whether the small difference they have observed for the tails of Rh^+ and Rh^{2+} can really be interpreted as evidence of the occurrence of post field ionization of Rh^+. If Rh^+ distribution can have a low energy tail produced by noise signals, then Rh^{2+} distribution, with twice the kinetic energy of the Rh^+ ions, should have larger noise signals and a more prominent low energy tail can be expected also.

With the advent of a high resolution pulsed-laser time-of-flight atom-probe,[79] it is now possible to derive routinely ion energy distributions in the field evaporation of various materials. Tsong[70] found that ion energy distributions of ions in low temperature field evaporation, when only ions of one charge state are formed, usually exhibit a very narrow distribution width, FWHM, of less than $0.3nF$ to $0.4nF$ eV, where n is the charge state of the ions and F is the field strength. This represents a spatial zone width of ion formation of less than 0.3 to 0.4 Å. A preliminary measurement with a greatly improved pulse-laser time-of-flight atom-probe indicates that the ionization zone for low temperature field evaporated ions, under the condition that only one charge state exists, should be about 0.2 to 0.3 Å in width, comparable to that of He^+ ions in field ionization or field desorption at or below the best image field.[80] Thus there is no evidence of post field ionization in the formation of these ions. An example of these measurements is shown in Fig. 2.14(*a*) and (*b*). Whereas Ir^{2+} ions show little low energy tail, Ir^{3+} ions show a long low energy tail. Considering also that the field strengths needed for field evaporation and the charge states of ions observed in low temperature field evaporation of all metals agree so well with the expected result from the evaporation field calculations, as listed in Table 2.2, evidence of post field ionization reported in many studies is very weak indeed. Post field ionization can occur if field evaporation is carried out under the conditions that ions of more than one charge state coexist. Under such conditions, Tsong[70] found the energy distributions of ions of the higher charge state become very wide, sometimes with the low energy tail extending to over 100 eV, as can be seen in Fig. 2.14(*b*). Ions in the low

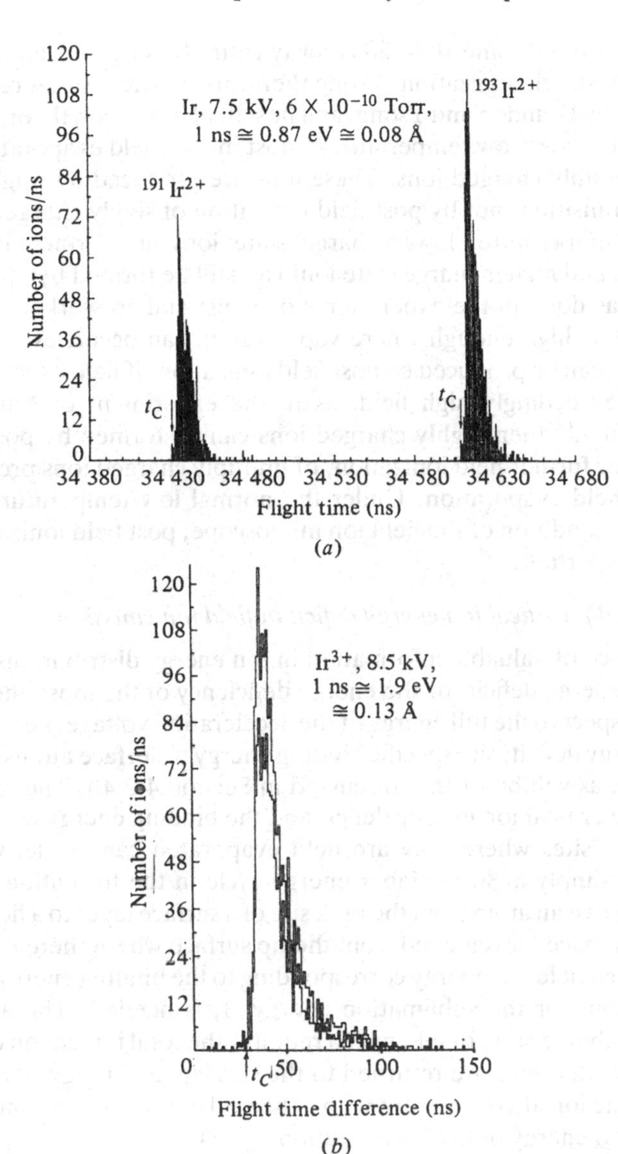

Fig. 2.14 (*a*) Ion energy distributions of the doubly charged Ir ions. The distribution width is very narrow without a low energy tail and therefore there is no evidence of post field ionization of field evaporated Ir^{2+} ions. Evaporation field calculations for low temperatures indicate that Ir will first field evaporate as Ir^{2+} ions when the field is gradually raised.

(*b*) A long low energy tail is found when the field is high enough so that a fraction of the ions formed are triply charged. Ir^{3+} ions in the tail section are obviously formed by post field ionization of Ir^{2+} ions far above the surface. Some of the ions are formed as far as 10 Å above the surface.

energy tail, formed some 10 to 20 Å away from the surface, can only be formed by post field ionization. Tsong therefore concludes that both the calculations by Brandon and Tsong, and post field ionization theories, are correct. Under very low temperatures, most metals field evaporate right away into doubly charged ions. These ions are produced by single step electronic transition, not by post field ionization of singly charged ions. At higher temperature, lower charge state ions are formed in field evaporation and higher charge state ions can still be formed by post field ionization, as done in the experiments of Ernst and co-workers. If the temperature is high enough where vaporization can occur, even singly charged ions can be produced by post field ionization. If field evaporation is done in exceedingly high field, as in the experiment of Müller & Krishnaswamy,[61] then highly charged ions can be formed by post field ionization, or further field ionization, of multiply charged ions produced initially in field evaporation. Under the normal low temperature field evaporation condition of the field ion microscope, post field ionization is not really important.

(*d*) *Critical ion energy deficit in field ion emission*

Another piece of valuable information in ion energy distributions is the critical ion energy deficit, or the energy deficiency of the most energetic ions with respect to the full energy of the acceleration voltage, neV. From this ion energy deficit, site specific binding energy of surface atoms can be determined, as will be further discussed in Section 4.2.4*b*. The relation between the critical ion energy deficit and the binding energy of surface atoms in the sites where they are field evaporated can be derived by considering simply a Born–Haber energy cycle in the formation of the ions. To remove an atom from the kink site of a surface layer to a field free space, i.e. a space far removed from the tip surface where there is a very strong applied field, an energy corresponding to the binding energy of the kink site atom, or the sublimation energy Λ, is needed. The atom is ionized to a charge state of $n+$, which requires the total ionization energy $\Sigma_i I_i$. When n electrons are returned to the surface, an energy of $n\phi_{\mathrm{em}}$ is regained. The ion also comes out with a thermal energy corresponding to the activation energy of field evaporation Q_n. Thus

$$\Delta E_{\mathrm{c}} = \Lambda + \Sigma_i I_i - n\phi_{\mathrm{em}} - Q_n. \tag{2.37}$$

A contact potential correction requires that in a retarding potential measurement the work function is replaced with that of the collector,[23] and that in a time-of-flight measurement it is replaced with the average work function of the flight tube.[30] In low temperature field evaporation, Q_n is only about 0.1 to 0.2 eV, or less than the uncertainty of most experimental measurements, and can be omitted for practical purposes.

It is important to recognize that the binding energy Λ in eq. (2.37) refers to the value in a field free space as all other parameters such as the maximum kinetic energy of ions and the ionization energies are all measured in a field free region also. Equation (2.37) is derived simply from a consideration of the total energy at two different states, an atom bound to the surface in a well-defined site in the absence of an applied field and the atom removed to a free space and being ionized there. Field evaporation occurs by deformation of the atomic and ionic potential energy curves in the high field region. However, as the total energy of the system is conserved, the amount of the potential energy change which occurs in the high field region will be transformed back to the kinetic energy of the field evaporated ions in the field free region, and it is this kinetic energy which is measured. From eq. (2.37) it is clear that for measuring the binding energy of surface atoms the measurement has to be done at very low temperature and very high field and rate, so that Q_n can be neglected from the equation. Q_n is a quantity which is difficult to calculate theoretically.

In a measurement of critical ion energy deficit in d.c. field evaporation of silver, which field evaporates as Ag^+, the ion energy deficit, when the field evaporation is done at 78 K, is found to agree with the simple formula, $\Delta E_c = \Lambda + I_1 - \phi_{coll}$, to within 0.02 eV if Λ is taken to be the cohesive energy of silver.[81] An important implication of this result, which has not been emphasized before, is that the binding energy of a kink site atom on Ag crystal surface, or the sublimation energy, is found to be equal to the cohesive energy of silver to within the accuracy of the measurement, or about ± 0.02 eV. If atomic interaction in a solid is pairwise additive, then the total potential energy of a crystal containing N atoms is given by

$$U_{tot} = \tfrac{1}{2}\Sigma_i\Sigma_j' U_{ij} = \tfrac{1}{2}N\Sigma_j' U_{ij}, \qquad (2.38)$$

where U_{ij} is the potential energy between the ith atom and the jth atom and Σ_j' represents a summation over all j except for $j = i$. Here we assume that the number of surface atoms is much smaller than the total number of atoms in the solid, or the net effect of the surface is small. Thus the binding energy of every atom in the solid is the same as the cohesive energy of the solid. The cohesive energy of solid Λ_{coh} should then be given by U_{tot}/N, or

$$\Lambda_{coh} = U_{tot}/N = \tfrac{1}{2}\Sigma_j' U_{ij}, \qquad (2.39)$$

which is exactly the same as the binding energy of a kink site atom Λ, or the sublimation energy, since a kink site atom has coordination numbers exactly one half those of a bulk atom. It is remarkable that even though atomic binding in metals is known to be not pairwise additive, yet the measured binding energy of a kink site atom, or the sublimation energy, is exactly equal to the cohesive energy of the solid, as would be expected from pairwise additive interactions of atoms in the solid. Further measurements of the critical energy deficits of low temperature field evaporated ions of various charge states, 1+, 2+ and 3+, for over 10 metals, including Cu, Be, Na, Fe, Ni, Rh, Mo, W, Pt, Ir and Au, etc., by Tsong *et al.*[79,80], using a high resolution pulsed-laser ToF atom-probe, confirm eq. (2.37) to within about 0.2 to 0.3 neV when the sublimation energy is replaced with the cohesive energy. Latest result with an improved pulsed-laser time-of-flight atom-probe gives the critical energy deficits for W^{3+} and Rh^{2+} as 45.36 eV and 22.46 eV respectively. The sublimation energy calculated from eq. (2.37) and values of I_i listed in Table 2.2 is 8.88 eV for W and 5.92 eV for Rh. These values agree with values of the cohesive energy derived from thermodynamic methods as also listed in the same table, which are 8.90 eV and 5.75 eV, respectively. These agreements can now be interpreted as follows: the binding energies of kink site atoms, or the sublimation energies, of over 10 metals, derived from the measurement of ion energy distributions in low temperature field evaporation of ions of different charge states, all agree with the cohesive energies of these metals to within the energy resolution of the measurements. On the practical side, it may be possible to measure the cohesive energy of materials by measuring the critical ion energy deficits in low temperature field evaporation. With the resolution of the atom-probe continuously being improved, it is now possible and should be most interesting to study the binding energy of adsorbed atoms which are deposited on crystal surfaces of different atomic structures, or to measure the binding energy of surface atoms with coordination numbers different from those of kink site atoms. For our purpose here, the critical energy deficits in low temperature field evaporation of metals can be calculated from eq. (2.37) by simply using the cohesive energy of metals for the sublimation energy term.

When field evaporation is done at very high temperature, the activation energy can be very high and thus can no longer be neglected from eq. (2.37). For example, Ernst[62] finds that in field evaporation of Rh, at presumably a nearly constant rate, when the tip temperature is changed from 100 to 600 K, the activation energy changes from 0.05 eV to about 1.5 eV, and the field strength drops from 4.1 to 1.7 V/Å. The activation energies for Rh^{+} and Rh^{2+} are found to be nearly equal. Ernst interprets this as evidence of having all the Rh atoms field evaporated as Rh^{+}, and Rh^{2+} ions are produced later by post field ionization. This is not really necessary. As both the + and 2+ Rh ions are produced at nearly the same rate at the same temperature, the activation energies of their formation must be nearly equal. The result indicates nothing more than this fact at the temperature and field strength of the experiment. Unfortunately, effects of field strength and surface temperature are not carefully isolated. As the activation energy changes, so will the critical distance x_c. Thus by assuming an explicit form of the field dependence of the ionic potential, the atomic potential can be mapped out. The limitation of this method lies in the very small range of the potential which can be mapped, and also the uncertainty of the effects of the field on the atomic and the ionic curves. The field strength changes continuously during the measurement. Further, unless it is certain that Rh^{2+} are produced by post field ionization, the charge state of field evaporating ions is not known, and the ionic curve is not certain either.

(e) *Atomic and ionic tunneling*

At very low temperature, the potential energy barrier in field desorption will become easier by tunneling than by overcoming it with thermal activation, especially if the atoms are very light. The possibility of field desorption by atomic and ionic tunneling was discussed by Gomer & Swanson,[50,51] and also by Tsong.[52] We will refer details to these original papers, and discuss here only some of the essential results. It can be shown that when atomic and ionic tunneling and thermal activation are both taken into account, the optimum field desorption rate is given by[52]

$$\varkappa_e \approx \left\{1 + A \exp\left[\frac{4}{27a^2(kT)^3}\right]\right\} \nu \exp\left(-\frac{Q_n}{kT}\right), \tag{2.40}$$

where A is a slow function of T and the barrier height and shape, and a is a barrier parameter depending also on the shape and the height of the potential barrier. An estimate has been given of the threshold temperature where the tunneling and thermal activation effects become equally probable.[16] According to this estimate, for light atoms such as Be, this temperature may be as high as 400 K, whereas for heavier atoms it should be in the teens. Of course, there is little reliable information available about the potential barrier in field desorption, and the originally estimated threshold temperatures may be a little too high. Nevertheless it is expected that when the tip temperature is in the tens the tunneling effect should be important, at least for light atoms such as H, D, and Be. The tunneling effect has been used[52] to explain a charge state shift in low temperature field evaporation of Be observed by Barofsky and Müller.[76a] One should be able to confirm the occurrence of a tunneling effect in field desorption rate. When the rate becomes nearly independent of the tip temperature, one is sure that the tunneling effect is becoming important.[76b] Despite the importance of this problem, so far no careful search of this effect in field evaporation of materials has been reported. It is worthwhile to determine experimentally the threshold temperature when the particle tunneling effect becomes as important as thermal activation for field evaporation of atoms of different materials. From such studies, at least, one can gain some information on the potential barrier in field evaporation.

2.2.5 Cluster ion formation

In low temperature field desorption, ions formed are almost always atomic ions. A very small percentage of cluster ions has occasionally been observed in the atom-probe, especially in field evaporation of compound semiconductors,[82] but the number is too small to make a good scientific study of their formation. Only with the advent of the pulsed-laser atom-probe has it been possible to produce and therefore study a significant fraction of relatively large cluster ions, up to about 10 atoms or slightly larger. While metal cluster ions[83] of large sizes are in general more difficult to obtain, cluster ions of covalent bond materials such as silicon,[84,85] compound semiconductors,[86] and carbon[87] can reach a size of over 10 atoms fairly easily. Formation of clusters and cluster ions is a subject of considerable scientific and technological interest.[88] It is realized that powder and particles are one of the most common forms of raw materials from which it is relatively easy to manufacture end products with good uniformity in their physical and chemical properties. Small particles are also known to be chemically and catalytically much more active. From a scientific point of view, small atomic clusters represent another form of matter which is intermediate between condensed matters, or solid states, and atomic and molecular states. In small atomic clusters, the fraction of atoms constituting surface atoms is exceptionally high. The most commonly employed techniques in atomic cluster studies use high temperature vaporization of atoms, either with an oven or by irradiation of a solid surface with high power laser pulses. The vaporized atoms are carried away by a carrier gas and expanded through a nozzle. Relatively large clusters can be formed in the nozzle expansion supercooled gas. In order to detect these neutral clusters they are ionized either by electron excitation or by resonance absorption of laser photons. There are many subjects of interest, such as magic numbers, critical numbers, ionization energies of clusters of different sizes, binding energies of atoms in clusters, structures of atomic clusters, electronic energy levels in clusters, chemical reactivities of clusters, etc. As cluster science is an emerging field of study, data on cluster properties are still very sketchy. Therefore pulsed-laser stimulated field desorption has been able to make valuable contributions.

Two scientific aspects in cluster science have been investigated using pulsed-laser field desorption technique with a considerable degree of success.[85] One is the critical numbers, which are the smallest numbers of atoms in multiply charged cluster ions when the ions can still resist being

'Coulomb exploded', or rather being dissociated by the repulsive Coulomb force of the like charges in the cluster ions. The other is the magic numbers, which are the numbers of atoms in clusters when they exhibit a greater relative stability. In pulsed-laser stimulated field desorption, ions produced are often multiply charged. Also, after the ions are desorbed, they no longer have any chance to interact with other energetic particles. Thus the abundances of different ion species represent how atomic clusters are desorbed from the surface, and therefore also the binding mechanisms of clusters, the relations to the binding energy of surface atoms in the crystal, and the absolute stability of the ion species. This method therefore focuses on the formation process of clusters from the surface. In Coulomb dissociation of multiply charged ions, there is no excitation of the ions by energetic particles, therefore the true stability of multiply charged ions in resisting Coulomb repulsive force induced fragmentation is studied. In contrast, cluster ions produced by other techniques are either singly charged or simply assumed to be singly charged, since in many atomic cluster studies the mass spectrometers used do not have enough mass resolution to separate, for example, A^+ from A_2^{2+}. Also, cluster ions can still interact with other atoms or clusters. If a thermal equilibrium can truly be reached in the carrier gas for these clusters, the abundance of clusters of different sizes represents the relative stability of these atomic clusters at the nozzle expanded gas temperature, provided that the method used for ionizing these clusters for the purpose of detecting them does not produce additional excitations or dissociations. Therefore the pulsed-laser field desorption method studies aspects and properties of cluster and cluster ions and their formation which differ slightly from those studied by the conventional techniques.

In pulsed-laser field desorption, larger cluster ions are more easily produced for covalent bond materials such as Si and C. When a Si tip is field evaporated with laser pulses of very low power density and a d.c. holding field close to the d.c. evaporation field, only Si^{2+} ions are formed. The energy distribution is comparable to those of metal ions, although the critical ion energy deficit does not agree with that expected from eq. (2.37). This is due to a photo-excitation effect, as will be discussed in the next section. When the laser power is increased so that the surface temperature can reach above ~1000 K, then in addition to doubly charged atomic ions, singly charged atomic ions and cluster ions of + and 2+ charges are formed, as seen in Fig. 2.15(*a*) to (*c*). It is found that under such conditions ions can be continuously produced without blunting the tip. Presumably Si atoms continue to migrate from the tip shank to the tip apex by temperature and field gradient induced surface diffusion. There are a few features of interest. First, most of the cluster ions are doubly

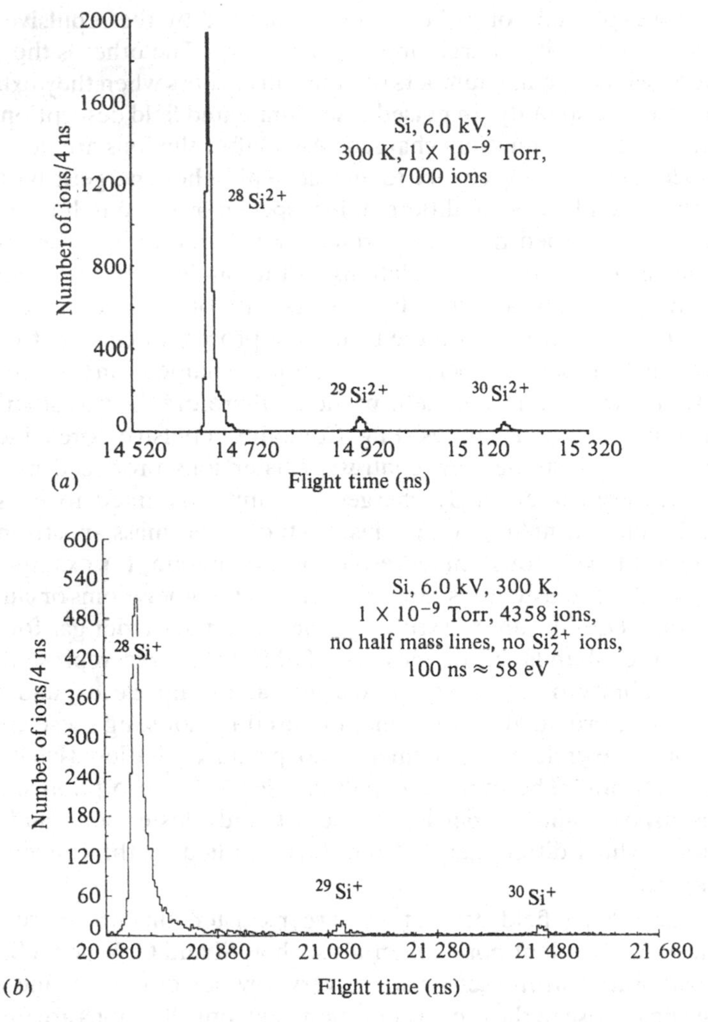

Fig. 2.15 (*a*) In low temperature pulsed-laser field evaporation of silicon, ions formed are all doubly charged. The energy distributions are very narrow, similar to those found in low temperature field evaporation of metals. However, the onset flight times are always shorter than the calculated values, indicating a photo-excitation effect as will be discussed in Sec.2.2.6.

(*b*) When the laser power is increased so that the tip is heated to over 1000 K, silicon field evaporates as 2+ and + atomic ions and cluster ions. Si^+ and cluster ions show long and persistent low energy tails, indicating that a lot of these ions are dissociation products of larger cluster ions. Si^{2+} distribution, on the other hand, remains sharp since they are not dissociation products of other ions.

(*c*) Cluster ions are mostly doubly charged, but Si_4^+ is also formed. The smallest doubly charged cluster ions are Si_3^{2+} and the most abundant ones are Si_4^{2+} and Si_6^{2+}.

(*d*) The situation is quite similar for carbon, except that now the most

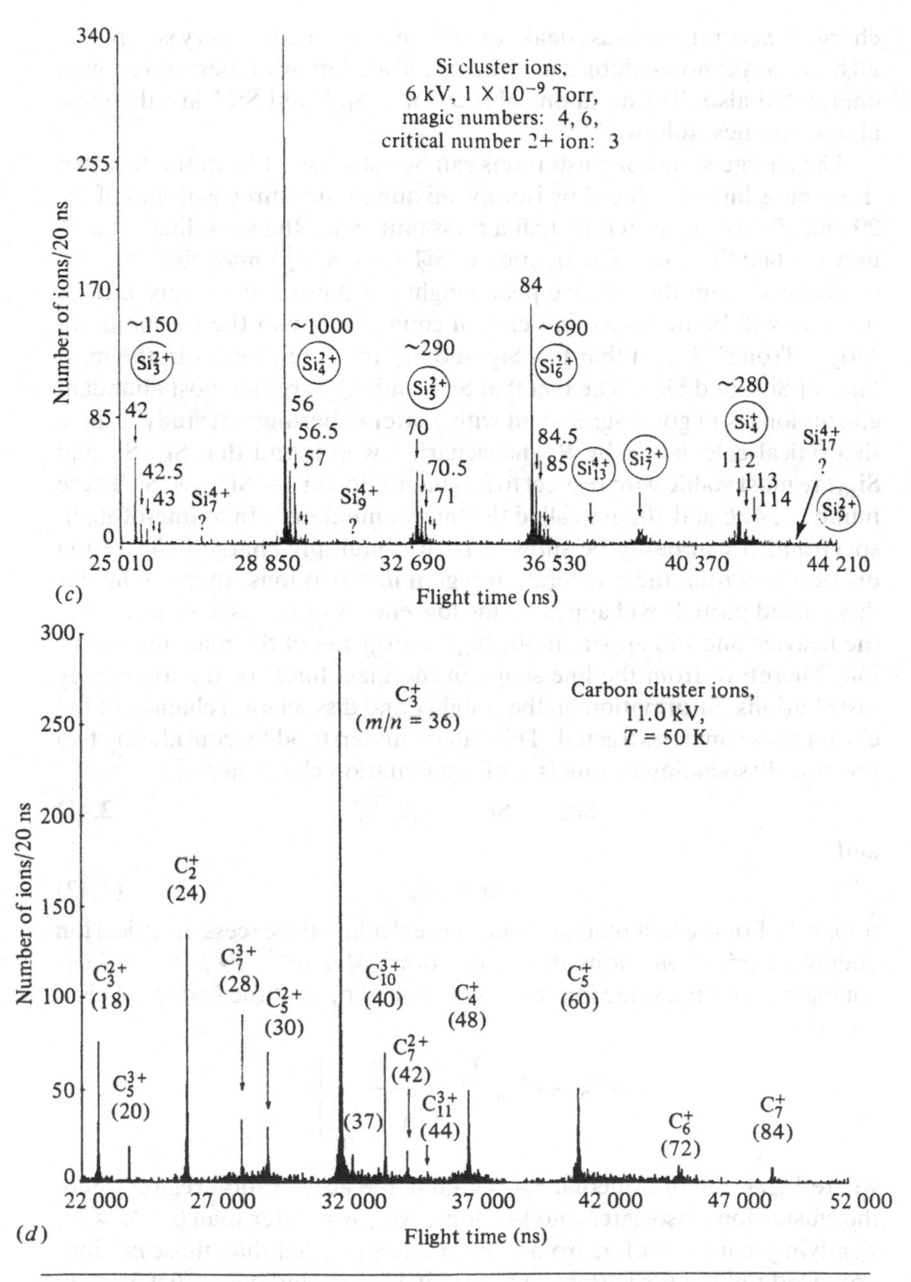

abundant charge state of the cluster ions is 1+, and the most abundant cluster ions are C_3^+ and C_5^+. Also there are a small number of doubly charged, odd numbered cluster ions and some triply charged cluster ions. For 3+ cluster ions, the larger the cluster, the larger the signal intensity, indicating the occurrence of Coulomb repulsive force induced dissociation of small 3+ clusters.

charged. Second, the mass peaks of Si^{2+} and Si_3^{2+} are still very sharp. But all other peaks now exhibit a low energy tail and many of them have a high energy tail also. Third, among cluster ions, Si_4^{2+} and Si_6^{2+} are the most abundant ones, followed by Si_5^{2+}.

The charge states of cluster ions can be established from the fact that those mass lines produced by isotope mixing (Si has three isotopes of 28, 29 and 30) are separated by half a mass unit. Still, the mass lines of Si_4^{2+} may contain Si_2^+ ions. The fraction of Si_2^+ ions in Si_4^{2+} mass lines may be established from the relative peak heights of these isotope mixed mass lines, as will be discussed in detail in connection with the formation of Mo_2^{2+}. Tsong[85] found that few Si_2^+ and Si_3^+ ions are present in the mass lines of Si_4^{2+} and Si_6^{2+}. The fact that Si_4^{2+} and Si_6^{2+} are the most abundant cluster ions is in good agreement with a later experimental study[89] and a theoretical calculation by Raghavachari,[90] who found that Si_4, Si_6 and Si_{10} are most stable with respect to fragmentation $Si_n \rightarrow Si_{n-1} + Si$. These numbers, 4, 6 and 10, are called the 'magic numbers'. In a time-of-flight spectrum, it can easily be shown that if a multiply charged cluster ion dissociates within the acceleration region into two ions, then the lighter dissociated particle will appear in the low energy tail of its mass line while the heavier one will appear in the high energy tail of the mass line of the ion. Therefore from the line shape of the mass lines, or the ion energy distributions, information on the stability and dissociation channel of the cluster ions can be extracted. This can be understood by considering two possible dissociation channels, or fragmentation channels,

$$Si_m^{n+} \rightarrow Si^+ + Si_{(m-1)}^{(n-1)+}, \tag{2.41}$$

and

$$Si_m^{n+} \rightarrow Si + Si_{(m-1)}^{n+}, \tag{2.42}$$

for $n = 2$. For the first reaction, one may calculate the excess in critical ion energy *deficit* of m^+ ions in the reaction $(M + m)^{2+} \rightarrow M^+ + m^+$, as compared with the same ion species produced right at the surface. This is given by

$$\Delta E_m = eV_{\text{int}} \left[1 - \frac{2}{\left(1 + \frac{M}{m}\right)}\right], \tag{2.43}$$

where V_{int} is the intermediate potential in the acceleration region where the cluster ion dissociates into two ions. ΔE_m is greater than 0 if $M > m$, signifying that m^+ suffers from a larger energy deficit than those m^+ ions produced right at the surface. $\Delta E_m = 0$ if $M = m$, and $\Delta E_m < 0$ if $M < m$. The last condition indicates that those dissociated heavier ions actually have a larger energy than the same ion species produced right at the surface. Thus they will appear in the high energy tail of the mass line. For dissociative reaction eq. (2.42), it is fairly obvious that both the dissociation products will have less energy than the same species produced right at the surface. Now let us examine Fig. 2.15. Both Si^{2+} and Si_3^{2+} exhibit sharp energy distributions without either a low energy tail nor a high energy tail. These ion species are therefore produced right at the surface and are not produced by dissociation of cluster ions. Si^+, on the other hand, shows a long and persistent low energy tail. Ions in the tail are definitely dissociation products of other cluster ions. In a time-of-flight spectrometer with a straight flight tube, dissociated neutral atoms, as long as they have already gained sufficient energies, can also be detected as Si^+. Thus the low energy tail of Si^+ mass line really contains both the dissociated Si^+ ions and the neutral Si atoms. As Si^+ is the lightest of all dissociated ion species, it cannot have a high energy tail as just discussed. Indeed, no high energy tail is found. All other cluster ion species show both high and low energy tails. When the dissociation channel is by emission of a neutral Si atom then both the Si atom and the other cluster ion will have a larger energy deficit, or both of them will appear in the low energy tails. The high energy tails, however, can be produced by dissociation of a multiply charged cluster ion into two ions of lesser charges and lesser masses. The heavier ion will then have excess kinetic energy. It is possible that the high energy tail of, for example, Si_4^{2+} really contains Si_2^+ ions produced by the dissociation reaction: $Si_3^{2+} \rightarrow Si_2^+ + Si^+$.

Formation of another covalent bond material, carbon, has also been extensively studied by many investigators using different techniques. Although silicon and carbon are both covalent bond materials, and therefore can be expected to behave similarly, both theoretical calculations and experimental observations seem to show considerable differences in their behaviors. For silicon, the most stable small clusters in resisting dissociation are 10-atom, 6-atom and 4-atom clusters. For carbon, they are 3-atom, 5-atom and 7-atom clusters. Their fragmentation energies, calculated by Raghavachari,[91] are shown in Fig. 2.16. In pulsed-laser stimulated field desorption, such differences can be observed. Tsong & Liu[84,87] found that, in general, in pulsed-laser stimulated field desorption of silicon and carbon, the abundances of cluster ions decrease exponentially with the cluster sizes. However, in the case of Si, Si_4^{2+} and Si_6^{2+} stand out, while in the case of carbon, singly charged and doubly charged C_3, C_5 and C_7 ions stand out, in excellent agreement with the theoretical calculations[91] and also with many other experimental observations.[92] In contrast to silicon where most cluster ions are doubly charged and no 3+ ions are found, Liu & Tsong found that carbon ions are mostly singly charged and a small fraction of them is either doubly or triply charged. Table 2.3 lists the relative abundances of cluster ions of

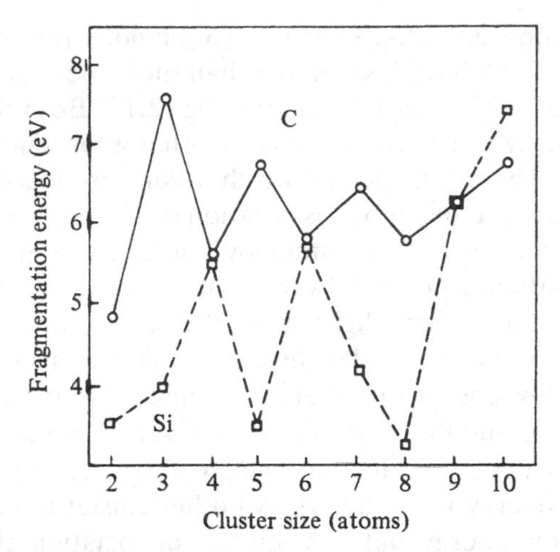

Fig. 2.16 Fragmentation energies for small silicon and carbon clusters as functions of cluster size, as calculated by Raghavachari. The dashed curve represents the fragmentation energies as a function of cluster size n for the reaction $Si_n \rightarrow Si_{n-1} + Si$. Note the most stable clusters are Si_{10}, Si_6 and Si_4. The solid curve indicates the fragmentation energies of small carbon clusters as a function of cluster size n for the reaction $C_n^+ \rightarrow C_{n-1}^+ + C$. Odd number clusters are more stable and have linear chain structures.

various sizes and charge states in pulsed-laser stimulated field desorption of Si and C obtained by Tsong & Liu.

Cluster ions are often observed in field evaporation of compound semiconductors such as GaAs and GaP etc. In high voltage pulse field evaporation of GaAs and GaP, only a few per cent of cluster ions are

Table 2.3 *Relative abundances of cluster ions in pulsed laser stimulated field evaporation*

Carbon			Silicon	
+	2+	3+	+	2+
C_3(1.00)	C_3(0.18)	C_{10}(0.13)	Si_4(0.08)	Si_4(1.00)
C_2(0.33)	C_5(0.09)	C_7(0.06)		Si_6(0.47)
C_5(0.25)	C_7(0.05)	C_5(0.04)		Si_5(0.15)
C_4(0.18)		C_{11}(<0.01)		Si_3(0.09)
C_7(0.03)				Si_7(0.06)
C_6(0.03)				
C_9(0.01)				
C_8(<0.01)				

observed for As and P.[93,94] On the other hand, in pulsed-laser stimulated field evaporation, a large amount of cluster ions can be formed; the abundances of cluster ions again depend on the laser intensity, or the degree of pulsed heating of the tip. The cluster ions can be as high as 30% of the total ions.[95] Ga atoms seem to field evaporate exclusively as singly charged atomic ions while As and P atoms can be field evaporated as 1+ and 2+ atomic and cluster ions. Cluster sizes of up to about 5 have been observed. The main object of these studies is the field evaporation behavior and the reliability in the compositional analysis of compound semiconductors, as will be discussed further in Chapter 4.

For multiply charged ions, there is the smallest cluster size below which the cluster ion will be 'Coulomb exploded'. The size has been found by the University of Konstanz group to be as large as 40 to 50 for van der Waal clusters and about 10 for metal clusters.[96] On the other hand, many other experiments, such as in liquid metal ion sources[97] and ion sputtering of solid surfaces,[98] quite often found doubly charged 3-atom cluster ions of metals. For silicon and carbon, the smallest 2+ cluster ions are also 3-atom clusters, as can be seen from Table 2.3. That Coulomb repulsive force can really induce dissociation, or fragmentation, of cluster ions is most evident in the relative abundances of 3+ carbon cluster ions. Whereas the relative intensities of singly and doubly charged cluster ions of different sizes in general decrease exponentially with the cluster sizes, for 3+ carbon cluster ions this trend is reversed, as can be seen from Table 2.3. Thus C_{10}^{3+} is more abundant than C_7^{3+}, which is in turn more abundant than C_5^{3+}. This reversed abundance is obviously due to Coulomb dissociation of small 3+ cluster ions.

For metals, the smallest doubly charged cluster ions found are diatomic cluster ions of Mo, as shown in Fig. 2.17.[99] In field evaporation, as the temperature of the tip surface is gradually increased and the field decreased, the charge states of the ions will shift toward lower ones. For Mo, before formation of Mo^+, Mo_2^{2+} is formed and dominates the spectrum. The tip temperature is around 1000 K and the field strength is around 2.5 to 3.0 V/Å. As Mo^+ and Mo_2^{2+} have the same mass-to-charge ratios, it is worthwhile explaining how these two ion species can be distinguished and can have their relative intensities established. Mo has 7 isotopes of masses 92, 94, 95, 96, 97, 98 and 100 u. If the mass spectrum contains only Mo^+ ions then the spectrum should contain seven mass lines of mass-to-charge ratios of 92 to 100 u. On the other hand, if the ion species is Mo_2^{2+}, then a random mixing of the 7 isotopes will result in formation of 15 mass lines of mass-to-charge ratios ranging from 92 to 100 at intervals of 0.5 u but with 92.5 and 99.5 absent. The time-of-flight mass spectrum clearly shows these 15 mass lines. This does not mean that there are no Mo^+ ions in these mass lines. To find the fractional abundances of

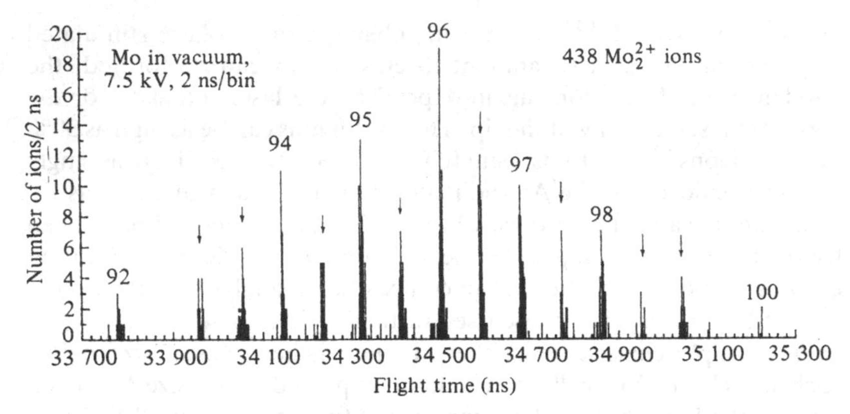

Fig. 2.17 Portion of a pulsed-laser time-of-flight spectrum showing the formation of doubly charged Mo diatomic cluster ions in pulsed-laser stimulated field evaporation of Mo. From a multinomial expansion analysis, it is concluded that few Mo^+ ions are formed. Although there are a total of only 438 ions in the entire spectrum, all the expected 15 mass lines due to an isotope mixing of the 7 Mo isotopes are present. A mass line, even when it contains only three ions such as that of $M/n = 100$, is clearly identifiable.

Mo^+ and Mo_2^{2+} one may resort to a multinomial expansion coefficient analysis. Let us represent the fractional abundances of the 7 Mo isotopes by a, b, c, d, e, f and g. The calculated and experimental fractional abundances of the 15 mass lines in Mo_2^{2+} are listed in Table 2.4. When the calculated abundances are compared with the experimental abundances, one reaches the conclusion that the spectrum shown in Fig. 2.17 contains few if any Mo^+ ions. If on the other hand, a fraction of the ions are Mo^+, say p, then the relative abundances of each mass line can also be calculated. For example, that of $M/n = 92$ should be $[pa + (1 - p)a^2]$, and that of $M/n = 94$ should be $[pb + (1 - p)(b^2 + 2ad)]$, etc. Thus the fraction p can be obtained by best fit of theoretical abundances and experimental abundances of different M/n mass lines.

The Mo_2^{2+} mass lines are extremely sharp, indicating that these ions are produced right at the surface, not by dissociation of other ions. Since there is no low energy tail in the seven mass lines (each of them is a high resolution ion kinetic energy distribution) corresponding to those of Mo^+, there is no dissociation of the cluster ions in the acceleration region. From this, it is concluded that the life time of Mo_2^{2+} ions is greater than 100 ns at a temperature of about 1000 K. Thus the barrier height for Coulomb dissociation is at least $kT \cdot \ln(100\text{ ns}/1 \times 10^{-13}\text{ s}) = 1.2\text{ eV}$. This is a very large barrier and the ion must be considered stable regardless of whether the binding state is in an absolute minimum or a relative minimum of the interatomic potential curve. Pulsed-laser field evapor-

Table 2.4 *Fractional abundances of ion species in* Mo_2^{2+}

M/n	Abundance	Calculated (%)	Experimental (%)
92	a^2	2.2	2.7 ± 0.7
93	$2ab$	2.8	3.4 ± 0.9
93.5	$2ac$	4.7	5.7 ± 1.1
94	$b^2 + 2ad$	5.8	6.8 ± 1.2
94.5	$2ae + 2bc$	5.8	5.3 ± 1.1
95	$c^2 + 2af + 2bd$	12.8	12.3 ± 1.6
95.5	$2be + 2cd$	7.1	7.8 ± 1.3
96	$d^2 + 2ag + 2bf + 2ce$	13.2	15.1 ± 1.7
96.5	$2cf + 2de$	10.9	10.0 ± 1.4
97	$e^2 + 2bg + 2df$	10.8	10.7 ± 1.7
97.5	$2cg + 2ef$	7.7	4.6 ± 1.0
98	$f^2 + 2dg$	9.0	7.8 ± 1.3
98.5	$2eg$	1.8	1.6 ± 0.9
99	$2fg$	4.6	3.4 ± 0.9
100	g^2	0.9	0.7 ± 4

ation experiments thus overthrow some of the earlier results on critical numbers in 'Coulomb explosion', although Coulomb dissociation of 3+ carbon cluster ions is now definitely established.

An attempt has also been made to derive the binding energy of atoms in clusters from a measurement of the critical energy deficit of cluster ions. For $n+$ cluster ions of m atom size, from consideration of Born–Haber energy cycle, the critical ion energy deficit can be easily shown to be given by[100]

$$\Delta E_c^{n+}(m) = m\Lambda + \Sigma_i I_i' - E_b(m) - n\phi_{av} - Q, \qquad (2.44)$$

where I_i' is the ith ionization energy of the neutral cluster and $E_b(m)$ is the total binding energy or the total cohesive energy of the cluster. Thus, in principle, if values of I_i' for different i's are known, the binding energy of atoms in clusters of different sizes can be derived from a measurement of the critical energy deficits of these cluster ions. At present, however, such data are incomplete. In addition, in pulsed-laser field desorption, clusters are formed only if the tip temperature is raised sufficiently. Thus Q cannot be neglected. Accurate determination of Q is not an easy problem. One may also eventually be able to find favorable conditions for producing cluster ions, and to predict their charge states using a calculation similar to those of Brandon and Tsong for atomic ions. On the basis of the image-hump model, the favorable cluster ion species should have the lowest evaporation field as given by[101]

$$F_e^{n+}(m) \simeq [\Delta E_c^{n+}(m)]^2/n^3e^3. \qquad (2.45)$$

At the present time, such calculations are impossible because of the lack of data for the needed parameters such as I_i' and $E_b(m)$, etc.

2.2.6 Photon stimulated field desorption

Field desorption from a tip surface can be greatly enhanced both by laser pulses and by a synchrotron radiation source. This is of course how a pulsed-laser time-of-flight atom-probe operates. The question is, what effect causes the enhancement in the field desorption rate. Two effects can be considered. The simplest possible effect is by heating of the emitter surface. Another effect is by direct photon induced electronic excitations. In their first development of a pulsed-laser imaging atom-probe, Kellogg & Tsong observed,[102] besides enhanced field desorption by laser pulses, a thermal and field induced surface migration of surface atoms, or the directional walks[103] of surface atoms if the laser power is high and the applied field low, so that the existence of temperature effects is obvious. In that paper, the tip temperature under the irradiation of laser pulses is estimated from the lowering of the evaporation voltage of the tip surface. For the normal operation of the pulsed-laser atom-probe, the temperature rise can be less than 100 K, if the d.c. applied field is sufficiently close to the desorption field at the tip temperature. Quantitative evidence of this simple thermal effect can be found in the ion energy distribution of pulsed-laser field desorbed inert gas atoms, in which field adsorbed atoms are thermally or flush desorbed in ns.[39] The ion energy distributions are identical to those found in ordinary field ionization in the critical ion energy deficits, the distribution widths, and the secondary resonance tunneling peak structures. There is no manifestation of photon energies in these ion energy distributions. Thus there is no effect of a direct electronic excitation by photons in pulsed-laser stimulated field desorption from metal surfaces if the photon intensity is not excessively high. This is, however, not the case when the tip is a semiconductor such as silicon or GaAs.

Using a high resolution pulsed-laser time-of-flight atom-probe, it is now possible to achieve 1 to 2 parts in 10^5 accuracy in the measurement of the critical ion energy deficit in field desorption, which corresponds to about 0.1 to 0.2 eV out of a total ion kinetic energy of about 10 keV, more than sufficient to detect ion energy changes by photo-excitation effects. It is found that, irrespective of the laser intensity, the ion kinetic energy distributions in pulsed-laser field desorption of semiconductors such as Si or GaAs always show smaller critical energy deficits than values calculated using eq. (2.37). This excess energy can be from several to over 10 eV even when the laser intensity is very low where field evaporation of metals will not show any discrepancy in critical energy deficits from the same equation. An example is shown in Fig. 2.18(*a*). Ions with excess energy can only be produced by direct photo-excitation. As no peaks corresponding to the photon energy have been found, the excitation is a solid state effect, perhaps involving a direct electronic excitation from the atomic state to vacant states in the conduction band of the surface as discussed in Fig. 2.8. In field desorption, when the desorbing atom reaches x_c by thermal vibration, the energy levels of atomic electrons are just sufficient to be transferred to the conduction band edge of the semiconductor surface. x_c in this case has to be calculated by replacing the work function ϕ_{em} with the electron affinity χ of the semiconductor surface. One may ask why the same process cannot be observed in field desorption from metal surfaces. The answer is similar to that discussed in

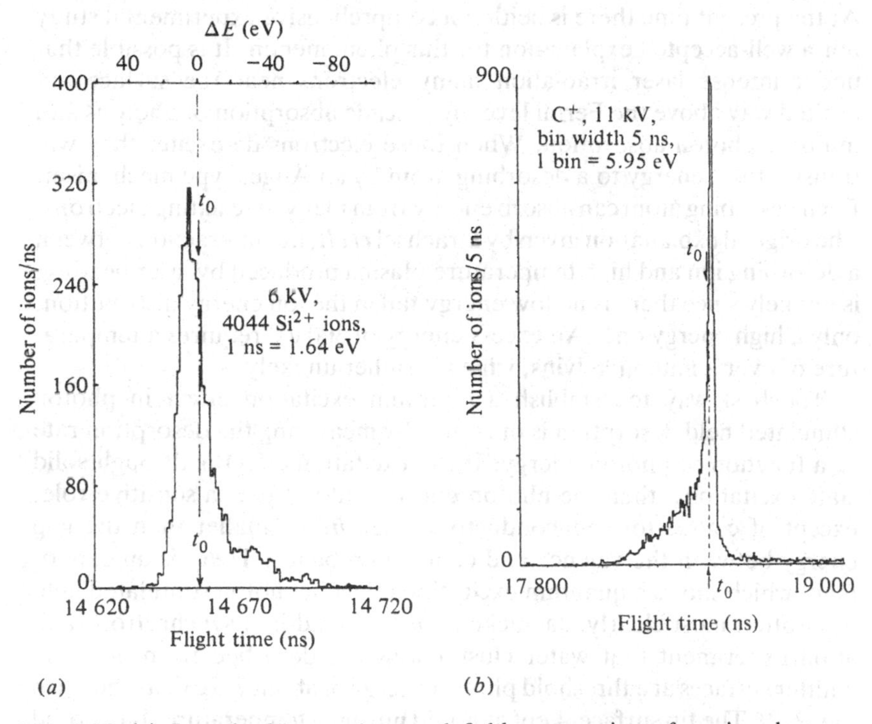

Fig. 2.18 (*a*) In pulsed-laser stimulated field evaporation of semiconductors, even when the laser power is very low, where metal ions will not show any excess kinetic energy, semiconductor ions always show excess kinetic energy of over several eV. An example is shown here for Si^{2+}. The onset flight time, when there is no photon-excitation, is indicated by t_0.

(*b*) If the laser power is very intense, then the excess energy can amount to over 300 eV irrespective of whether it is a semiconductor or a metal. An example is shown here for C^+.

connection with photon stimulated field ionization. Even though photon excitation may be able to produce an ion, it will be reneutralized right away before it can be accelerated away from the surface.

Another strange photo-excitation effect has been observed. Drachsel *et al.*[83] find that if the laser power is sufficiently high then the ion energy distribution of pulsed-laser stimulated field desorbed metal ions contains a high energy peak which is well separated from the regular field desorption peak. The excess energy can be as large as a few hundred eV, and this excess energy increases with laser power intensity. Tsong[30,70] did not find a well separated peak, but he did find a high energy tail. Again, the tail can extend over a few hundred eV when the laser intensity is increased. These observations literally mean that under intense laser irradiation, ions are ejected out of the surface with an initial kinetic energy of a few hundred eV. An example is shown in Fig. 2.18(*b*) for C^+. At the present time there is neither a comprehensive experimental study nor a well-accepted explanation for this phenomenon. It is possible that under intense laser irradiation many electrons near the surface are excited way above the Fermi level by cascade absorption of photons and multiple photon absorption. When these electrons de-excite, they will transfer their energy to a desorbing atom by an Auger type mechanism. Each desorbing atom can absorb energy from many de-exciting electrons. The original explanation given by Drachsel *et al.*, i.e. interaction between a desorbing ion and high temperature plasma produced by laser heating, is unlikely since there is no low energy tail in the ion energy distribution, only a high energy one. An excess energy of 300 eV requires a temperature of over a million kelvins, which is rather unlikely.

The best way to establish a quantum excitation effect in photon stimulated field desorption is of course by measuring the desorption rate as a function of photon energy. If the excitation effect is through solid state excitations, then the photon energy will not play a sensitive role, except of course for semiconductors when $h\nu$ is smaller than the gap energy between the valence and conduction bands. There is an experiment which shows a quantum excitation effect in photon stimulated field desorption most clearly. Jaenicke *et al.*[104] found in a synchrotron radiation experiment that water cluster ions are desorbed from field ion emitter surfaces at a threshold photon energy of about 7.5 eV, as shown in Fig. 2.19. The tip surface, kept at liquid nitrogen temperature, is exposed to about 10^{-9} Torr of water, and the desorbed ions are found to be protonated water cluster ions. The emitter surface is presumably covered with a thick ice layer of several tens to a few hundred Å in thickness, thus no effect of the tip material is found. The threshold energy is found to agree, within experimental uncertainties, with the appearance energies of forming protonated water cluster ions in d.c. field desorption and

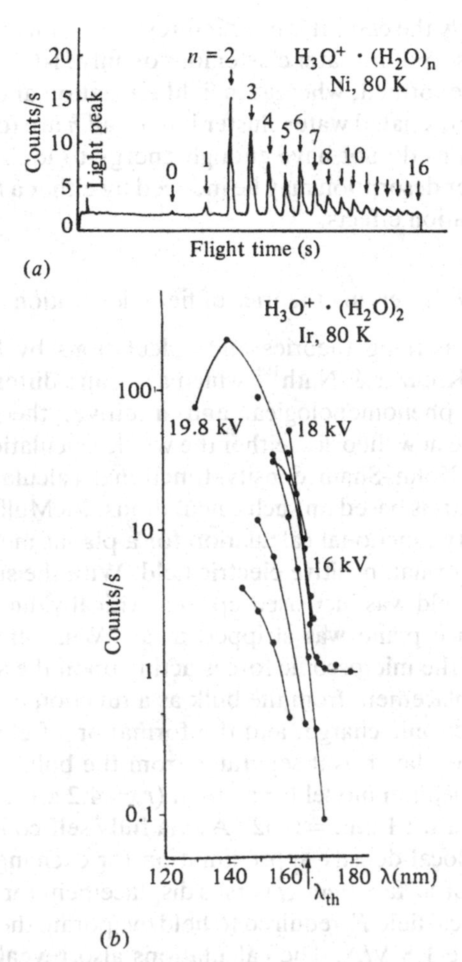

Fig. 2.19 (*a*) Time-of-flight spectrum in photon stimulated field desorption of water using synchrotron radiation, obtained by Jaenike *et al.*[104]

(*b*) Measurement of the threshold photon energy in photon stimulated field desorption of water, or rather of ice. The threshold photon wavelength is found to be ~160 nm or $h\nu \simeq 7$ eV.

pulsed-laser stimulated field desorption.[100,101] The appearance energy is the sum of the critical ion energy deficit and the work function term. This energy is thus equivalent to the ionization energy if no rearrangement of atoms occurs in the adsorption–desorption process. This agreement strongly suggests that photon stimulated field desorption is involved with a direct electronic transition between the ground state of the molecule and a vacant state at the conduction band edge of the surface. If it were not for the rearrangement of atoms of the adsorbed and desorbed species, this would be exactly the case. It is most interesting to note that in photon stimulated field desorption, water cluster ions of only $H_3O^+ \cdot (H_2O)_n$ with n greater than 2 are formed, whereas in field ionization and pulsed-laser field desorption, protonated water cluster ions of all n are formed. In this regard, laser photons do not have enough energy to ionize water molecules, and the laser desorption may be induced by either a heating effect or solid state excitation effects.

2.2.7 Emerging theories of field desorption

There are some emerging theories and calculations by McMullen & Perdew[105] and by Kreuzer & Nath[106] which are quite different from the accepted, though phenomenological and intuitive, theories of field desorption. In these new theories, either the whole calculations are based on self-consistent Kohn–Sham density–functional calculations, or the field penetration part is based on such calculations. McMullen & Perdew performed a density functional calculation for a planar metal surface in the presence of a normal, positive electric field. With the subsurface ion layers frozen, the field was increased up to a critical value at which the entire surface lattice plane was stripped away. What emerged was a detailed picture of the microscopic forces acting upon the surface layer, its equilibrium displacement from the bulk as a function of the field, the distribution of electronic charge, and the formation of electronic resonances in the surface layer as it separates from the bulk. These effects were studied for a jellium model for sodium ($r_s = 4.2$ a.u., surface layer thickness $d = 5.65$ a.u.; 1 a.u. = 0.529 Å) via fully self-consistent calculations within the local-density approximation for exchange and correlation. From plots of surface energy versus displacement for several fields of interest, the critical field F_c required to field evaporate the rigid surface layer is found to be 1.8 V/Å. The calculations also reveal information about electronic-charge distribution, electronic resonances which develop with increasing separation of the surface layer from the bulk, and various components of the surface-layer-bulk binding force. A similar calculation for F_c with parameters appropriate for A1 (111) surface finds it to be 4.5 V/Å. These calculations are very complicated, and although based on the rigorous formalism of the jellium model, they give results in poor agreement with experimental observations, at least for the evaporation field. Experimentally, aluminum field evaporates around or slightly lower than the best image field of neon, or less than 3.3 V/Å, in also poor agreement with the simple calculations of Brandon, and Tsong; they give a field of 1.9 V/Å in one case and 1.61 V/Å in another case. For Na there are no data available, but, realistically, the evaporation field is unlikely to be greater than 1 V/Å. The simple calculations give a field of 1.1 V/Å.

Also so far, the new theory does not consider the possibility of field evaporation as multiply charged ions. The advantage of the new method lies in the possibility of investigating how electronic-charge distribution changes as the atoms are field evaporated from the surface, and also how the binding force changes as the atoms are removed from the surface.

Kreuzer & Nath[106] have calculated the electronic properties of the metal within a tight binding cluster approach on the basis of the ASED-MO (atom superposition and electron delocalization molecular orbital) method,[107] with local electric fields taken from self-consistent jellium calculations.[108] These calculations use density-functional methods, treating positive ion cores as a structureless jellium of uniform positive charge to which an electron gas reacts within the local density approximation. Different metals are represented by different values of r_s, the radius of the Wigner–Seitz cell. Using an embedded 4-atom cluster with the evaporating atom at the apex of a tetrahedral structure, the potential energy as a function of distance in the electric field is calculated. A field evaporation field, F_{ev}, is defined as the field where the atomic potential curve no longer has an activation barrier. Values of F_{ev} obtained for Fe, Nb, Rh and W for different charge states are calculated and are listed in Table 2.5 together with cohesive energies calculated by this method for these metals. A simple scaling law for the evaporation field strength is also given to be

$$F_{ev} = \frac{3V_c}{2ne\left(\dfrac{9\pi}{4}\right)^{1/3} r_s^{1/2}}, \tag{2.46}$$

where V_c is the cohesive energy of the metal. Values calculated from this simple equation are also listed in Table 2.5. In general, except for Al, the agreement with experimental value is no better than the simple calculations of Brandon and Tsong. On the other hand, the field dependence of activation energy obtained with these calculations agrees with the experimental result of Kellogg[109] much better than that obtained with the simple image-hump model.

2.3 Field adsorption

In an applied field of the order of a few V/Å, gas atoms and molecules, which normally will not adsorb on a surface at a given temperature, may adsorb on the surface by an effect of the applied field. Field adsorption can occur at a temperature much higher than that in ordinary adsorption. For example, He and Ne can be field adsorbed on tungsten surfaces at a temperature as high as 100 K with a probability of near one, whereas in ordinary physisorption the probability is nearly zero at such a tempera-

Table 2.5 *Evaporation fields based on cluster calculation (cc) and from the scaling law (sl)*

Ion species	r_s	F_{ev} (cc) V/Å	F_{ev} (sl) V/Å	Experimental value V/Å	n for scaling law
Be^{2+}	1.87		3.2	3.4	1
Al^{+}	2.07		3.5	~3.3	1
Fe^{2+}	1.33	5.4	4.3	3.5	2
Fe^{2+}	3.00	2.6	2.9		1
Cu^{+}	2.67		2.8	~3.0	1
Zn^{2+}	2.3		1.33		1
Nb^{2+}	3.07	3.4	3.8	3.5	2
$Mo^{2+,3+}$	1.61		5.2	4.6	2
Ru^{2+}	1.41		4.2	4.0	2
$Rh^{2+,3+}$	2.81	2.9	4.1	4.8	1
Ag^{+}	3.02		2.5	2.5	1
W^{3+}	1.5	8.0	6.5	5.7	2
	2.06	6.0	5.6		2
	3.0	4.0	4.6		2
Re^{3+}	1.5		4.7	4.8	2
$Ir^{3+,2+}$	1.42		6.4	5.3	2

ture. Field adsorption was invoked to explain a time-of-flight atom-probe experiment by Müller *et al.*[110] They found that when a tungsten tip was field evaporated in the presence of an image gas, the mass spectrum also contained ions of the image gas atoms. As the high voltage pulses used in field evaporation were only about 10 ns in width, the chance of detecting a hopping image gas atom should be negligibly small. The image gas ions could come only from atoms in the adsorbed state. McKinney & Brenner[111] found that image gas ions could be detected in an image gas pressure as low as 10^{-9} Torr, further supporting the idea of field adsorption. Initially, the adsorption was thought to arise from the polarization binding term, $-\frac{1}{2}\alpha F_0^2$. However, it was soon realized that this energy will produce only a hopping motion of image gas atoms unless the temperature is exceptionally low, when field induced condensation will occur. Field adsorption was explained when Tsong & Müller[112] showed that the short range interaction needed for field adsorption could arise from a field induced dipole–dipole interaction between an image gas atom and a surface atom, and the degree of coverage of field adsorption, or equivalently the probability of field adsorption of a surface adsorption site, exhibited a behavior identical to that of the Langmuir adsorption isotherms, as will be discussed in the next sections.

2.3.1 Dynamics of field adsorption

Let us consider first a phenomenological theory of field adsorption. A physical surface is not a completely flat surface but rather has atomic structures, atomic steps, etc. When an external electric field is applied, the field is by no means uniform everywhere as a jellium model of metal surface would imply. Instead, the field is enhanced near a protruding atom, which may be either a kink site atom or an atom in the high index plane. A good approximation of the field distribution above such an atom is to consider the atom a small conducting sphere embedded in a positive and negative jellium, exposing only about half of the sphere outside the surface. If one uses the classical picture of a metal surface, or neglects considering the spilling of negative charge jellium into the vacuum as needed to minimize the potential energy of the surface and also the suppressing of negative jellium into the bulk for producing the applied positive field, then the field distribution above the surface atom is exactly that of a dipole field superimposed on a uniform field. The field induced dipole moment of a surface atom, regardless of whether it is a kink site atom, a ledge site atom, or a surface atom sitting in the middle of a plane, should be proportional to the applied field strength at least to the first order approximation. This proportionality may be called the 'effective' polarizability, α_M, of the surface atom. Its value should depend on the adsorption site of the atom, and in principle can be derived from a measurement of the field dependence of the field desorption rate as discussed earlier. As to adatoms, their effective polarizability can also be derived from a directional walk experiment as will be discussed in Chapter 4. This model is schematically shown in Fig. 2.20. If we consider only one surface atom, in what is usually referred to as an isolated-atom model, then the binding energy H is given by[112]

$$H \simeq \tfrac{1}{2}\alpha_A(f_A - 1)F^2, \tag{2.47}$$

where

$$f_A = \left[\frac{1 + \dfrac{2\alpha_M}{d^3}}{1 - \dfrac{4\alpha_A\alpha_A}{d^6}}\right]^2, \tag{2.48}$$

α_A is the polarizability of the gas atom, values of which are accurately known, α_M is the effective polarizability of the surface atom, and d is their separation. Using the polarizability data obtained for the kink site atoms from a measurement of the field dependence of the field evaporation rate of the W (110) layer, one finds that the short range binding energy of He on W surfaces at the best image field is about 0.2 to 0.3 eV, which is in fair agreement with experimental measurement, as will be discussed in

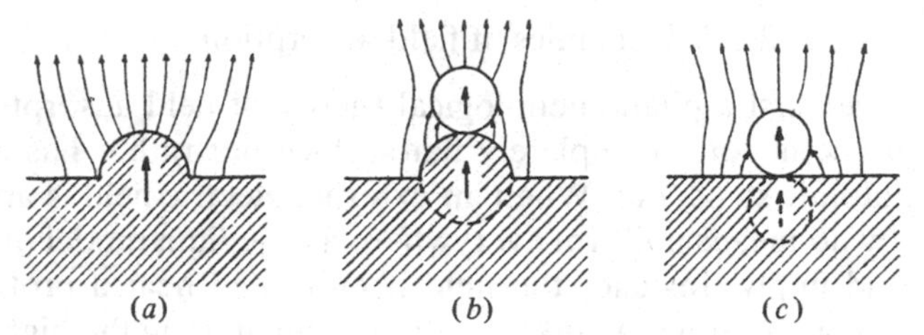

Fig. 2.20 (*a*) Schematic diagram showing the field distribution above a protruding surface atom. This distribution is the superposition of a uniform applied field and a dipole field.
(*b*) A short range binding energy in field adsorption arises from a field induced dipole–dipole interaction between an image gas atom and a surface atom.
(*c*) A polarized gas atom can also interact with its own image on a plane.

greater detail in the next section. This model predicts the field adsorption sites to be the apex of atoms in the more protruding positions, in agreement with pulsed-laser field desorption images of helium taken with a pulsed-laser imaging atom-probe.[113] The field desorption image is found to be in an image spot to image spot correspondence with the field ion image. This is possible only if He ions are desorbed from the atop sites.

A real surface is composed of an array of atoms.[60] Both the discreteness of the charge distribution, or the monopole effect, and the depolarization effect due to the charge array have to be considered.[114] Equations (2.47) and (2.48) overestimate the binding energy of field adsorption by a few per cents for He atoms and by as much as 40% for image gas atoms with large polarizability, such as Ar, on a non-loosely packed surface. On the other hand, there is another force which will contribute about the same amount of energy to field adsorption. Even on a completely flat jellium surface, a polarized atom can interact with its own image, as shown in Fig. 2.20(*c*). This contribution depends on the polarizability of the image gas atom and the applied field strength. For He in a field of 4.5 V/Å, this additional binding energy is about 0.015 eV, or about 10% of the field induced dipole–dipole interaction. Thus the effect of the charge array is partly offset by the image interaction.[115]

More sophisticated theoretical calculations have been reported by Nath *et al.*[116] In their model, the originally rather deep ground state of the field adsorbed inert gas atom is lifted close to the Fermi level by the applied field. Interaction between the field adsorbed atom and the surface occurs similar to that in ordinary chemisorption. The surface is represented by atom clusters of 4 to 14 atoms which are immersed in a jellium of positive charges and negative charges in an applied field. The variation of field strength in the near surface region, or the effect of field penetration, is derived from self-consistent density functional calculations. The field dependence of the binding energy derived with this tight binding cluster calculation, after some adjustment, is in good agreement with experimental data obtained by Tsong & Müller,[112] by Kinkus & Tsong[117] and by Ernst *et al.*[42] using a variety of methods as shown in Fig. 2.21. One must be aware that the accuracy of the data is still very limited. The latter authors have also obtained a good fit of their data to the theoretical curve when the distance between the image plane and the nuclear plane of the surface layer is properly adjusted. All these experimental data also fit well to the isolated-atom model with an effective polarizability of surface atoms of about 4 to 5 Å^3. Field adsorption of hydrogen has also been studied in detail.[118,119] Kellogg finds that the field adsorption of hydrogen on a tungsten surface depends on the second power of the field strength, as would be expected from the isolated-atom model. Of course the charge-array model will also show the same field dependence. The data collected by Kinkus & Tsong for field adsorption in the middle of the W (112) plane clearly shows that the charge-array model is superior to the isolated atom model, especially for highly polarizable inert gas atoms such as argon field adsorbed on atoms in a flat plane. On the other hand, the isolated-atom model may be adequate for describing field adsorption at plane edges of large facets such as the bcc (110), and the fcc (111) and (001) surfaces. One notices that the field adsorption energy is of the order of 0.2 to 0.3 eV near the best image field of inert gas atoms. This is almost two orders of magnitude larger than the physisorption energy. In an earlier work based on electron stimulated field desorption, Ernst *et al.*,[120] using an electron stimulated field desorp-

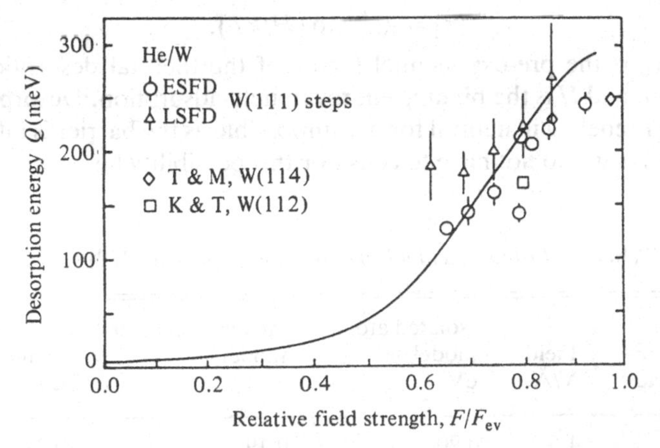

Fig. 2.21 Binding energy of field adsorption, or the desorption energy, of He as a function of field derived by different investigators using various methods. The solid line is a theoretical curve given by Ernst *et al.*[42].

tion method, found the field adsorption energy to be only 0.030 eV for He at a field of 4.4 V/Å, about an order of magnitude smaller than the values obtained earlier by Tsong & Müller. However, data obtained later by the same authors now agree with those obtained by Tsong, Kinkus & Müller. Unfortunately, all available data are no more accurate than ±10 to 15%. It is often difficult to tell from these data how different models are compared with each other. Additional data for Ne and Ar are listed in Table 2.6.

2.3.2 Kinetics of field adsorption

The binding energy in field adsorption can be derived from consideration of the kinetics of field adsorption. Specifically, it can be determined from a temperature dependence of the probability of field adsorption on an adsorption site, or the degree of coverage of field adsorption on a plane. As will be shown, a consideration of the probability of field adsorption based on adsorption time and desorption time leads to an equation equivalent to the Langmuir adsorption isotherm, but specific to the problem of field adsorption.[112,115] Let us focus on one surface atom. The average time it takes to have an image gas atom field adsorbed on the surface atom, τ_a, is

$$\tau_a = 1/z_s\eta \tag{2.49}$$

where z_s is the number of gas atoms arriving at the surface atom per unit time and η is the sticking coefficient at zero coverage. An adsorbed atom can be thermally desorbed. The average life time of adsorption at temperature T, or the mean time of desorption at T, τ_d, is given by

$$\tau_d = \nu_0^{-1} \exp(H/kT), \tag{2.50}$$

where ν_0 is the pre-exponential factor of the thermal desorption rate equation, and H is the binding energy in field adsorption. Desorption by particle tunneling in neutral form is impossible as the barrier is infinite in width, thus we do not have to consider this possibility here.

Table 2.6 *Primary field adsorption energy on a W (112) surface*

Gas species	Field V/Å	Isolated atom model eV	Adsorption layer model eV	Experimental value eV
He	4.5	0.20	0.19	0.17
Ne	3.9	0.19	0.16	0.14
Ar	2.5	0.24	0.13	0.12

Let us represent the probability of field adsorption at any instant of time t by $p(t)$. This function has to satisfy the difference equation,

$$p(t) = p(t)\left(1 - \frac{dt}{\tau_d}\right) + \frac{dt}{\tau_a}[1 - p(t)] = p(t) + \frac{dp}{dt}dt. \qquad (2.51)$$

This equation simply states that in dt, $p(t)$ will be reduced by $p(t)\,dt/\tau_d$, and increased by $dt[1 - p(t)]\tau_a$. For the case where the surface atom is not field adsorbed with an image gas atom at $t = 0$, we have $p(0) = 0$. The solution satisfying this initial condition is

$$p(t) = \left(\frac{\tau_d}{\tau_a + \tau_d}\right)\left\{1 - \exp\left[-\left(\frac{1}{\tau_a} + \frac{1}{\tau_d}\right)t\right]\right\}. \qquad (2.52)$$

The equilibrium probability of field adsorption is therefore given by

$$p(\infty) = \tau_d/(\tau_a + \tau_d). \qquad (2.53)$$

The problem now reduces to finding how τ_a is related to the experimental parameters. The gas supply functions for different tip geometry have already been discussed in Section 2.1.2 and they are given by eqs (2.9), (2.11) and (2.12). z_s is the total gas supply function Z multiplied by s_A/A where s_A is the cross-section of a surface atom in capturing an incoming gas atom, and A is the total area of the gas supply function. For a large field enhancement factor, $\phi = \alpha F^2/2kT$, z_s can be approximated by

$$z_s = C' F_0^i p_g / T^j, \qquad (2.54)$$

where $i = 1$ and $j = 1$ for spherical and cylindrical emitters, and $i \sim 2$ and $j \sim 3/2$ for a conically shaped emitter; p_g is the gas pressure. C' is a constant which may be called the gas supply constant. Thus the adsorption time is

$$\tau_a = CT^j/p_g F_0^i, \qquad (2.55)$$

where $C = 1/C'\eta$ is a constant which may be determined from an experiment. By combining these equations, one obtains

$$p(\infty) = \left[1 + \frac{\nu_0 CT^j}{p_g F_0^i}\exp\left(-\frac{H}{kT}\right)\right]^{-1}. \qquad (2.56)$$

This equation for $i = j = 1$ was originally derived by Tsong & Müller. It was recognized by Rendulic[121] to be a form of Langmuir isotherm specific to field adsorption. The probability of adsorption on a surface atom is of course equivalent to the degree of coverage of the surface if all the atoms in the surface have the same field strength. In the present equation, the field enhancement as well as the time-dependent behavior have been taken into account. For analyzing experimental data, eq. (2.56) is best rearranged in the following form:[115]

$$\ln\left[\left(\frac{1}{p} - 1\right)\Big/ T^j\right] = \ln\left(\frac{\nu_0 C}{p_g F_0^i}\right) - \frac{H}{kT}. \qquad (2.57)$$

Thus a plot of $\ln[(1/p - 1)/T^{i}]$ versus $1/kT$ will give a straight line of slope $-H$ and an intercept $\ln(\nu_0 C/p_g F_0^i)$. From this plot, C and H can be derived. Once C and H are determined, τ_a and τ_d can be calculated. The time dependence of p is then determined.

In most experimental measurements, the probability of field adsorption on a surface atom, or the degree of coverage of a surface, is determined by desorbing field adsorbed atoms by either pulsed-laser stimulated field desorption, or by electron shower induced field desorption. A probe-hole covering many atoms is used, and each signal detected is counted as detection of one desorbed atom. This is not really correct, and a correction accounting for signal overlap has to be made. It can be shown that if the probability of detecting one ion signal, with the probe-hole covering n atoms is denoted by P, then

$$p = [1 - (1 - P)^{1/n}]/e_d, \qquad (2.58)$$

where e_d is the detection efficiency of the particle detector.

Analysis of data using eq. (2.56) was reported by Tsong & Müller, and later similar analysis was also used for data analysis by Ernst *et al.*[120] It is found that without correction of the possible signal overlap as given by eq. (2.57), little inaccuracy is introduced into the value of H, although value of the intercept of the plot can be affected. Kinkus & Tsong analyzed their data using eq. (2.57); an example is shown in Fig. 2.22. Their data indeed fit well into the linear plot at high temperatures, but do not fit well at low temperatures. It can be shown that if there is another adsorption state with a lower adsorption energy, then deviation from a linear plot similar to the data shown will result. Unfortunately, statistical fluctuations of the data are too large to determine the binding energy of the second adsorption state accurately. The second adsorption state may be explained in terms of an interaction of the field induced dipole of an image gas atom with its own image. Another possibility is that theoretically, in the quantum well of field adsorption potential produced by all possible interactions, there is more than one energy level if the well is deep enough. Image gas atoms will occupy these energy levels with probabilities given by the Boltzmann factor. Atoms can be field desorbed from any of these energy levels. It will be most interesting to study these energy levels both theoretically and experimentally in the near future.

2.4 Field dissociation by atomic tunneling

A compound ion may dissociate in a high applied electric field into a neutral atom and an ion. This dissociation process, generally known as field dissociation, was theoretically treated by Hiskes[123] in 1961 as an atomic tunneling phenomenon. It is similar to field ionization of an atom

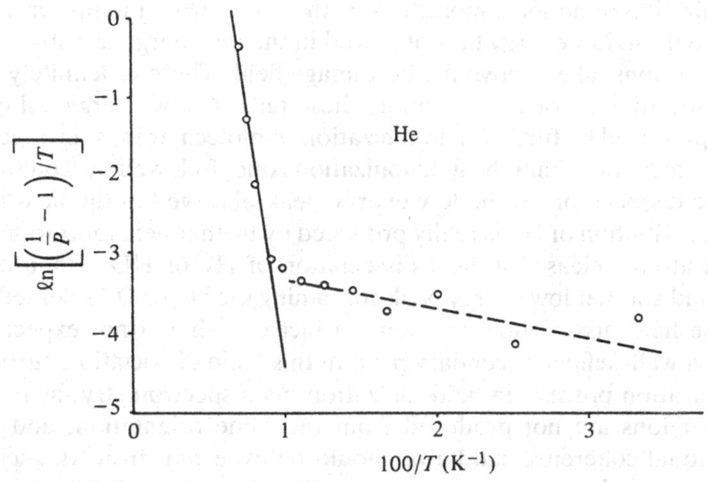

Fig. 2.22 ln [$1/P - 1)/T$] vs. $1/T$ plot for field adsorption of He on the W (112) surface, obtained from a pulsed-laser stimulated field desorption experiment. In the high temperature region, the data points fit well into a straight line of $H = 0.17$ eV, as expected from a theoretical analysis. In the low temperature region, the data points deviate significantly from the linear plot of the solid line. This deviation indicates that there may be another binding state with a much smaller binding energy of ~0.08 eV.

in a high electric field by tunneling of an atomic electron, first theoretically treated by Oppenheimer. It appears that the first report of a successful search of field dissociation was made by Riviera & Sweetman[124] when they observed H^+ ions by passing a beam of H_2^+ ions through a region with an electric field greater than 10^5 V/Å. Hiskes' calculations showed that such a field was needed to field dissociate H_2^+ at an observable rate from their highly excited vibrational states. The data presented was not of good quality, and there seems to have been no comprehensive follow-up work. In field ion mass spectrometry, Beckey and co-workers[125] interpreted the low energy tail often observed in the mass lines of large organic molecular ions as being produced by field dissociation. Although Beckey[126] discussed theories of field dissociation in great detail, these were essentially what Hiskes had already discussed earlier in his paper, and none of the experimental data presented showed clear evidence of field dissociation. Hanson[127] also interpreted low-energy peaks observed in field ion energy distributions of H^+ as being produced by further field ionization of field dissociated neutral H from H_2^+ and HD^+. All the experimental evidence relies heavily on interpretations of data. For example, a low energy tail in the field ion energy distribution can be produced by at least two other mechanisms. One is field ionization of hopping molecules further away from the tip surface, similar to those low energy tails observed in the ion energy distributions of inert gas ions taken above the best image field. There is definitely no field dissociation involved in forming these tails. A low energy tail can also be produced by further field ionization of molecular ions, produced by field ionization within the field ionization zone, followed by Coulomb repulsive dissociation. If the low energy peak observed in the field ion energy distribution of H^+ is really produced by further field ionization of those H atoms released in field dissociation of H_2^+ or HD^+, then one should find another low energy peak containing the H^+ or D^+ released in the same field dissociation reaction. In fact one should not expect to observe a well-defined secondary peak in this 'field dissociation–further field ionization process' in field ionization mass spectrometry, as these molecular ions are not produced from the same orientation, and no orientational coherence can be or should be expected. Instead, such a process can produce only a low energy tail in the energy distribution of H^+.

There exist some compound ions which are ideally suited for studying field dissociation, particularly for studying tunneling, vibrational and rotational effects of the field dissociation reaction. These are metal helide and possibly also metal hydride ions. The recent studies[128,129] of field dissociation of $HeRh^{2+}$ are especially interesting since the data obtained are very well defined, and the interpretation is most direct, as will be discussed in Section 2.4.2. The problem of field dissociation is fundamentally interesting for the reason that the potential barrier is mass-ratio and ion orientation dependent. It is one of the more interesting quantum mechanical tunneling phenomena which exhibits many novel features.

2.4.1 Basic theory

Let us consider a compound ion which consists of a light 'neutral atom' and a 'heavy ion of charge $n+$'. The two particles are bound together either by a polarization force or by a chemical bond. This clear separation of charges is obviously possible only if the ionization energy of the light atom is greater than the n-th ionization energy of the heavy atom. The vector connecting the two nuclei and pointing from the center of the light atom toward the center of the heavy ion is represented by $\mathbf{r}_n$. The applied field direction is assigned as the z-direction. For simplicity we will first assume that $\mathbf{r}_n$ lines up in the field direction. The equations of motion of the two particles are then given respectively by

$$M\ddot{R} = -f(r_n) + neF, \qquad \text{where } \mathbf{r}_n = \mathbf{R} - \mathbf{r}, \tag{2.59}$$

$$m\ddot{r} = +f(r_n), \tag{2.60}$$

where **R** is the position vector of the heavy ion with mass M, **r** is the position vector of the light atom with mass $m, f(r_n)$ is the interaction force between the two particles, and neF is the electric force of the applied field acting on the ion. By combining these two equations, one obtains

$$\mu\ddot{r}_n = -f(r_n) + neF/(1 + M/m) = -\mathrm{d}V(r_n)/\mathrm{d}r_n. \quad (2.61)$$

Thus

$$V(r_n) = U(r_n) - neFr_n/(1 + M/m), \quad (2.62)$$

where $U(r_n)$ is the potential energy of the two particles in the absence of the applied field. It is now clear that the Schroedinger equation governing the relative motion of the two particles, after converting it to the three-dimensional case, is given by

$$-\frac{\hbar^2}{2\mu}\nabla^2\psi(\mathbf{r}_n) + \left[U(r_n) - \frac{2e\mathbf{F}\cdot\mathbf{r}_n}{\left(1 + \frac{M}{m}\right)}\right]\psi(\mathbf{r}_n) = E\psi(\mathbf{r}_n). \quad (2.63)$$

The reason that a compound ion can be field dissociated can be easily understood from a potential energy diagram as shown in Fig. 2.23. When $\mathbf{r}_n$ is in the same direction as **F**, the potential energy curve with respect to the center of mass, $V(r_n)$ is reduced by the field. Thus the potential barrier width is now finite, and the vibrating particles can dissociate from one another by quantum mechanical tunneling effect. Rigorously speaking, it

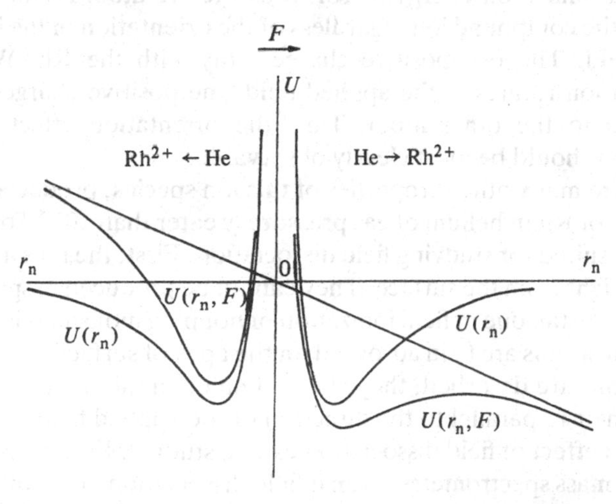

Fig. 2.23 When $\mathbf{r}_n$, which points from the neutral atom to the ion, lines up in the same direction as the applied field **F**, the potential energy of the system is reduced on one side by the field. Thus the compound ion, in one of its vibrational states, can dissociate by particle tunneling. If $\mathbf{r}_n$ is anti-parallel to **F**, then the potential energy bends upward, and field dissociation becomes impossible. The direction of $\mathbf{r}_n$ is denoted by the arrow of the He → Rh^{2+} bond.

is not the light atom which is doing all the tunneling, but rather both particles are doing the tunneling. However, the reduced mass is most effectively reduced by the particle with the small mass, and the tunneling effect is possible because of the light particle; therefore one may say that the particle with a lighter mass is doing the tunneling. On the other hand, when $\mathbf{r}_n$ is in the opposite direction to $\mathbf{F}$, the potential barrier increases with particle separation, and no field dissociation can occur. Thus the field dissociation rate depends on the orientation of the compound ion with respect to the direction of the applied field. This orientation effect can be directly observed in an experiment which will be described below.

2.4.2 Experimental observation with $HeRh^{2+}$

The theory of field dissociation was originally developed for molecular hydrogen ions. However, conceptually it is more complicated to explain how the theory works in this system. In a positive applied field, the remaining electron in the H_2^+ tends to stay at the back side, i.e. with respect to the applied field direction, of the molecular ion. The orientation of the ion is effectively reversed just by rotation of the molecular ion through 180°, not 360° as normally would according to the theory. Conceptually, theory of field dissociation manifests in a most clear way for $HeRh^{2+}$. First, the ionization energy of He is at least 6 eV higher than the second ionization energy of Rh; thus the He atom remains almost neutral in the compound ion regardless of the orientation of the ion in the applied field. The two positive charges stay with the Rh. When the compound ion rotates in the applied field, the positive charges are not transferred to the other atom. Thus the orientation effects of field dissociation should be most clearly observable.

There are many other properties of this ion species, produced by field desorption of Rh in helium of gas pressure greater than 10^{-8} Torr, which are ideally suited for studying field dissociation. First, the ions are always desorbed right from the surface. They cannot be produced in space. Thus no low energy tail due to field ionization of hopping molecules is possible. Second, He atoms are field adsorbed on the apex of surface atoms. When $HeRh^{2+}$ ions are desorbed, they always line up in the same orientation, i.e. with their $\mathbf{r}_n$ parallel to the direction of the applied field $\mathbf{F}$. Thus the orientation effect of field dissociation can be studied. In contrast, in field ionization mass spectrometry, even if field dissociation of H_2^+ or HD^+ can occur, these molecular ions are produced in random orientations, and no orientational coherence effects of field dissociation can be expected. Third, the reduced mass of $HeRh^{2+}$ is still very small, and particle tunneling effects can still be important.

In pulsed-laser field desorption of helium from rhodium surfaces, if the experiment is done without field evaporating the surface then the ion energy distribution of He^+ ions is essentially the same as that in ordinary field ionization. He atoms are thermally desorbed from the surface, and are then field ionized when they pass through the ionization zone of field ionization. Also if pulsed-laser field evaporation of rhodium is done in UHV at a field of 4.6 to 4.9 V/Å using a very small laser power then the ion energy distribution is very narrow, with a very well defined peak without having either a secondary peak or a low energy tail. However, when the pulsed-laser field evaporation is done in the presence of about 1×10^{-8} Torr of He, but otherwise under identical conditions, then three mass lines, each of them a high resolution ion energy distribution, are found. This is shown in Fig. 2.24. The He^+ line is identical to that found in field desorption without field evaporating the emitter surface, the Rh^{2+} line now shows a secondary peak, otherwise the main peak is identical to field evaporation in UHV, and there is a $HeRh^{2+}$ line. For a tip of radius 420 Å, the secondary peak is found to have an additional ion energy deficit of 51 eV, indicating that these ions are produced further away from the surface. The secondary peak, in contrast to a low energy tail, indicates the existence of a field dissociation zone, similar to the ionization zone in

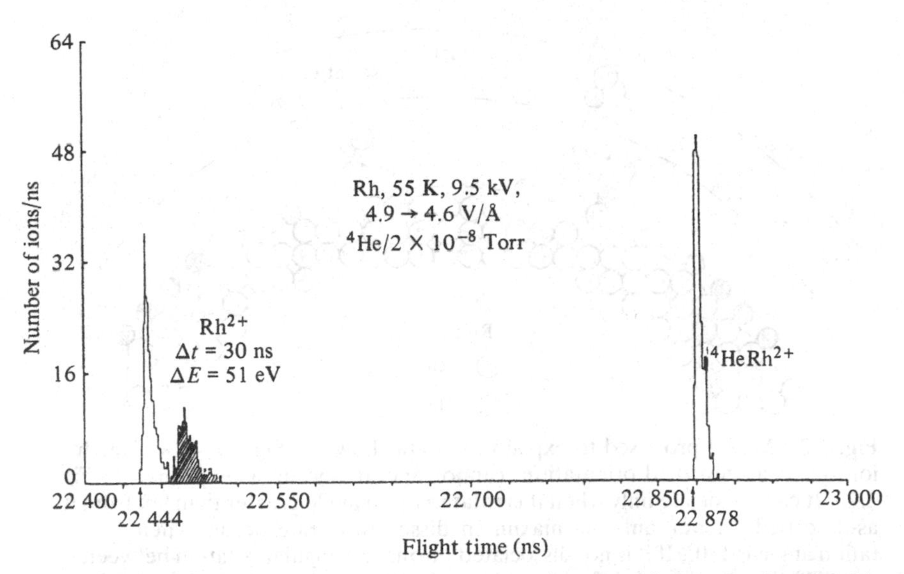

Fig. 2.24 Portion of a time-of-flight spectrum in pulsed-laser stimulated field evaporation of a Rh tip in 2×10^{-8} Torr of ^4He. Besides the formation of $^4HeRh^{2+}$, the Rh^{2+} line now shows a low energy peak of ~51 eV additional energy deficit (shaded). Rh^{2+} ions in this secondary peak are produced by field dissociation of $^4HeRh^{2+}$ in the field dissociation zone which is about 150 Å in width and is centered at ~220 Å above the tip surface.

field ionization. A model for field dissociation in a well-defined spatial zone was therefore proposed by Tsong, as shown in Fig. 2.25. The reason that field dissociation can occur only in a well-defined spatial zone can be seen from the Schroedinger equation governing the relative motion of Rh^{2+} and He in the compound ion as given by eq. (2.63). Field dissociation can occur only if $\mathbf{F}\cdot\mathbf{r}_n$ is positive so that the potential energy curve can be bent downward. When a $HeRh^{2+}$ is field desorbed from the surface, $\mathbf{r}_n$ is antiparallel to $\mathbf{F}$. Field dissociation cannot occur. It can occur only if the ion rotates by at least 90°. Maximum rate of field dissociation occurs when the ion is rotated by 180°. When the rotation angle exceeds 270°, no field dissociation can occur either. The next time the compound ion lines up in the right orientation for field dissociation to be possible again, the ion is already too far away from the surface and the field strength is already too low to field dissociate the compound ion, thus only one secondary peak is observed.

Using the field distribution of eq. (2.13), one can show that the peak position of the secondary peak is given by,

$$x = \frac{r_t}{2}\left\{\exp\left[4k\left(1 + \frac{M}{m}\right)\frac{\Delta t}{t_0} - 1\right]\right\}, \qquad (2.64)$$

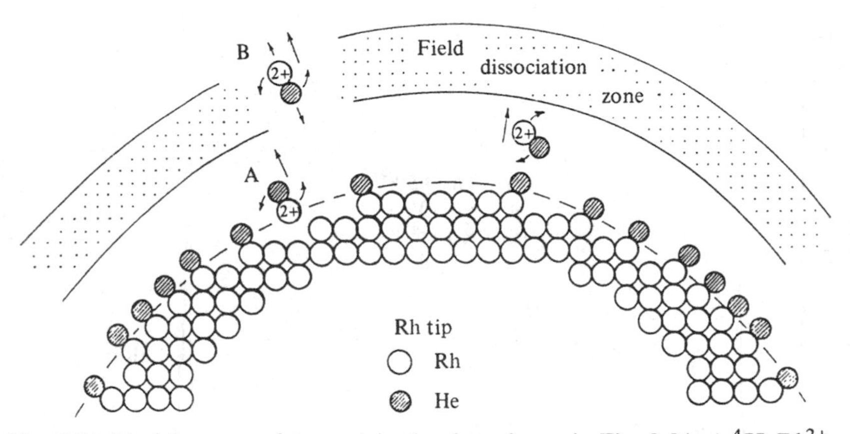

Fig. 2.25 Model proposed to explain the data shown in Fig. 2.24. A $^4HeRh^{2+}$ ion, in its as-desorbed orientation, cannot field dissociate as explained in Fig. 2.23. It can dissociate only when it is rotated by an angle greater than 90° from its as-desorbed orientation. The maximum dissociation rate occurs when the rotation angle is 180°. If it is not dissociated during the angular rotation between 90 and 270° then it has to wait for the right orientation again. By that time, the ion is too far away from the surface and the field is too low for field dissociation to be possible. Thus a well-defined 'field dissociation zone' exists. From the additional energy deficit of the secondary Rh^{2+} peaks and the field distribution above the surface, the dissociation time and the 180° rotation time of the compound ion is calculated to be 790 ± 21 fs.

where k is defined by $F_0 = V/kr_t$ which have a value of about 5, M is the mass of Rh, m is the mass of He, $\Delta t = 30$ ns is the difference in flight time of the main and secondary peaks, and $t_0 = 22\,444$ ns is the onset flight time of the main peak. Using these data Tsong & Cole[130] conclude that the field dissociation zone of $HeRh^{2+}$ for a tip radius, r_t, of 420 Å is about 150 Å in width which is centered at 220 Å from the tip surface. The flight time of the ion in traveling a distance of 220 Å can be calculated and it is 790 femtoseconds. Thus the rotation time by 180° of this compound ion in a field of about 4.8 V/Å is 790 ± 21 fs. Rotation of compound ions under the large torque of the applied electric field, where the rotational quantum number changes by at least several within half a revolution is a problem still poorly understood. One should also ask whether field desorbed $HeRh^{2+}$ ions are coherent in their vibrational motion with respect to the instant they are field desorbed. It is hard to imagine that this is not the case. As field desorption should be most favored when a Rh atom vibrates away from the surface plane, $HeRh^{2+}$ ions formed can be expected to be coherent in their vibrational motion. If this is the case, then the secondary Rh^{2+} peak should exhibit fine structures of the vibrational motion of the compound ion. Miskovsky & Tsong[131] have made a detailed calculation based on this argument. They conclude that the secondary peak should contain four fine structure peaks arising from the vibrational motion. Even though the pulsed-laser time-of-flight atom-probe used for this experiment is already one of the highest resolution time-of-flight systems available, it is not quite enough to resolve these peaks even if they exist. A search is underway to find these vibrational features. Discussions of the question of rotation and vibration of $HeRh^{2+}$ ions will be referred to the original papers.[130,131]

2.4.3 Isotope effects and tunneling probabilities

Any atomic tunneling phenomenon should exhibit a strong isotope effect. In field dissociation of $HeRh^{2+}$, this effect has indeed been investigated by Tsong,[121] who replaces ^4He with ^3He. He found that under identical experimental conditions, except the replacement of He isotopes, $^3HeRh^{2+}$ no longer field dissociate, as shown in Fig. 2.26. This observation, at first glance, is most surprising since one would expect the tunneling probability to be much larger for a particle of smaller reduced mass. It can however be understood from the Schroedinger equation, eq. (2.63), governing the relative motion of the two particles in the applied field. The potential barrier reduction term, $2e\mathbf{F}\cdot\mathbf{r}_n/(1 + M/m)$, is much smaller in magnitude for ^3He than for ^4He. Apparently the effect of the barrier change is more important than the effect of the reduced mass. This is indeed found to be true in a detailed theoretical calculation,[122] as will be discussed in the next paragraph. The physical reason that

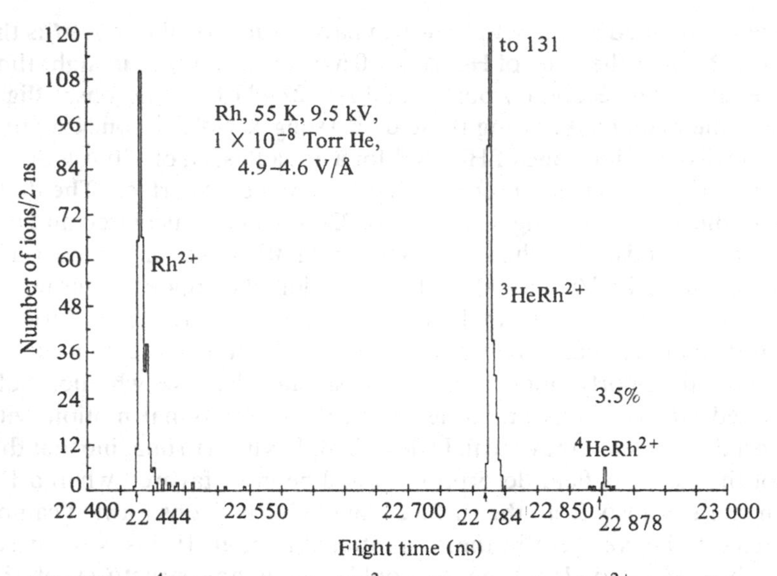

Fig. 2.26 When ^4He is replaced with ^3He, the secondary Rh^{2+} peak disappears even though $^3HeRh^{2+}$ ions are still formed. At first glance, this strong isotope effect is most surprising since one would expect $^3HeRh^{2+}$ to field dissociate more easily than $^4HeRh^{2+}$ because of its smaller reduced mass. This 'peculiar' isotope effect is the result of a center of mass transformation in the applied field, as can be understood from the Schroedinger equation of eq. (2.63), already explained in the text.

$^3HeRh^{2+}$ is more difficult to field dissociate than $^4HeRh^{2+}$ is because of a center of mass transformation. This transformation, as seen in eq. (2.62), changes the potential of the ion in the applied field.

To interpret field dissociation data of $HeRh^{2+}$ properly, one must know the interaction $U(\mathbf{r}_n)$ between a He atom and a Rh^{2+} ion. Such information is of basic interest in its own right. Although bonding in metal helide ions with three-dimensional transition metal atoms has been studied theoretically using a complete active space self-consistent field-level *ab initio* and Dirac–Fock expansion methods by Hotokka *et al.*,[132] the calculations are too involved for $HeRh^{2+}$. Thus a computational method based on an effective medium theory has been carried out by Tsong & Cole.[122] The effective medium theory has of course been extensively used in calculating physical adsorption interactions.[133] The interaction between a He atom and a Rh^{2+} ion in an applied field $\mathbf{F}$ is found to be

$$U(r, \mathbf{F}) = \frac{8.395 \times 10^3}{r^2} \exp(-qr) - f_4(qr)\frac{5.8733}{r^4} - f_6(qr)\frac{6.0046}{r^6}$$

$$-f_8(qr)\frac{25.52}{r^8} - \frac{2\mathbf{F}\cdot\mathbf{r}}{\left[1+\dfrac{M}{m}\right]}\,\mathrm{eV}, \qquad q = 4.7732, \tag{2.65}$$

for r in Å and F in V/Å. $f_n(x)$ is defined by

$$f_n(x) = 1 - \mathrm{e}^{-x}\sum_{k=0}^{n}\frac{x^k}{k!}. \tag{2.66}$$

Figure 2.27(a) and (b) show $U(r, F)$ as a function of $\mathbf{r}$ in various applied fields for $^4HeRh^{2+}$ and $^3HeRh^{2+}$ when $\mathbf{r}$ is in the same direction as $\mathbf{F}$. When $F = 0$, the dissociation energy is found to be ~0.34 eV, a value similar to those obtained by Hotokka *et al.* for 3-d metal helide ions. The barrier penetration probability as a function of applied field has been calculated using the WKB approximation with a numerical method for tunneling from a vibrational state which is assumed to be 300 and 500 K above the bottom of the well, and the result is shown in Fig. 2.28. It is seen that under the same applied field, the field dissociation rate for $^4HeRh^{2+}$ is at least three to four orders of magnitude greater than that for $^3HeRh^{2+}$, in good agreement with the experimental observations. A more detailed numerical calculation, at different flight times and with the vibrational levels carefully identified, gives an even better agreement with the experimental result. The transmission coefficient is a maximum of 1.73×10^{-2} at $x = 211$ Å and the zone width extends from ~140 to 300 Å. Thus the field dissociation zone is about 160 Å in width and it is centered at about 208 Å above the surface. This should be compared with the field ionization zone which is ~4 Å above the surface and is only about 0.2 to 0.3 Å in width. Of course, in field ionization the tunneling particle is an electron, whereas in field dissociation the tunneling particle is a neutral atom.

2.5 Field ion image formation

2.5.1 Mechanisms of field ion image formation

With all the basic physical processes now discussed in the earlier sections, we are now ready to summarize the mechanisms of image formation in the field ion microscope. In the FIM, the sample is a sharp tip usually prepared by electrochemical etching of a piece of thin wire. The surface of a tip so prepared is in general not atomically smooth, but rather contains a lot of rough regions and asperities. Such an irregular surface cannot be used as a projecting surface for field ions to form a uniformly magnified image of the surface at the screen. Once the microscope is carfully evacuated to UHV, an image gas is admitted to a pressure of

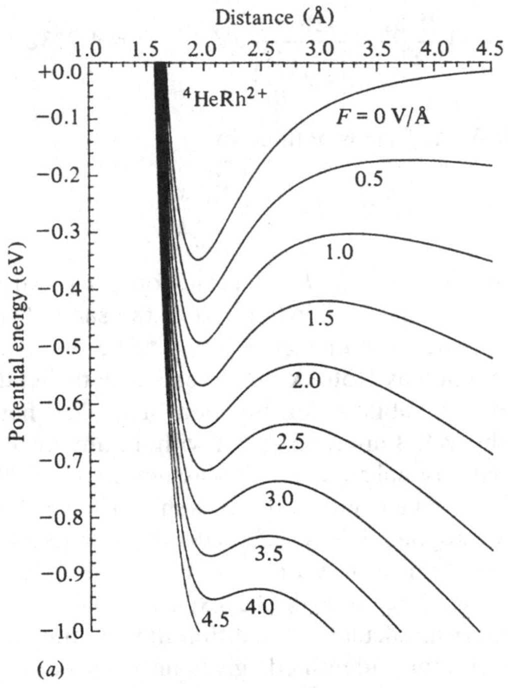
Distance (Å)
1.0 1.5 2.0 2.5 3.0 3.5 4.0 4.5
+0.0
−0.1
−0.2
−0.3
−0.4
−0.5
−0.6
−0.7
−0.8
−0.9
−1.0
Potential energy (eV)
^{4}HeRh^{2+}
F = 0 V/Å
0.5
1.0
1.5
2.0
2.5
3.0
3.5
4.0
4.5
(a)

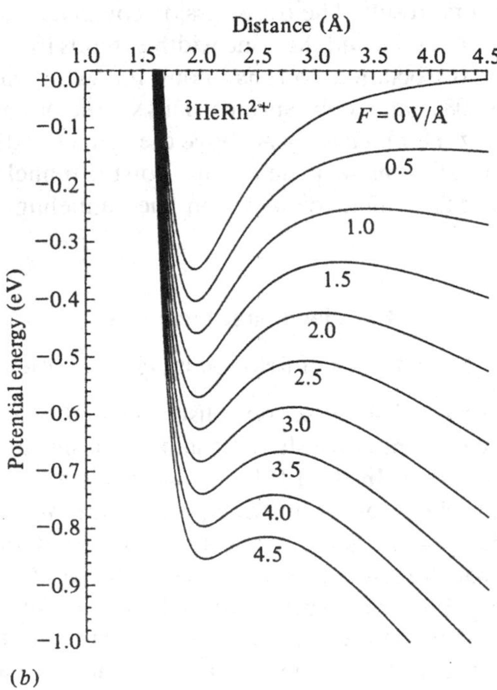
Distance (Å)
1.0 1.5 2.0 2.5 3.0 3.5 4.0 4.5
+0.0
−0.1
−0.2
−0.3
−0.4
−0.5
−0.6
−0.7
−0.8
−0.9
−1.0
Potential energy (eV)
^{3}HeRh^{2+}
F = 0 V/Å
0.5
1.0
1.5
2.0
2.5
3.0
3.5
4.0
4.5
(b)

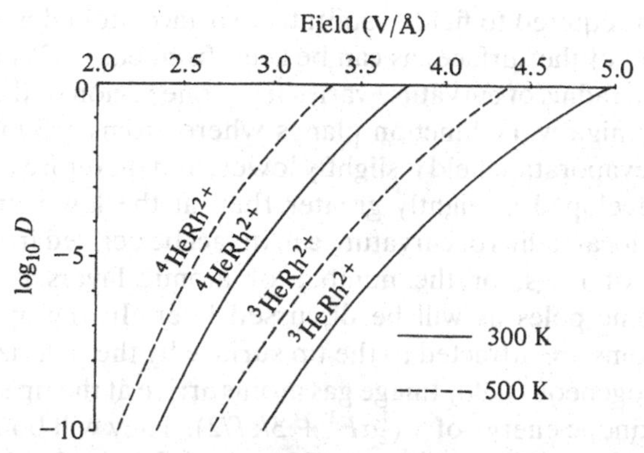

Fig. 2.28 Particle penetration probability in field dissociation of $^4HeRh^{2+}$ and $^3HeRh^{2+}$ from a vibrational state 300 K and 500 K above the bottom of the potential energy curve. At the same field, the particle barrier penetration probability for $^3HeRh^{2+}$ is three to four orders of magnitude smaller than that for $^4HeRh^{2+}$, in good agreement with the experiment.

about 1×10^{-4} Torr and the tip is then cooled down to a temperature as low as the image gas permits, i.e. without condensation, or to the resolution of the microscope which one desires. As will be discussed in the next section, the lower the tip temperature, the better will be the resolution of the microscope. The tip is connected to a positive d.c. high voltage power supply and the voltage is gradually raised. When the local field strength above the asperities is close to the evaporation field of the tip materials, atoms at the surface of these asperities start to field evaporate, and the sharp asperities are therefore gradually removed. The emitter surface eventually develops into an 'atomically smooth' surface with small facets formed at different crystallographic poles, as can be seen most clearly in the ball model shown in Fig. 1.3(*a*). Each atomic layer forms a small ring, or each ring-shaped image represents a nearly circularly shaped atomic layer and its lattice step to the next layer. The

---

Fig. 2.27 (*a*) Theoretical potential energy curves for $^4HeRh^{2+}$ in different applied field strengths. At zero field, the binding energy of the system is about 0.34 eV, and the equilibrium separation is about 2 Å. In an applied field of over 4 V/Å, the potential barrier reduces rapidly, and dissociation by particle tunneling becomes possible.

(*b*) Similar curves for $^3HeRh^{2+}$. Note that under the same applied field, the potential barrier for field dissociation is much higher and wider than that for $^4HeRh^{2+}$.

field strength required to field evaporate a surface atom depends on the work function of the surface, as can be seen from eqs (2.35) and (2.36). Therefore the radius of curvature varies from one region of the surface to another. For high work function planes where atoms are more closely packed, the evaporation field is slightly lower, and therefore the radius of curvature developed is slightly greater than at the low work function regions. The local radius of curvature can in fact be derived from counting the number of rings, or the number of atomic layers between two crystallographic poles as will be discussed later. In the applied field, image gas atoms are attracted to the tip surface by the polarization force in the inhomogeneous field. Image gas atoms arrive at the tip surface with an average kinetic energy of $\sim(\frac{1}{2}\alpha F^2 + 3kT/2)$. They will bounce off the surface when they hit it, losing a small fraction of their kinetic energy. A fraction of the arriving image gas atoms will be trapped to the emitter surface region, and will hop around until they lose a large fraction of their kinetic energy by thermal accommodation. Nearly a monolayer of image gas atoms will be field adsorbed on the apex of the more protruding surface atoms, which are also the imaged atoms as field desorption images exhibit a one-to-one correspondence with the field ion images. These hopping image gas atoms, during their hopping motion, may be field ionized by having one of their electrons tunneling into the metal. The ionization occurs most readily in disks about 0.2 to 0.3 Å thick which are located beyond the critical distance of field ionization ~4 Å above the surface. These disks are located above those surface atoms which are field adsorbed with an image gas atom. The size of the disks is not known accurately, but should depend on the overlapping integral of the atomic wavefunction of the image gas atom and the wavefunction of the surface atom. It is perhaps of the order of half the sum of the diameters of the image gas atom and the surface atom. An image gas atom, once it has been field ionized, will be accelerated away from the surface in nearly the radial direction of the nearly hemispherical tip surface toward the screen assembly. Every second, about 1000 to 10 000 ions originate from the same ionization disk. Thus the field ion image is a map of the field ionization disks above surface atoms. A schematic diagram showing this mechanism of image formation has already been shown in Fig. 1.2.

To be able to give rise to a field ion image, the evaporation field of the tip material has to be higher than the image field of the image gas used. Experimental observation with tungsten indicates that the best image field of He is about 4.4 V/Å, of Ne is about 3.7 V/Å and of Ar and of H_2 is about 2.2. V/Å. One should then choose an image gas with the best image field slightly lower than the evaporation field of the material. On the other hand, the applied field has the function of protecting a low temperature field evaporated clean emitter surface from being contaminated by low ionization energy, chemically reactive residual gases in the field ion microscope. Thus by using a low ionization energy inert gas for imaging the emitter surface, one automatically sacrifices the protection of the strong applied field against surface contamination, and therefore a more stringent requirement is needed for the vacuum condition as well as the purity of the image gas.

An intrinsically interesting question is how the field ion image contrast is produced and how field adsorption affects field ionization. The image contrast may arise from an enhancement in field ionization by field adsorption of image gas atoms. Schmidt *et al.*[134] find that when helium image gas is mixed with a small amount of Ne, scintillations can be observed of the field ion image. A comprehensive study of this effect has been carried out by Rendulic.[135] The scintillations arise from temporary replacement of field adsorbed helium atoms by neon atoms as under the same field strength the field adsorption energy of Ne should be considerably larger because of its larger polarizability. Ne atoms either enhance, or suppress much less, the field ionization of He atoms. From the study with He–Ne mixed gases, Rendulic concludes that the field adsorbed He atom suppresses field ionization whereas the field adsorbed Ne atom enhances field ionization. Rendulic's experimental result was later confirmed by the theoretical calculations of Nolan & Herman[136] and of Iwasaki & Nakamura.[137] On the other hand, McLane *et al.*[138] reported a much more direct measurement (though a somewhat qualitative one in this preliminary study) of the effect of field adsorption on field ionization rate, which contradicts all these studies. They find that the rate of helium field ions detected increases with time after the field emitter surface is pulse field evaporated to quickly deplete all the field adsorbed atoms. This observation can indicate only that field ionization is enhanced by the field adsorption of helium atoms. Sweeney & Tsong[139] made a comprehensive study of the enhancement effect with a similar method using an imaging atom-probe.

In the experiment, field adsorbed atoms on the field emitter surface are quickly removed by pulsed-field evaporation of the surface with a high voltage pulse of 20 ns width. The rate of field ionization as a function of time is then measured from the entire field ion emitter surface using the single ion detection sensitivity of the Chevron channel plate of the imaging atom-probe. The field ionization rate from a freshly created surface, without field adsorbed atoms, is found to be much smaller. As the surface is gradually field adsorbed again with the image gas atoms, the field ionization rate increases with time and eventually returns to a saturation value. Examples of these recovery curves are shown in Fig. 2.29. It is quite clear that field ionization is greatly enhanced by field adsorption of image gas atoms regardless of whether the field adsorbed

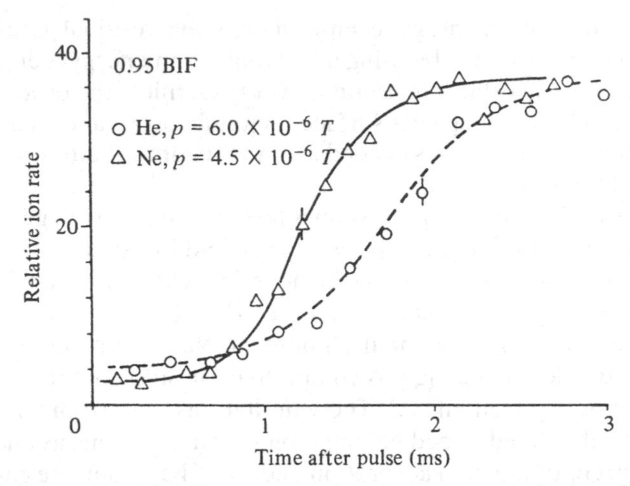

Fig. 2.29 Recovery of field ion current after the tungsten emitter surface is pulse field evaporated to deplete all the field adsorbed image gas atoms at the surface. Zero time refers to the time when the surface is completely depleted of field adsorbed atoms.

atoms are He or Ne. From many similar measurements, the enhancement factor is found to depend on the field strength, with a value which is as large as 40 for He and over 100 for Ne at a field way below the best image field, and approaches one when the field is way above the best image field. Thus Sweeney & Tsong's experimental result, based on a much more direct method, is in disagreement with the earlier experimental result of Rendulic, and also with the theoretical calculations of Nolan & Herman and Iwasaki & Nakamura. Field ionization enhancement can arise from an increase in the electronic transition rate, a change in the gas supply, or an increase in the dwelling time of an image gas atom in the ionization zone. The dwelling time is expected to be increased drastically by lowering the tip temperature, or by increasing the applied field. However, when the temperature is lowered, or the field is increased, the enhancement factor is found to decrease. Thus the theory of a possible dwelling time change can be ruled out. The effects of gas supply and electronic transition rate changes are difficult to isolate. Nevertheless, a reasonable estimate made by these authors indicates that the electronic transition rate is enhanced by field adsorption of both He and Ne.

After the above discussion the field ion image contrast can now be easily understood. According to the calculations of Fonash, Sharma and Schrenk,[140] who considered carefully the variation in the surface potential right above a surface atom, and an atom between two surface atoms, an image intensity variation of about 10% can be expected on a surface plane. This small variation obviously cannot explain why the field images show a contrast of at least 1000%. Since according to the measurement of Sweeney & Tsong, field adsorption of He or Ne can greatly enhance field ionization, the observed image contrast must be due to the effect of field adsorption. There are at least two experimental facts supporting this interpretation. First, there is a one-to-one correspondence between the field ion image and the field desorption image. Second, image contrast is much higher at low field strength and decreases drastically at high field, in complete agreement with the field ionization enhancement factor measured by Sweeney & Tsong. At the moment there is no new theory to explain these experimental observations.

2.5.2 Resolution of the field ion microscope

In many respects, a theory of resolution for the field ion microscope is very similar to that for the field emission microscope.[15] There are at least three factors which will limit the resolution of the FIM.[16] First, when one tries to confine a particle within a very small spatial region, then according to Heisenberg's uncertainty principle the momentum of the particle in the lateral direction of the surface can no longer be precisely defined. The spread in the lateral velocity component of the image gas ions will then limit the size of the smallest image spot obtainable on the screen, or, equivalently, limit the resolution of the FIM. Second, field ions are formed from hopping image gas atoms. These atoms, before being field ionized, have a velocity component parallel to the emitter surface. This component depends on the kinetic energy of the gas atoms before they are field ionized, and cannot be smaller than the thermal energy of the tip temperature. When image gas atoms arrive at the tip surface, they have a kinetic energy the sum of the polarization energy and the thermal energy of the FIM chamber temperature. The better the gas atoms are thermally accommodated to the tip temperature, the better will be the resolution of the FIM. Third, field ionization can occur within a cross-sectional area where the wavefunction of the image gas atom and the wavefunction of the surface atom overlap. This intrinsic limitation cannot be improved by an aritifical method.

In a point projection microscope, the image magnification is given by

$$\eta = R/\beta r_t, \tag{2.67}$$

where R is the tip-to-screen distance, and β is an image compression factor which takes into account the fact that field ions do not follow exactly the radial direction of the nearly hemispherical tip surface because of the tip shank and other supporting leads. For simple field emitter geometries, β can be calculated by determining the ion trajectories.[141] It can always be found from field ion images where angular separations among different crystallographic poles are well known. In normal tip geometry β has a value between 1.5 and 1.8. The ion beam broadens on its way to the screen by the slightly compressed radial projection, so that a size of Δy_t on the tip surface appears as $\eta\ \Delta y_s$ at the screen (see Fig. 2.30). Beside the broadening due to the radial projection of the ions, the tangential velocity of the ions, v_t, will further broaden the spot size. To find the effect of the tangential velocity component v_t, we approximate the actual tip geometry by a freely suspended sphere, so that the force acting on the ions is inversely proportional to the second power of the distance, and it is also a central force. Conservation of angular momentum of the central force problem must be satisfied. Thus

$$p_\theta = Mr^2\dot{\theta}, \tag{2.68}$$

where r is the distance from the center of the sphere and θ is the angle between r and the radial vector. At the tip surface where $r = r_t$,

$$p_\theta = Mr_t v_t. \tag{2.69}$$

It is now only a few steps to show that

$$\dot{r} = \left\{\frac{2e[V_0 - V(r)]}{M}\right\}^{1/2} = \left(\frac{2eV_0}{M}\right)^{1/2}\left(1 - \frac{r_t}{r}\right)^{1/2}, \tag{2.70}$$

$$\theta_0 = \int_{r_t}^{R} \frac{v_t r_t}{r^2}\frac{\mathrm{d}r}{\dot{r}} \approx 2v_t\left(\frac{M}{2eV_0}\right)^{1/2}, \tag{2.71}$$

$$t = \int_{r_t}^{R} \frac{\mathrm{d}r}{\dot{r}} \approx R\left(\frac{M}{2eV_0}\right)^{1/2}. \tag{2.72}$$

The spot size at the screen due to the lateral motion in two opposite directions, $\pm v_t$, of the ions for a tip of spherical shape is therefore

$$\Delta y_s(v_t) = 2R\theta_0 = 4v_t t. \tag{2.73}$$

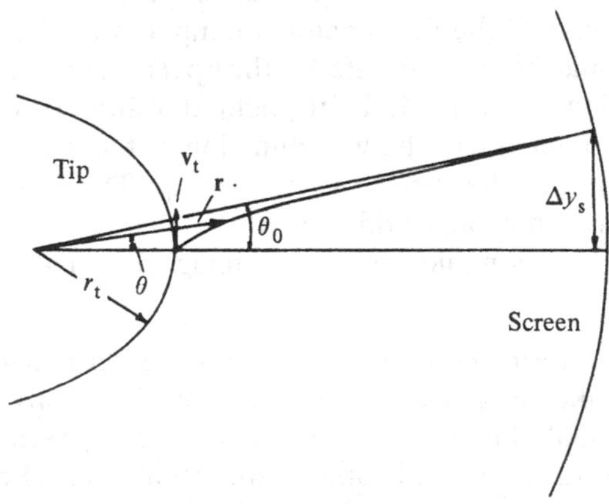

Fig. 2.30 A diagram explaining the trajectory of an ion emitted with a large tangential velocity from the apex of the tip surface. In reality, θ_0 may be as small as a few minutes of the arc. The spot size at the screen is twice Δy_s.

Let us now consider the increase in the spot size due to the effect of Heisenberg's uncertainty. When a particle is confined to pass through a small space of width Δy_t at the tip, the uncertainty in the tangential component of the momentum of the particle is of the order of $\hbar/2\,\Delta y_t$, and the corresponding velocity component is $\hbar/2M\,\Delta y_t$. Thus the spread of the spot size at the screen by this uncertainty alone is

$$\Delta y_{s,u} = 4t \cdot (\hbar/2M\,\Delta y_t), \tag{2.74}$$

therefore

$$\Delta y_s = \{(\eta\,\Delta y_t)^2 + (2\hbar t/M\,\Delta y_t)^2\}^{1/2}. \tag{2.75}$$

We then minimize the spot size by letting $d(\Delta y_s)/d(\Delta y_t) = 0$; this gives $\Delta y_s = 2(\eta\hbar t/M)^{1/2}$. Therefore the optimum resolution after replacing V_0 by $\varkappa F_0 r_t$, is

$$\delta_u = \Delta y_s/\eta = 2(\beta^2\hbar^2 r_t/2\varkappa M e F_0)^{1/4}. \tag{2.76}$$

To estimate the broadening of spot size by the thermal velocity of the ions, let us assume that the image gas atoms immediately before ionization have a Maxwell–Boltzmann velocity distribution with an effective temperature T which is very close to the tip temperature T_t. If $n(y)\,dy$ represents the number of ions arriving at the screen between y and $y + dy$, we have

$$n(y)\,dy = N\left(\frac{M}{2\pi MkT}\right)^{3/2}\int_0^\infty \exp\left(-\frac{Mv_x^2}{2kT}\right) \times dv_x \int_0^\infty \exp\left(-\frac{Mv_z^2}{2kT}\right) dv_z \cdot \exp\left(-\frac{Mv_v^2}{2kT}\right) dv_y. \tag{2.77}$$

But

$$y = R\theta_0 = 2v_t t = 2R(M/2eV_0)^{1/2} v_y. \tag{2.78}$$

Carrying out the integration and replacing v_y with y according to the last equation one obtains

$$n(y_t) = K\exp\{-eV_0 y^2/4kTR^2), \tag{2.79}$$

where K is a constant. The spreading of the spot size by thermal lateral velocity, y_t, can be obtained by imposing the criterion that $n(y_t) = n(0)/e$, where e is the base of the natural logarithm. With this criterion one obtains

$$\delta_t = 2y_t/\eta = 4\{kT\beta^2 r_t/\varkappa e F_0\}^{1/2}. \tag{2.80}$$

The ionization disk size has not been discussed in detail. It can be obtained from a calculation similar to that of Sharma & Schrenk by imposing a similar criterion where the ionization rate falls to $1/e$, the maximum value, as in our discussion of δ_t. Lacking such a calculation, we may simply assume that the resolution limited by the ionization disk size, δ_0, is of the order of the radius of the image gas atom. As all these effects

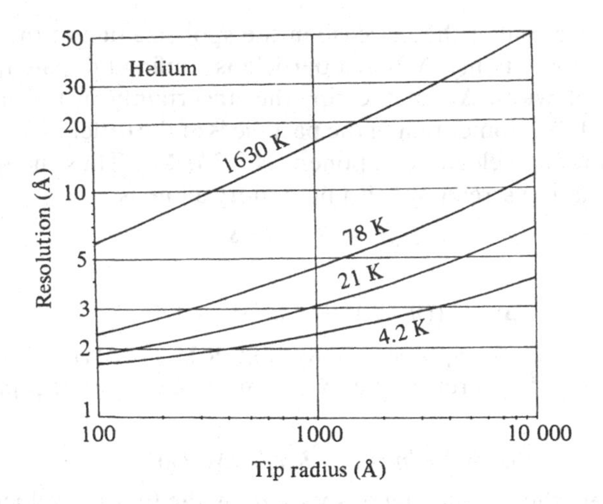

(a)

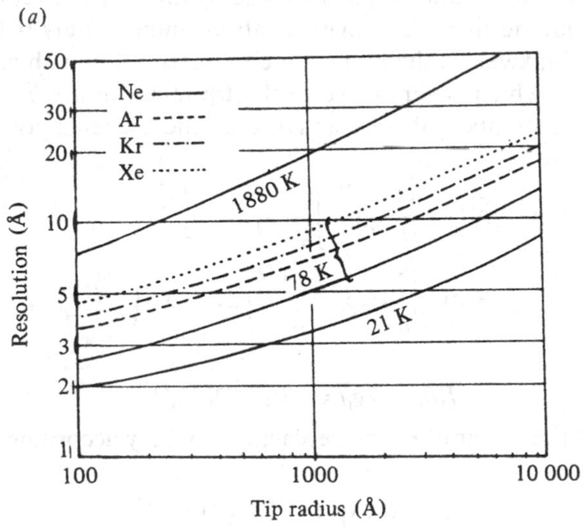

(b)

Fig. 2.31 (*a*) Calculated resolution of the field ion microscope using He as the image gas. The resolution depends on the tip radius as well as the equilibrium temperature of the gas atoms just prior to field ionization.
(*b*) Calculated resolution of the FIM using a few other commonly used image gases.

are statistical in nature, the resolution of the microscope is

$$\delta = \left\{\delta_0^2 + 4\left(\frac{\beta^2 r_t \hbar^2}{2\varkappa e M F_0}\right)^{1/2} + 16\left(\frac{\beta^2 k T r_t}{\varkappa e F_0}\right)\right\}^{1/2}. \tag{2.81}$$

The final step is a minor modification made by Chen & Seidman[142] of the derivation given by Müller & Tsong.[16] The best resolution of the FIM can be achieved by working with the smallest possible tip radius r_t, by reducing the thermal equivalent temperature of the ionized gas atoms, and by using an image gas with the highest ionization potential so that F_0 will be the highest, and with the smallest size image gas atoms. Helium satisfies all these requirements while Ne and H_2 with cooling below 20 K are the second best choices. Figure 2.31(*a*) and (*b*) give some idea of the expected resolution of the FIM.

In practice, the resolution of the FIM also depends on the size of the surface plane. A smaller plane always gives an atomically better resolved image. The resolution, under normal good operating conditions, is about 2.5 to 3.0 Å, and nearest neighbor atoms in most materials can be resolved if the tip radius and the facet size are sufficiently small. In the book by Müller & Tsong,[16] atomically resolved He ion images are shown of the chains of W atoms (2.74 Å spacing) across the [111] zone of a tungsten tip of 840 Å average radius, taken at $V_t = 24.1$ kV and 12 K, and a fully resolved (001) plane of Pt, 2.77 Å spacing, from a tip of $r_t = 164$ Å taken at 21 K and 5.6 kV. In Fig. 2.32 we show a few atomically resolved

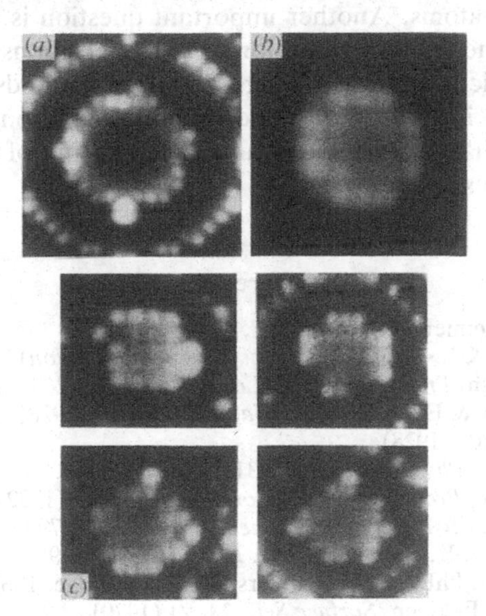

Fig. 2.32 (*a*) He field ion image of an Ir (001) plane. As the size of the plane is too large, atoms are not resolved.

(*b*) He field ion image of a Pt (001) plane. Atoms, which are separated by a distance of 2.80 Å, are fully resolved in this image.

(*c*) He field ion image of a few non-reconstructed, (1 × 1), W (001) surfaces. Atoms are separated by a distance of 3.16 Å. For metals, the structures of surfaces prepared by low temperature field evaporation almost always have the unreconstructed (1 × 1) structures.

images of the non-reconstructed Pt (001) and W (001) surfaces. The atomic spacing in the Pt (001) plane is about 2.8 Å and that in the W (001) plane 3.16 Å.

The resolution we have discussed so far refers to the smallest distance between two surface atoms where the two atoms can still be seen in the FIM as separate atoms. There is another kind of resolution which has considerable utility. When an atom is deposited on a surface, one will want to know how accurate the position of the adatom can be mapped. What is the minimum distance between two adatom positions which can be distinguished? It has been shown by an image mapping method that the accuracy of mapping the position of an adatom can be better than 0.5 Å.[143] With the availability of computers to accurately digitize the positions of adatoms or surface atoms this resolution is probably now much better of ~0.3 Å. There must be a limit of this resolution, although there has been no theoretical or experimental discussion of this kind of resolution. With careful mapping of the positions of an adatom during its random walk diffusion, information on adsorption sites can be derived. We will discuss this point in connection with studies of the behavior of single adsorbed atoms. Another important question is how accurately one can determine the two-dimensional relative positions of atoms in two successive atomic layers using image digitizing methods. The field ion microscope can, in principle, provide such information. This problem seems not yet to have been studied carefully at the time of this writing, but may soon be investigated.

References

1 J. R. Oppenheimer, *Phys. Rev.*, **31**, 67 (1928).
2 R. H. Fowler & L. Nordheim, *Proc. Roy. Soc. (London)*, **A119**, 173 (1928); L. Nordheim, *Proc. Roy. Soc. (London)*, **A121**, 626 (1928).
3 R. W. Gurney & E. U. Condon, *Nature*, **122**, 439 (1928); G. Gamow, *Zeit. Phys.*, **51**, 204 (1928).
4 C. Lanczos, *Z. Physik*, **68**, 204 (1931).
5 R. W. Gurney, *Proc. Roy. Soc. (London)*, **134**, 137 (1932).
6 M. G. Inghram & R. Gomer, *J. Chem. Phys.*, **22**, 1279 (1954).
7 E. W. Müller & K. Bahadur, *Phys. Rev.*, **102**, 624 (1956).
8 M. J. Southon, PhD Thesis, University of Cambridge, 1963.
9 H. A. M. van Eekelen, *Surface Sci.*, **21**, 21 (1970).
10 T. T. Tsong & E. W. Müller, *J. Appl. Phys.* **37**, 3065 (1966).
11 M. Knudsen, *Ann. Physik*, **34**, 593 (1911).
12 B. Baule, *Ann. Physik*, **44**, 145 (1914).
13 H. Y. Wachman, Role of the Surface in Thermal Accommodation, in *Dynamics on Manned Lifting Entry*, John Wiley, New York, 1963.
14 O. Nishikawa & E. W. Müller, *J. Appl. Phys.*, **35**, 2806 (1964).
15 R. Gomer, *Field Emission and Field Ionization*, Harvard University Press, 1961.

16 E. W. Müller & T. T. Tsong, *Field Ion Microscopy, Principles and Applications*, Elsevier, New York, 1969.
17 E. W. Müller, *Air Force Report*, AFOSR-TN-55-307 (1955).
18 M. K. Southon & D. G. Brandon, *Phil. Mag.*, **8**, 579 (1963).
19 H. D. Beckey, J. Dahmen & H. Knoppel, *Z. Naturforschung*, **21a**, 141 (1966).
20 T. T. Tsong & E. W. Müller, *J. Chem. Phys.*, **41**, 3279 (1964).
21 G. R. Hanson & M. G. Inghram, *Surface Sci.*, **55**, 29 (1976).
22 A. J. Jason, R. P. Burns & M. G. Inghram, *J. Chem. Phys.*, **44**, 4351 (1965); A. J. Jason, R. P. Burns, A. C. Parr & M. G. Inghram, *J. Chem. Phys.*, **44**, 4351 (1966).
23 A. J. Jason, *Phys. Rev.*, **156**, 266 (1967).
24 E. W. Müller & S. V. Krishnaswamy, *Surface Sci.*, **36**, 29 (1973).
25 T. Utsumi & N. V. Smith, *Phys. Rev. Lett.*, **33**, 1294 (1974).
26 N. Ernst, G. Bozdech & J. H. Block, *Int. J. Mass Spectrom. Ion Phys.*, **28**, 33 (1978).
27 H. J. Heinen, F. W. Röllgen & H. D. Beckey, *Z. Naturforsch.*, **29a**, 773 (1974).
28 A. R. Anway, *J. Chem. Phys.*, **50**, 2012 (1969).
29 T. T. Tsong, & T. J. Kinkus, *Phys. Rev.*, **B29**, 529 (1984).
30 T. T. Tsong, *Phys. Rev.*, **B30**, 4946 (1984); *Surface Sci.*, **177**, 593 (1986).
31 D. S. Boudreaux & P. H. Cutler, *Phys. Rev.*, **149**, 170 (1966); *Surface Sci.*, **5**, 230 (1966).
32 R. Haydock & D. R. Kingham, *Surface Sci.*, **103**, 239 (1981).
33 T. T. Tsong, T. J. Kinkus & C. F. Ai, *J. Chem. Phys.*, **78**, 4763 (1983).
34 M. E. Alferieff & C. B. Duke, *J. Chem. Phys.*, **46**, 938 (1967).
35 J. A. Appelbaum & E. G. McRea, *Surface Sci.*, **47**, 445 (1975).
36 R. A. Armstrong, *Surface Sci.*, **47**, 666 (1975).
37 A. A. Lucas, *Phys. Rev. Lett.*, **26**, 813 (1971); *Phys. Rev.*, **B4**, 2939 (1971).
38 A. A. Lucas & M. Sunjic, *J. Vac. Sci. Technol.*, **9**, 725 (1972).
39 T. T. Tsong, T. J. Kinkus & S. B. McLane, *J. Chem. Phys.*, **78**, 7497 (1983).
40 E. W. Müller & T. T. Tsong, *Progr. Surface Sci.*, **4**, 1 (1973).
41 T. Sakurai & E. W. Müller, *Surface Sci.*, **49**, 497 (1975).
42 N. Ernst, W. Drachsel, Y. Li, J. H. Block & H. J. Kreuzer, *Phys. Rev. Lett.*, **57**, 2686 (1986).
43 T. T. Tsong, J. H. Block, M. Nagasaka & B. Viswanathan, *J. Chem. Phys.*, **65**, 2469 (1976).
44 B. H. C. Niu, J. R. Beacham & P. J. Bryant, *J. Chem. Phys.*, **67**, 2039 (1977).
45 B. Viswanathan, W. Drachsel, J. H. Block & T. T. Tsong, *J. Chem. Phys.*, **70**, 2582 (1979).
46 E. W. Müller, *Phys. Rev.*, **102**, 618 (1956).
47 W. Schottky, *Z. Physik*, **14**, 63 (1923).
48 T. T. Tsong, *J. Chem. Phys.*, **54**, 4205 (1971); *J. Phys. F, Metal Phys.*, **8**, 1349 (1978).
49 R. G. Forbes, R. K. Biswas & K. Chibane, *Surface Sci.*, **114**, 498 (1982); R. G. Forbes, *Surface Sci.*, **116**, L195 (1982).
50 R. Gomer, *J. Chem. Phys.*, **31**, 341 (1959).
51 R. Gomer and L. W. Swanson, *J. Chem. Phys.*, **38**, 1613 (1963).
52 T. T. Tsong, *Surface Sci.*, **10**, 102 (1968).
53 D. G. Brandon, *Surface Sci.*, **3**, 1 (1965).
54 A. J. Bennet & L. M. Falicov, *Phys. Rev.*, **151**, 512 (1966); J. W. Gadzuk, *Surface Sci.*, **6**, 133 (1967).

55 T. T. Tsong & E. W. Müller, *Phys. Rev.*, **181**, 530 (1969).
56 T. T. Tsong, *Surface Sci.*, **81**, 28 (1979); **85**, 1 (1979).
57 D. M. Newns, *J. Chem. Phys.*, **50**, 4572 (1969).
58 S. Nakamura & T. Kuroda, *Surface Sci.*, **70**, 452 (1978).
59 E. W. Müller, in *IV. Intern. Kongress f. Elektronenmikroskopie, Berlin*, 1958, Springer, Berlin, 1960,Vol. 1, 820.
60 T. T. Tsong, *Surface Sci.*, **70**, 211 (1978).
61 E. W. Müller & S. V. Krishnaswamy, *Phys. Rev. Letter*, **37**, 1011 (1976).
62 N. Ernst, *Surface Sci.*, **87**, 469 (1979).
63 R. Haylock & D. R. Kingham, *Phys. Rev. Letter*, **44**, 1520 (1980).
64 D. R. Kingham, *Surface Sci.*, **116**, 273 (1982).
65 R. S. Chamber, M. Vesely & G. Ehrlich, 17*th Field Emission Symp.*, Yale University, 1970.
66 D. M. Taylor, PhD Thesis, University of Cambridge, 1970.
67 N. Ernst & Th. Jentsch, *Phys. Rev.*, **B24**, 6234 (1981).
68 G. L. Kellogg, *Surface Sci.*, **111**, 205 (1981); *Phys. Rev.*, **B24**, 1848 (1981).
69 M. Konishi, M. Wada & O. Nishikawa, *Surface Sci.*, **107**, 63 (1981); M. Leisch, *Surface Sci.*, **159**, L445 (1985); K. Murakami, Ph.D Thesis, Osaka Univ. (1985).
70 T. T. Tsong, *Surface Sci.*, **177**, 593 (1986).
71 D. G. Brandon, *Brit. J. Appl. Phys.*, **16**, 683 (1965); *Phil. Mag.*, **14**, 803 (1966).
72 T. T. Tsong & E. W. Müller, *Phys. Stat. Solidi*, **a1**, 513 (1970).
73 R. G. Forbes, *Surface Sci.*, **64**, 367 (1977); **70**, 239 (1978).
74 S. V. Krishnaswamy & E. W. Müller, *Z. Phys. Chem. Neue Folge*, **104**, 121 (1977).
75 R. J. Walko & E. W. Müller, *Phys. Stat. Solidi*, **a9**, K9 (1972).
76 a. D. F. Barofsky & E. W. Müller, *Surface Sci.*, **10**, 177 (1968); b. A. Menand & D. R. Kingham, *J. Phys. C: Solid State Phys.* **18**, 4539 (1985).
77 D. McKinstry, *Surface Sci.*, **29**, 37 (1972).
78 A. R. Waugh & M. J. Southon, *Surface Sci.*, **68**, 79 (1977).
79 T. T. Tsong, S. B. McLane & T. J. Kinkus, *Rev. Sci. Instrum.*, **35**, 1442 (1982); T. T. Tsong, Y. Liou & S. B. McLane, *Rev. Sci. Instrum.*, **55**, 1246 (1984).
80 Y. Liou, J. Liu & T. T. Tsong, to be published.
81 T. T. Tsong, W. A. Schmidt & O. Frank, *Surface Sci.*, **65**, 109 (1977).
82 T. T. Tsong, Y. S. Ng & A. J. Melmed, *Surface Sci.*, **77**, L187 (1977).
83 W. Drachsel, Th. Jentsch & J. H. Block, *Proc. 29th Intern. Field Emission Symp.*, Almqvist & Wiksell, Stockholm, 1982, 299.
84 G. L. Kellogg, *Appl. Surface Sci.*, **11/12**, 186 (1982).
85 T. T. Tsong, *Appl. Phys. Letters*, **45**, 1149 (1984); *Phys. Rev.*, **B30**, 4946 (1984).
86 For formation of cluster ions in pulsed-laser field evaporation of GaAs, GaP and InP etc. see: A. Cerezo, C. R. M. Grovenor & G. D. W. Smith, *Appl. Phys. Lett.*, **46**, 567 (1985); O. Nishikawa, E. Nomura, M. Yanagisawa & M. Nagai, *J. de Physique Coll.*, **C2**, 303 (1986); M. Tomita & T. Kuroda, *Surface Sci.*, **201**, 385 (1988). These studies are mainly concerned with the field evaporation behavior of compound semiconductors and the formation of cluster ions in pulsed-laser stimulated field evaporation and the relative intensities of these ions under different laser power intensities, also with the compositional analysis of these materials. Not for the purpose of studying science of clusters and cluster ions.

87 J. Liu & T. T. Tsong, *Phys. Rev.*, **B38**, 8490 (1988).
88 See for example: A. Herrman, E. Schumacher & L. Worste, *J. Chem. Phys.*, **68**, 2327 (1978); P. Jena, B. K. Ras & S. N. Kana (eds), Physics and chemistry of small clusters, *NATO ASI*, **B158**, Plenum NY (1987) and references therein; W. R. Brown, R. R. Freeman, K. Raghavachari & M. Schlütter, *Science*, **234**, 860 (1987) and references therein.
89 L. A. Bloomfield, R. R. Freeman & W. L. Brown, *Phys. Rev. Lett.*, **54**, 2246 (1985).
90 K. Raghavachari, *J. Chem. Phys.*, **84**, 5672 (1986).
91 K. Raghavachari, *J. Chem. Phys.*, **87**, 2191 (1987) and many other references therein.
92 See for examples E. A. Rohlfing, D. M. Cox & A. Kaldor, *J. Chem. Phys.*, **81**, 3322 (1984); H. W. Kroto, J. R. Heath, S. C. O'Brien, R. F. Curl & R. E. Smally, *Nature*, **318**, 162 (1985).
93 Y. Ohno, T. Kuroda & S. Nakamura, *Surface Sci.*, **75**, 689 (1978).
94 T. T. Tsong, Y. S. Ng & A. J. Melmed, *Surface Sci.*, **77**, L187 (1978).
95 O. Nishikawa, E. Nomura, M. Yanagisawa & M. Nagai, *J. de Physique, Coll.*, **C2**, 303 (1985).
96 K. Sattler, S. Mühlback, O. Echt, P. Pfau & E. Reckenagel, *Phys. Rev. Lett.*, **47**, 160 (1981); P. Pfau, K. Sattler, P. Pflaum & E. Reckenagle, *Phys. Lett.*, **A104**, 262 (1984).
97 See compilation of data by W. Drachsel, Th. Jensch, K. Gingerich & J. H. Block, *Surface Sci.*, **156**, 173 (1985).
98 P. Joyes & J. van der Walle, *J. Phys.*, **B18**, 3805 (1985).
99 T. T. Tsong, *J. Chem. Phys.*, **85**, 639 (1986).
100 T. T. Tsong & Y. Liou. *Phys. Rev.*, **B32**, 4340 (1985).
101 T. T. Tsong, *J. Vac. Sci. Technol.*, **B3**, 1425 (1985).
102 G. L. Kellogg & T. T. Tsong, *J. Appl. Phys.*, **51**, 1184 (1980).
103 T. T. Tsong & R. J. Walko, *Physica Status Solidi*, **a12**, 111 (1972); G. L. Kellogg, *Phys. Rev.*, **B12**, 1343 (1975).
104 S. Jaenicke, A. Ciszewski, W. Drachsel, U. Wigman, T. T. Tsong, J. R. Pitts, J. R. Block & D. Menzel, *J. de Physique, Coll.*, **C7**, 343 (1986).
105 E. R. McMullen & J. R. Perdew, *Solid State Commun.*, **44**, 945 (1982); *Phys. Rev.*, **B36**, 2598 (1987).
106 H. J. Kreuzer & K. Nath, *Surface Sci.*, **183**, 591 (1987).
107 A. B. Anderson, *J. Chem. Phys.*, **63**, 4430 (1975) and references therein.
108 P. Gies & R. R. Gerhardts, *Phys. Rev.*, **B33**, 982 (1986).
109 G. L. Kellogg, *Phys. Rev.*, **B29**, 4304 (1984).
110 E. W. Müller, S. B. McLane & J. A. Panitz, *Surface Sci.*, **17**, 430 (1969); E. W. Müller, S. V. Krishnaswamy & S. B. McLane, *Surface Sci.*, **23**, 112 (1970).
111 J. T. McKinney & S. S. Brenner, *16th Field Emission Symp.*, Pittsburgh, PA, 1969.
112 T. T. Tsong, & E. W. Müller, *Phys. Rev. Letters*, **25**, 911 (1970); *J. Chem. Phys.*, **55**, 2884 (1971).
113 G. L. Kellogg & T. T. Tsong, *Surface Sci.*, **110**, L559 (1981).
114 R. G. Forbes & M. K. Wafi, *Surface Sci.*, **93**, 192 (1980).
115 T. T. Tsong, *Surface Sci.*, **140**, 377 (1984).
116 K. Nath, H. J. Kreuzer & A. B. Anderson, *Surface Sci.*, **176**, 261 (1986).
117 T. J. Kinkus & T. T. Tsong, *J. Vac. Sci. Technol.*, **A3**, 1521 (1985).
118 G. L. Kellogg, *J. Chem. Phys.*, **74**, 1479 (1981).

119 M. Hellsing & B. Hellsing, *Surface Sci.*, **176**, 249 (1986).
120 N. Ernst, G. Bozdesch & J. H. Block, *Proc. 27th Intern. Field Emission Symp., Tokyo*, 1980, 151.
121 K. D. Rendulic, *Surface Sci.*, **28**, 285 (1971).
122 M. W. Cole & T. T. Tsong, *Surface Sci.*, **79**, 325 (1977).
123 R. Hiskes, *Phys. Rev.*, **122**, 1207 (1961).
124 A. C. Riviera & D. R. Sweetman, *Phys. Rev. Lett.*, **5**, 560 (1960).
125 H. D. Beckey & H. Knoppel, *Z. Naturforschg.*, **21A**, 1920 (1966).
126 H. D. Becky, *Field Ionization Mass Spectrometry*, Pergamon Press, New York, 1971.
127 G. R. Hanson, *J. Chem. Phys.*, **62**, 1161 (1975).
128 T. T. Tsong & Y. Liou, *Phys. Rev. Lett.*, **55**, 2180 (1985); T. T. Tsong, *J. Appl. Phys.*, **58**, 2404 (1985).
129 T. T. Tsong, *Phys. Rev. Lett.*, **55**, 2826 (1985).
130 T. T. Tsong & M. W. Cole, *Phys. Rev.*, **B35**, 66 (1987).
131 N. Miskovsky & T. T. Tsong, *Phys. Rev.*, **B38**, 11188 (1988).
132 M. Hotokka, T. Kindstedt, P. Pyykko & B. O. Roos, *Mol. Phys.*, **52**, 23 (1984).
133 J. K. Norskov, *Phys. Rev.*, **B26**, 2785 (1982); N. D. Lang & J. K. Norskov, *Phys. Rev.*, **B27**, 4612 (1983); A. Frigo, F. Toigo, M. W. Cole & F. O. Goodman, *Phys. Rev.*, **B33**, 4184 (1986).
134 W. Schmidt, Th. Reisner & E. Krautz, *Surface Sci.*, **26**, 297 (1971).
135 K. D. Redulic, *Surface Sci.*, **34**, 581 (1973).
136 D. A. Nolan & R. M. Herman, *Phys. Rev.*, **B8**, 4088 (1973); **B10**, 50 (1974).
137 H. Iwasaki & S. Nakamura, *Surface Sci.*, **49**, 664 (1975); **52**, 588 (1975).
138 S. B. McLane, S. V. Krishnaswamy & E. W. Müller, *Surface Sci.*, **27**, 367 (1971).
139 J. H. Sweeney & T. T. Tsong, *Surface Sci.*, **104**, L179 (1981).
140 S. J. Fonash & G. L. Schrenk, *Phys. Rev.*, **180**, 649 (1969); S. P. Sharma & G. L. Schrenk, *Phys. Rev.*, **B2**, 598 (1970).
141 A. M. Russell, *J. Appl. Phys.*, **33**, 970 (1962).
142 Y. C. Chen & D. N. Seidman, *Surface Sci.*, **27**, 231 (1972).
143 T. T. Tsong, *Phys. Rev.*, **B6**, 417 (1972); P. L. Cowan & T. T. Tsong, *Surface Sci.*, **67**, 158 (1977).

3

Instrumentations and techniques

3.1 Introduction

The field ion microscope is perhaps the simplest of all atomic resolution microscopes as far as mechanical and electrical designs are concerned. The atomic resolution microscopes, at the present time, include also different types of electron microscopes,[1] the scanning tunneling microscope (STM)[2] and the atomic force microscope (AFM)[2] Before we discuss the general design features of the field ion microscope it is perhaps worthwhile to describe the first field ion microscope,[3] and a very simple FIM[4] which can be constructed in almost any laboratory. The first field ion microscope, shown in Fig. 3.1, is essentially a field emission microscope[5] except that it is now equipped with a palladium tube with

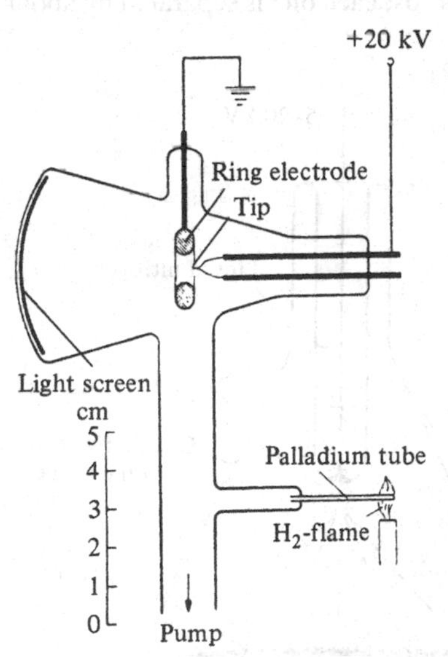

Fig. 3.1 The first field ion microscope. Hydrogen could be introduced into the microscope by heating a palladium tube. There was no provision for cooling the tip.

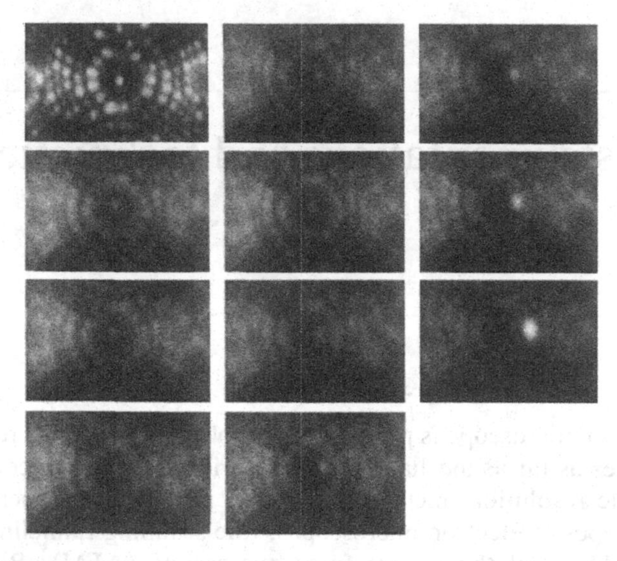

Fig. 3.2 The first picture at the upper left-hand corner is a 78 K He field ion micrograph of a W (112) surface with a W adatom on it. From top to bottom and then from left to right: the rest of the images are ~290 K He field ion micrographs of the same surface where diffusion of the adatom can be clearly seen. When the adatom is near the center of the plane, it performs a random walk. However, when it is slightly off the center it is driven toward the plane edge by a field gradient induced driving force, as will be discussed in Chapter 4. These are real time photos; each one is separated by about 5 s.

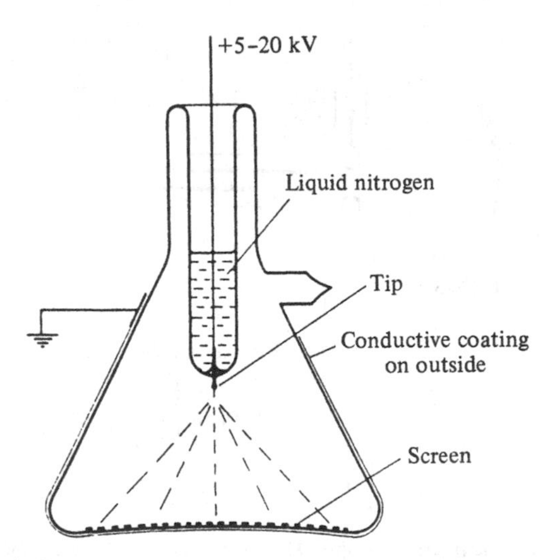

Fig. 3.3 Simple Pyrex glass field ion microscope which can be constructed in most laboratories.

which hydrogen of a few mTorr can be admitted to the microscope by simply heating it with a hydrogen flame, for example with an alcohol lamp. The tip is connected to a high positive voltage power supply. Hydrogen was chosen as the image gas because of its low sputtering efficiency; thus the microscope can be operated in the field electron emission mode by just reversing the polarity of the tip voltage. Hydrogen requires an image field of about 2.2 V/Å, which is at least several times higher than the field strength needed for the field emission microscope. In the first FIM, no provision was available for cooling the tip, as the need of cooling the tip to achieve better image resolution was realized only much later. Room temperature field ion images not only have a very poor resolution, but even worse, are also lacking in image contrast. However, it was later realized that even with the tip kept at the room temperature, the resolution of the FIM is sufficient to see individual adsorbed metal atoms on a large plane, especially if He or Ne is used as an image gas.[6] In fact, the drifting motion of an adatom under the influence of a driving force, resulting from a field gradient existing across a plane, can be seen at room temperature in real time, as shown in Fig. 3.2.

If one strives for the elegance of simplicity, then a very simple design which requires no more than the facility of most high school or undergraduate laboratories to construct can be considered. As shown in Fig. 3.3, the FIM chamber is made of a Pyrex glass, say a 500 ml Erlenmeyer flask, into which a test-tube cold finger with a single tungsten lead-through is sealed after a phosphor screen has been deposited on the bottom of the flask. A tungsten tip, prepared by electrochemical polishing using a method described in Section 3.1.2, is mounted on the cold finger. The finished tube is pumped down to vacuum in the 10^{-8} Torr range by using any conventional pump system and providing a light bake-out at about 250°C or slightly higher. The tube is either back-filled with helium to a pressure of about 5 mTorr and then sealed off, or is simply sealed off under high vacuum if no gas handling facility is available. In such a case, filling of helium can be done by diffusion through the glass wall at room temperature. For this purpose, the sealed-off tube is placed under a bell jar or a plastic bag which is flashed with steel cylinder quality helium at atmospheric pressure. Depending on the thickness of the glass wall, the diffusion takes one to a few days. For operating the field ion microscope, one has only to fill the cold finger with liquid nitrogen and apply a continuously adjustable, high voltage of some +5 to +20 kV to the tip. The ground electrode is provided by coating the lower outside wall of the Erlenmeyer flask with a thin film of hydroscopic liquid, such as a solution of calcium chloride in glycerine. The conductivity of the glass, having a resistance of about 10^{12} Ω, is sufficient to maintain the inner walls of the tube near ground potential. At a typical ion current of 10^{-9} A, the voltage drop is only on the order of 1 kV, and there are enough secondary electrons and slow ions drifting in the tube to equilibrate the wall and phosphorescent screen charges, even without a conductive coating of the inner wall. As there is no image intensification device in this simple field ion microscope, the intensity of the ion image is very weak, and observation must be made in complete darkness with very well dark-adapted eyes. When the voltage is gradually raised, image spots will start to appear at a few kV applied voltage. By raising the voltage very slowly, one should be able to see field evaporation of surface atoms, and the surface will eventually develop to a good field ion pattern. It is however possible that the tip may rupture by field stress before the surface is completely developed into an end form and a new tip will then have to be installed by repeating the whole preparation steps again. For a tungsten tip, the chance of having an end form developed should be better than 90%. The limitation and inconvenience of such a simple FIM for scientific studies is quite obvious.

3.1.1 Basic design of the field ion microscope

The field ion microscope is very simple in mechanical and electrical designs. The basic requirements are that the tip must be able to be effectively cooled down to the desired low temperature, and the electrical insulation must be good enough to sustain a tip voltage of about 20 kV or higher. The lower the tip temperature, the better the resolution of the field ion microscope. The higher the voltage the tip mounting assembly can sustain, the larger the radius of the tip one can use for FIM examinations. As the field strength in V/Å is approximately given by $F = V/5r$, when r is expressed in Å, the tip radius can be as large as 900 Å for a 20 kV image voltage when He is used and about 1800 Å when Ar or H_2 is used. Further increasing the tip voltage is not desirable for two reasons. First, the resolution deteriorates for larger tip radius. Second, the X-ray created by ion bombardment of the counter electrode may become hazardous to the health of the observer. Although high-voltage FIM can easily be developed simply by employing better electrical insulators, it is not generally used.

In the early days, field ion microscopes were almost always constructed out of Pyrex glasses, and no convenient internal image intensification device was available. Cooling of the tip was always done by filling the cold finger with liquid nitrogen, liquid neon or liquid hydrogen.[4] With the development of stainless steel vacuum components, few field ion microscopes nowadays use such designs. Almost all field ion microscopes now use stainless steel chamber and ultra-high-vacuum components. Tip cooling can still be done with a cold finger filled with liquid nitrogen, or a

Displex closed-cycle helium refrigerator, which has the advantage of being able to adjust the tip temperature all the way from about 10 K to room temperature. For the purpose of image intensification, a micro-channel plate can be used. Figure 3.4(*a*) is a schematic diagram of a field ion microscope which uses a helium Displex refrigerator for the tip cooling. The temperature of the tip can be adjusted all the way from

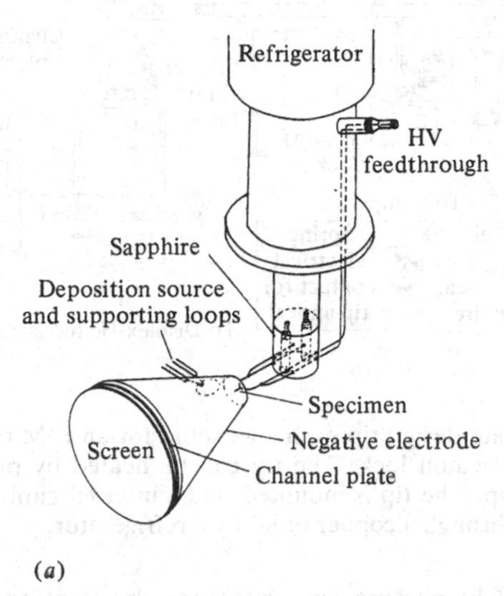

(*a*)

Fig. 3.4 (*a*) Simple metal field ion microscope which uses a Helium Displex Refrigerator for tip cooling. A tip is mounted on a 5 mil Pt wire heating loop with two 1 mil Pt potential leads spot welded onto it. An adatom deposition source, a coil supported by two degassable loops, is also shown. Vacuum is easier to achieve with this simple design.

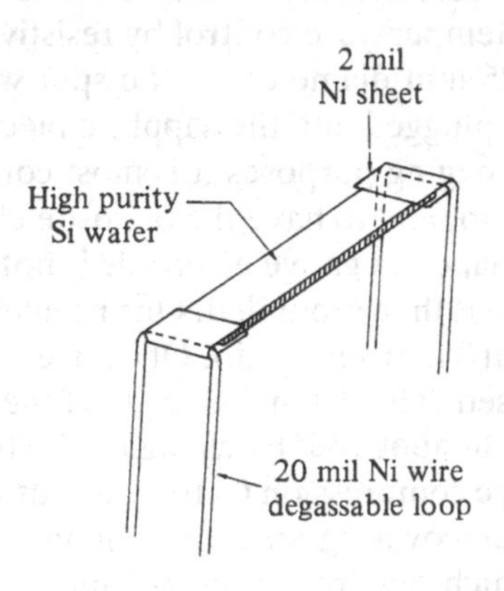

(*b*)

Fig. 3.4 (*b*) Deposition source for silicon adatoms.

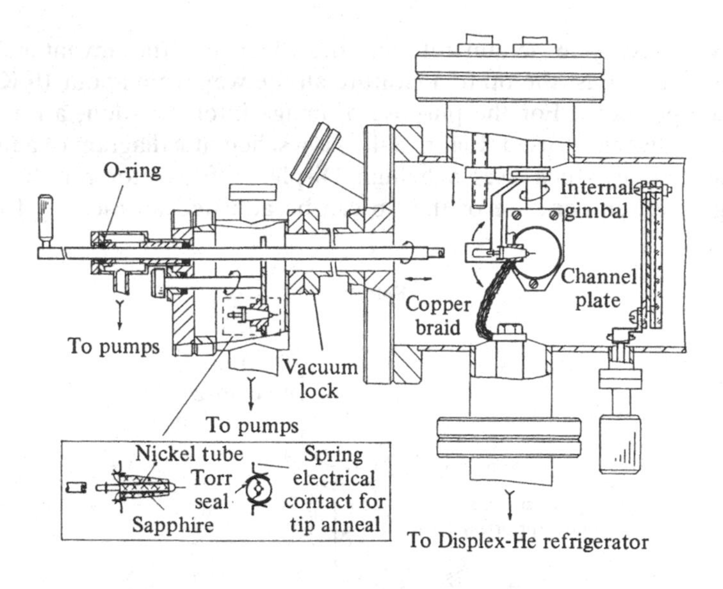

(*c*)

Fig. 3.4 (*c*) Penn State design tip exchange setup for an FIM or a pulsed-laser atom-probe with a vacuum lock. The tip can be heated by passing a current through the wire loop. The tip is mounted on an internal gimbal and is cooled through a copper braid by a refrigerator.

about 10 K to 300 K by passing a current through a heating element of the refrigerator. A sapphire piece is tightly attached to the copper base of the refrigerator tip. Sapphire is an excellent thermal conductor at low temperature and an excellent electrical insulator. The tip is spot welded onto a tip heating loop which can be a wire of Pt, Mo or W of about 0.1 mm diameter. For tip temperature control by resistivity heating, two Pt potential leads of 0.025 mm diameter can be spot welded to the loop. These wires are tightly plugged into the sapphire piece by wedge shaped plugs. For most FIM imaging purposes it is most convenient to apply a positive voltage to the tip, and to have the opposite electrode grounded. In that case the cone shaped negative electrode is not needed. The front face of the channel plate is then grounded. One usually uses a 3 in channel plate and places it about 3 in from the tip. Of course, a smaller or a larger channel plate can be used. The extension angle of the channel plate with the tip can be adjusted to about 60° by changing the tip-to-channel-plate distance. With the image compression factor of about 1.5, one will be able to image the tip surface covering an extension angle of about 90°. For certain experiments which require heating or annealing of the tip to high temperatures, accurate monitoring of the tip temperature is easier if the tip is kept near the ground potential. In that case, a negative imaging voltage is applied to the front face of the channel plate and a cone shaped electrode. The cone shaped electrode is used to avoid the excessively high voltage which would otherwise be needed to establish the field strength required for the imaging.

One can also use a simple cold finger instead of a refrigerator, in which case liquid nitrogen cooling is most convenient. In early days, liquid hydrogen was used, but it is expensive and dangerous and has now been replaced by the refrigerator. Liquid helium can be used, but the heat capacity is very small. Again, a sapphire piece has to be tightly attached to the cold finger to make sure of a good thermal contact. Good thermal contact is a basic requirement for achieving the best resolution of the field ion microscope, and anyone who wants to build a FIM should make sure of having a good thermal contact between the tip and the cold finger.

For the purpose of image intensification, a microchannel plate is most convenient. A gain of 10^3 to 10^4 is sufficient. Without a microchannel plate, the image intensity is so low that one must try to use as high pressure as possible for the image gas, and 2 to 5×10^{-3} Torr is commonly used. When the gas pressure is too high the mean free path becomes too short and scattering of field ions can occur, which will deteriorate the image contrast and even more seriously may induce gas discharges. With the availability of microchannel plate for image intensification, image gas pressure in the 10^{-5} Torr range is now sufficient. For a typical tip size of a few hundred Å radius, imaged at around 10 kV, the exposure time using a typical camera with an *f* number of about 1.4 and a fast film of ISO (ASA) rating around 400, is only 1 to a few seconds.

Additional setups for tip manipulations can be installed in the FIM. For example, it is often desirable to deposit atoms of various kinds onto a well developed field ion tip surface, either for studying the behavior of single atoms on the surface or for forming alloy layers on the surface. For such purposes, a vapor deposition source can be installed. The simplest deposition source is a coil of wire which can be heated by simply passing an electric current. It is however very important that both the supporting leads of the deposition coil and the coil are thoroughly degassed before the deposition. For deposition of non-metallic materials such as Si, different designs will have to be used. An example of a source for the deposition of Si atoms is shown in Fig. 3.4(*b*).

It is often desirable to have a separate chamber with a vacuum lock for the purpose of tip exchange and tip manipulation without the need of exposing the FIM chamber to atmosphere. There are many different designs used by different investigators. One employed in the author's laboratory, designed by G. L. Xiao, is shown in Fig. 3.4(*c*). A tip is mounted on a cone-shaped sapphire piece with a cross-shaped stainless steel thin rod attached at the back end. The tip mounting sapphire piece can be plugged into a copper seat of the cold finger or can be removed from the seat using a cylindrical transfer rod with two L-shaped cut-outs, as shown in the figure. When a tip is to be changed, the rod is moved into the FIM chamber, and is inserted into the back end of the tip-mount-sapphire piece, rotated through a small angle and then withdrawn into the tip exchange chamber. In the tip exchange chamber, there is a socket-assembly where several tip-mount-sapphire pieces are seated. To exchange a tip, the following steps are taken.

1. Withdraw the rod from the socket assembly through a cut-out.
2. Rotate the socket assembly until an empty socket is lined up with the transfer rod axis.
3. Insert the 'old' tip-mount-sapphire piece into the empty socket and withdraw the transfer rod.
4. Rotate the socket assembly until a socket with a new tip-mount-sapphire piece is lined up with the transfer rod.
5. Remove the new tip-mount-sapphire piece from the socket assembly.
6. Rotate the socket assembly until the cut-out is lined up with the transfer rod axis.
7. The new tip-mount-sapphire piece is transferred into the cold finger copper seat, and the rod is then withdrawn. The tip exchange is now complete and finally the UHV valve between the tip exchange chamber and the FIM chamber is closed.

3.1.2 Specimen preparation

The most convenient and effective method for preparing a tip specimen is by electrochemical polishing of a piece of thin wire of 0.05–0.2 mm diameter. Usually the methods developed for electropolishing thin film specimens in transmission electron microscopes are also applicable for polishing field ion microscope tips.[7] In Table 3.1 some of the commonly used emitter polishing solutions and conditions for the polishing are listed for various materials.[8]

For chemically reactive metals such as Fe, Ni, Mo and W, etc., tip polishing is in general very simple. As shown in Fig. 3.5, a beaker is filled with three quarters of the recommended polishing solution. A piece of thin wire is mounted on a mechanical manipulator so that the wire can be dipped into the solution to a desirable depth, and can also be lifted out of the solution. Usually a section of about 5 to 8 mm should be immersed in the solution. A counter-electrode, either a piece of Pt foil or simply a piece of tungsten wire, can be a loop or simply a straight piece of foil or wire. It is essential that the tip specimen wire is held in the vertical position so that the convection of the solution during the polishing can be

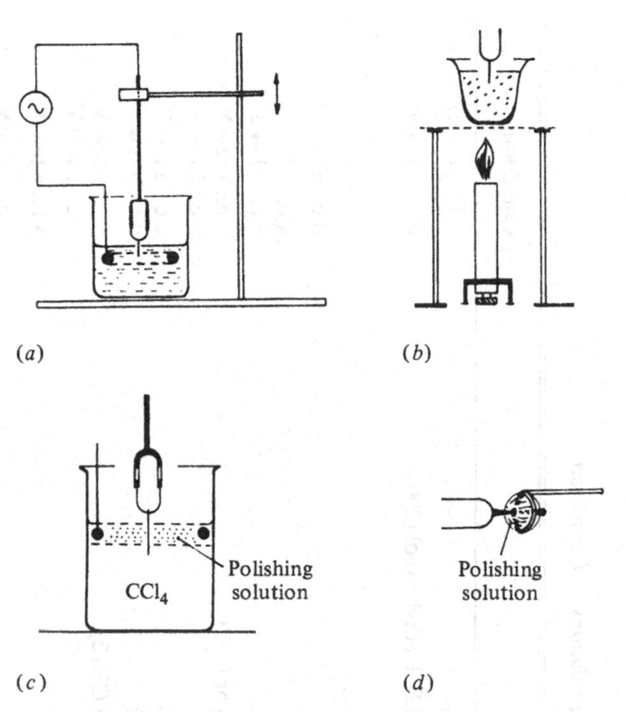

Fig. 3.5 (*a*) Simple tip polishing setup. A small piece of thin wire of tip material is spot welded onto a tip supporting loop which is usually made of 4 to 5 mil Mo or Pt wire. The loop is plugged into a Ni tube, which can be raised or lowered into a polishing solution. The polishing can be done with either a d.c. or an a.c. voltage of a few volts.

(*b*) In a molten salt method, a Pt or Fe beaker serves as an electrode. The salt is heated at a temperature just above the melting point of the salt.

(*c*) Floating thin layer tip polishing method.

(*d*) A tip can be prepared from a fine whisker using a drop of polishing solution supported on a Pt wire loop. Tip motion can be controlled with a micromanipulator. The etching part can be observed with a low magnification microscope.

symmetrical with respect to the wire and a symmetrically shaped tip can be more easily formed. By applying the recommended a.c. or d.c. voltage to the tip specimen and the counter electrode, a tip can usually be formed in from a few seconds to less than a minute. When a d.c. voltage is used, the tip should be connected to the positive polarity. Once in a while, the tip can be lifted for visual or microscopic inspection. In an optical microscope, the very end of a sharp tip cannot be resolved, as a sharp tip usually has a radius of the order of 50 Å or less, which is much less than the wavelength of the visible light. Thus it is not possible to tell how good a tip is by examining it in an optical microscope. When the tip surface appears

Table 3.1 *Tip polishing solutions and conditions*

Materials	Polishing solution	Conditions
Aluminum	10% perchloric aid in methanol	a.c.: a few V
Beryllium	$HCl(4) + H_2SO_4(4) + HNO_3(2)$ + ethyl glycol (200)	d.c.: a few V
Carbon (graphite)	Heat in oxidizing flame	
	Same as molybdenum	
Copper	Conc. H_3PO_4	d.c.: a few V
Gold	HCl (50%) + HNO_3 (50%)	a.c.: a few V
	20% KCN aq. sol.	a.c.: $10 \rightarrow 3$ V
Iridium	Molten $NaNO_3(3) + KOH(7)$ or NaCl	a.c.: a few V
	20% KCN aq. sol.	a.c.: $10 \rightarrow 3$ V
Iron and steel	Dilute HCl (1–10%)	a.c.: a few V
	$HNO_3(1) + HCl(1) + H_2O(2)$	d.c.: a few V
Molybdenum	NH_4OH aq. sol. (8) + conc. KOH(2)	a.c.: a few V
	Conc. KOH	d.c.: a few V
Nickel	40% HCl	a.c.: a few V
Nichrome alloys	10% HNO_3 or HCl	a.c.: a few V
Niobium	Molten $NaNO_3$	a.c.: a few V
	$HNO_3(9) + HF(1)$	d.c.: a few V
	$H_2SO_4(1) + HF(1) + H_2O(4)$	d.c.: a few V

Niobium nitrate	$HF(1) + HNO_3(3)$	d.c.: 10 → 3 V
Platinum	Molten $NaNO_3(4) + NaCl(1)$	d.c.: a few V
	20% KCN aq. sol.	a.c.: 15 → 3 V
Platinum alloys	Same as platinum	
Rhenium	12N H_2SO_4 or Conc. HNO_3	d.c.: ~10 V
Rhodium	Same as platinum	
Ruthenium	Same as platinum; aq. sol. KCl or KOH (25% of sat.)	a.c.: 1 to 30 V
Silicon	Conc. HNO_3(45%) + 48% HF(42%) + CH_3COOH(13%)	Dip into solution
Tantalum	Molten $NaNO_2$ or KOH	a.c.: ~8 V
	$HF(2) + CH_3COOH(0.5) + H_3PO_4(1) + H_2SO_4(1)$	d.c.: 5–15 V
Tin	HF	a.c.: a few V
Titanium	HF	d.c.: 4 to 12 V
Titanium nitrite	HF	a.c. or d.c.: a few V
Tungsten	Same as molybdenum	
Tungsten alloys	Same as molybdenum	
Zinc	Conc. KOH	d.c.: 10 to 15 V
Zirconium	10% HF	Dip into solution
GaAs	$H_2SO_4(4) + H_2O_2(1) + H_2O(1)$	Dip into hot solution
GaP	$HNO_3(3) + HCN(1)$	Dip into hot solution
LaB_6	40% HNO_3	a.c.: a few V

smooth and the tip end cannot be resolved in an optical microscope, in nine cases out of ten it is a good tip. Field ion images can of course reveal immediately whether a tip is good or not.

For noble metals such as Pt, Ir, Rh, Ru and their alloys, a molten salt tip polishing method is found to be most effective. The recommended salt mixture is put into a crucible of Pt or more simply of Fe to about three quarters capacity. The salt mixture is melted with a gas burner and the tip polishing can then proceed as before. The crucible can be used as the counter electrode. Usually one finds the best result by keeping the salt just near its melting point, which is around 400 to 500 °C.

Many other more sophisticated methods have been developed to handle special specimens. One of the methods is the floating thin layer method shown in Fig. 3.5(*c*).[9] A thin layer of polishing solution of a few mm depth is floated on the surface of an insulating and immiscible liquid of higher density, such as CCl_4. In this method, it is important that the section of the wire dipped inside the insulating liquid is kept as short as possible. If this section is too long, its weight is enough to create lattice defects when it fractures and drops off as the etched part of the wire becomes very thin. This method is good for producing lattice defects but not particularly useful if one is looking for a tip with a good surface! There was a period of a few years when many papers reported observations of streaky field ion images.[10] Such images are produced by sharp edges at the fractured part of the tip surface, produced during the dropping off of the unetched section. The thin floating layer method is no longer popular, although such tips may be suited for STM use.

For etching a very small specimen such as a very thin whisker, a drop of the polishing solution may be supported on a small wire loop, as shown in Fig. 3.5(*d*). The whisker is then dipped into the solution. All the electrochemical polishing may be done under a low power optical microscope. This tip polishing method has been used very successfully for polishing silicon and other semiconductor tips from thin whiskers, and also for polishing high temperature superconductor tips out of small, finely shaped chips.[11]

For a solid piece of specimen, thin rods of the sample can be prepared by cutting with a diamond saw or by a photo-resist etching method,[12] followed by ordinary electrochemical polishing. The rod can be inserted into a Pt or Ni tube, which is then spot welded onto a heating wire loop. Often it is necessary to fill the tube with silver paint to have good electric contact. In such a case, heating of the sample by passing a current through the loop may not be advisable. Instead, the tip may have to be annealed by using a properly focused CW laser beam.[13]

With the development of the STM and the achievement of atomic resolution with this microscope, the preparation and characterization of a probing sharp tip, ideally with the size of a single atom, is becoming a more and more important task. In field ion microscopy, a sharp tip of much less than 100 Å radius, sometimes perhaps a few atoms, can be easily obtained by electrochemical etching of a tip to a very slender shape. Such tips have too few atoms and too low an image voltage for FIM observations, and are usually made much duller by field evaporation. The most convenient tip radii in FIM studies range from about 200 to 500 Å. A slender tip, unfortunately, is mechanically unstable and thus unsuitable for STM use. An STM tip has to have a very large cone angle, but is very sharp right at the tip section, preferably with a concave outward profile. Two methods have been proposed for producing sharp tips for STM use, both of them using the (111) oriented tungsten tips. One employs ion sputtering of a tungsten tip by reversing the polarity of the tip to field emission mode in the presence of Ar or Ne gas. Gas atoms are ionized near the tip surface by the bombardment of field electrons. The tip surface is sputtered by these ions and can be sharpened to a radius so small as to consist of a (111) facet of three atoms. An adatom can be added to the top of this facet by vapor deposition, so that the final facet is a single atom.[14]

Another method is by electric field gradient induced surface diffusion, or build up of a tip. A (111) oriented tungsten tip is heated to about 800 K and simultaneously a negative field of about 0.2 V/Å is applied to the tip surface. Under the high temperature and the high field gradient existing at the tip surface, atoms diffuse toward the (111) pole where the electric field is the highest. A sharp protrusion can be built up at the (111) pole of the tip surface.[15]

3.1.3 Image interpretation

The crystal structure and lattice parameters of all materials suitable for field ion microscopy are already known from X-ray crystallography. Identification and indexing of crystal planes of a field ion image can be done most easily by comparing the micrograph with the expected symmetry and the surface structure of the lattice. In the early days, ball models were built for comparing with field ion images, as has been shown already in Fig. 1.3. By assuming that field ion images are formed only for atoms in the more protruding positions of the emitter surface, the ball model appears very similar to the field ion image. This resemblance suggests that the indexing of a micrograph can be done by comparing it with a geometrical projection of the crystal lattice. It is now generally agreed that field ion images can be represented approximately by a stereographic projection image of the nearly hemispherical emitter surface.[16] The field ion micrographs are of course, not uniformly magni-

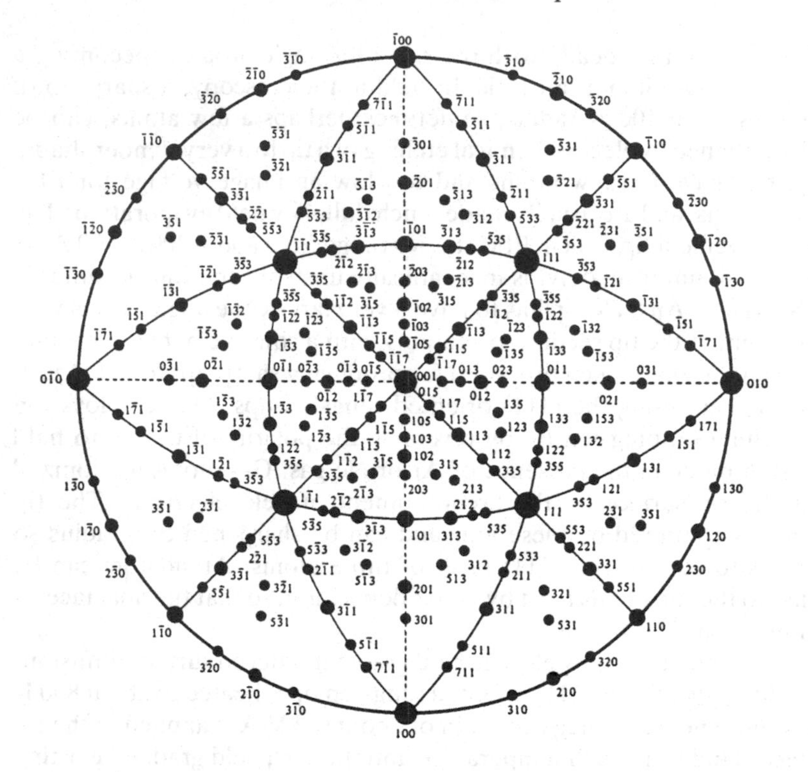

Fig. 3.6 (*a*) Stereographic projection map of a (001) oriented cubic crystal. The facet sizes reflect approximately those of the field ion image of a fcc crystal.

fied, owing to variations in the evaporation field of surface atoms at different crystal planes. The evaporation field depends on the work function of the surface, which varies from one plane to another and also from one region to another. Thus there are small changes in the local radius of curvature and in the image magnification as well. With the development of computer simulation methods for field ion images,[17] identification of crystal planes can now be done much more reliably than with ball models, and these models are now used only in illustrations for general audiences. Nevertheless, it is still very convenient to have stereographic projection maps which can be used for a quick and routine identification of crystal facets in the field ion micrograph. Figure 3.6(*a*) and (*b*) show stereographic projections of a (001) and a (110) oriented cubic lattice. The facet sizes reflect approximately the sizes of facets developed on field ion emitter surfaces of fcc metals for the (001) orientation map, and on those of bcc metals for the (110) orientation map. For a perfectly spherical emitter surface, the sizes of facets should be larger for planes with larger atomic step heights. In real field ion

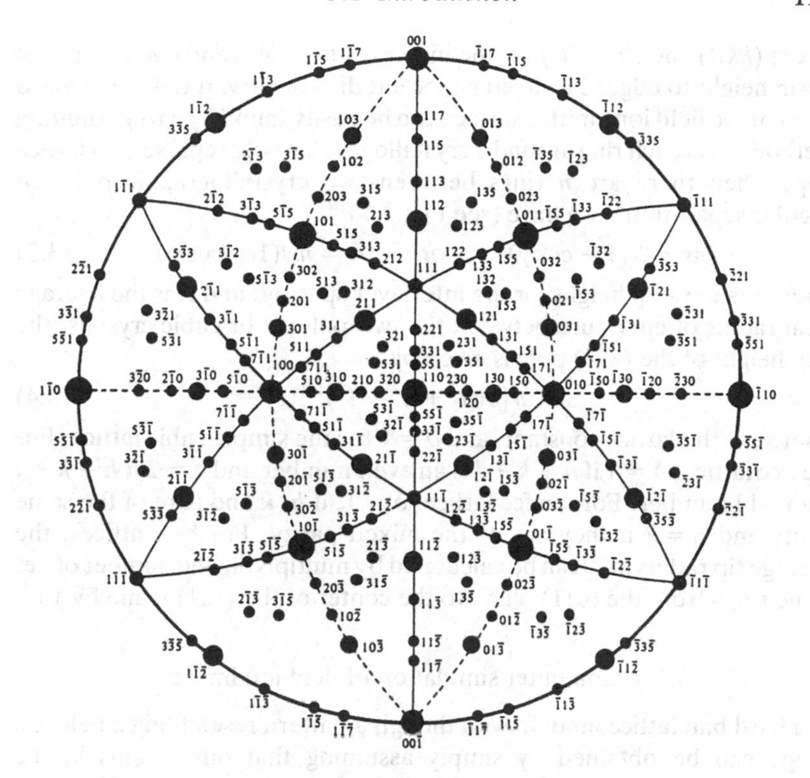

Fig. 3.6 (*b*) Stereographic projection map of a (110) oriented cubic crystal. The facet sizes reflect approximately those of the field ion image of a bcc crystal.

images the sizes of high atom density planes are further accentuated by field evaporation process. High atom density planes are planes with larger values of the work function. According to eqs (2.35) and (2.36), their evaporation fields should be lower, thus the radii of curvature of these planes should be larger for a field evaporation end form of the emitter surface and their facet sizes should also be larger.

It is often necessary to find the angle between two crystallographic poles of known Miller indices (hkl) and $(h'k'l')$. For the cubic lattice, it is given by

$$\cos\gamma = \frac{hh' + kk' + ll'}{(h^2 + k^2 + l^2)^{1/2}(h'^2 + k'^2 + l'^2)^{1/2}}. \tag{3.1}$$

For the hexagonal crystal, it is given by

$$\cos\gamma = \frac{hh' + kk' + \frac{1}{2}(kh' + hk') + \frac{3}{4}\frac{a^2}{c^2}ll'}{\left\{\left[h^2 + hk + k^2 + \frac{3}{4}\frac{a^2}{c^2}l^2\right]\left[h'^2 + h'k' + k'^2 + \frac{3}{4}\frac{a^2}{c^2}l'^2\right]\right\}^{1/2}}, \tag{3.2}$$

where $(hkil)$ and $(h'k'i'l')$ are the indices of the poles and c/a the ratio of prism height to edge. The average local radius of curvature between two poles of the field ion emitter surface can be easily found by a ring counting method.[18] As each ring around a crystallographic pole represents a lattice step, when there are n rings between two crystallographic poles of angular separation γ we have (see Fig. 3.7)[19]

$$ns = R(1 - \cos\gamma), \qquad \text{or} \qquad R = ns/(1 - \cos\gamma), \tag{3.3}$$

where s is the step height, or the interlayer spacing, and R is the average local radius of curvature between the two poles. For cubic crystals, the step height of the (hkl) pole is given by

$$s = a/\{\delta(h^2 + k^2 + l^2)^{1/2}\}, \tag{3.4}$$

where a is the lattice constant, and $\delta = 1$ for the simple cubic lattice. For the bcc lattice, $\delta = 1$ if $h + k + l$ is an even number and $\delta = 2$ if $h + k + l$ is an odd number. For the fcc lattice, $\delta = 1$, if h, k and l are of the same parity and $\delta = 2$ if they are of the mixed parity. For bcc lattices, the average tip radius in Å can be calculated by multiplying the number of net plane rings from the (011) plane to the center of the (121) plane by 16.

3.1.4 Computer simulation of field ion images

In a hard ball lattice model, even though a pattern resembling a field ion image can be obtained by simply assuming that only atoms in the protruding positions are imaged, it is difficult to be quantitative about the degree of protrusion required for an atom to be imaged in the field ion microscope. The hard ball model is cumbersome and inflexible. It is also expensive to construct a large model capable of showing high index planes. These difficulties can be easily solved by a computer simulation of field ion images, introduced by Moore in 1962.[17]

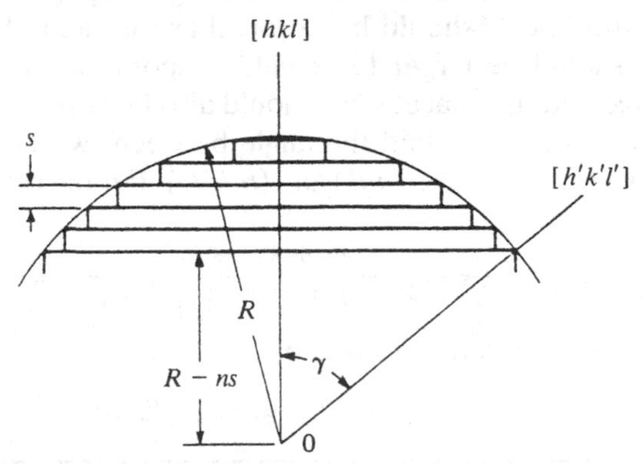

Fig. 3.7 Relation between the cap height and the step height of atomic layers on a spherically shaped crystal lattice.

For simplicity, Moore assumed that an emitter represents an ideal, spherical surface, which consists of atoms of an infinite crystal lattice with their centers lying inside a spherical surface. In the field ion microscope, not all surface atoms are imaged. In fact, not even all kink site atoms are imaged. It is assumed that only atoms with their centers lying sufficiently close to the spherical surface will give rise to field ion image spots. The degree of protrusion of an individual atom can now be quantitatively represented by the shortest distance p from the center of the atom to the spherical surface. In other words, in the thin shell model, it is postulated that in order for the atom to give rise to a field ion image, the distance p must be smaller than a critical value p_0 or else the center of the atom must lie within a thin shell of thickness p_0 below the spherical surface. The physical significance of this criterion can be interpreted as being equivalent to the assumption that the surface electronic charges of an ideal, spherical but stepped surface smooth out completely to form a geometrically nearly perfect, spherical surface. Thus the distance p for a given atom is a measure of the degree to which that atom is depleted of electronic charges by the application of the high voltage. This degree of charge depletion reflects the local electric field strength as given by Gauss' law. The field ionization rate is related to the electric field and hence also the distance p. An atom lying at a certain distance p_0 below the reference sphere will not field ionize enough to form a field ion image. This thin shell model, while lacking in details such as variations in the local radius of curvature of a field evaporated surface and possible effects of chemical bonds and field adsorption of image gas atoms, has been very successful in producing simulation images closely resembling field ion images.

The image simulation procedure is fairly straightforward in principle. Let us consider the simulation of an fcc lattice. The first step is to find the location of all surface atoms having their bonds broken by the spherical surface. An appropriate tip axis, most commonly [001] or [111], is chosen according to the tip orientation of the field ion image one intends to simulate. The center of the sphere coincides with the center of an atom, and the radius of the sphere is an integral multiple of the lattice parameter. As the result of this choice the reference sphere is made tangential to a (001) plane, or a (111) plane, at the tip apex; hence all atoms within that plane are outside of the spherical surface. Now all the atoms with their centers lying within a spherical thin shell of thickness p_0 are found and their coordinates identified. These are atoms which will give rise to field ion image spots. It is now a matter of choosing which of the projections one wants to use for forming a simulation image. As discussed already in the last section, field ion images can best be represented by a stereographic projection.

It is quite obvious that the number of simulated image points for a given lattice depends on the thickness of the thin shell one chooses for the computation. The shell thickness p_0 can be varied until the total number of image points agrees with that of an actual field ion image of the same average tip radius. Moore & Ranganathan[20] found that the best fit shell thickness depends on the radius of the tip, or the radius of the spherical surface. It ranges from 0.219*a* to 0.046*a* for a tip radius in the range of 20*a* to 250*a* where *a* is the lattice parameter. The number of planes appearing in the simulation image also depends on the tip radius as does the actual field ion image. This thin shell model can be easily extended to simulate field ion images of tips with lattice defects, such as various types of dislocations, grain boundaries, twin boundaries and domains in ordered alloys, etc. An example is shown in Fig. 3.8 where a [111] oriented Ir tip has a twin slice, created perhaps by the field stress of the image field. For some of the crystallographic poles, the atomic rings of the matrix match perfectly with those of the slice, a characteristic of the twin boundary. Since in this simulation, no image brightness variation is taken into account, the boundaries are more difficult to see than the real·field ion image. It is of course possible to add brightness variations into the simulated images.

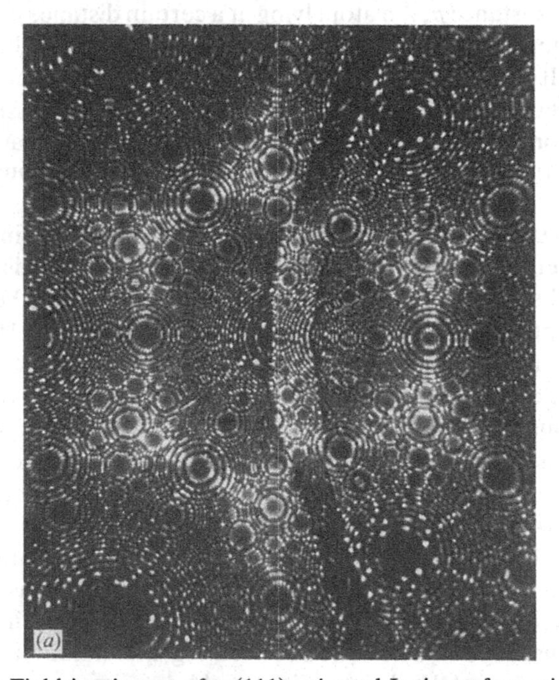

Fig. 3.8 (*a*) Field ion image of a (111) oriented Ir tip surface with a twin slice passing across the surface, obtained by Müller in the 1960s.

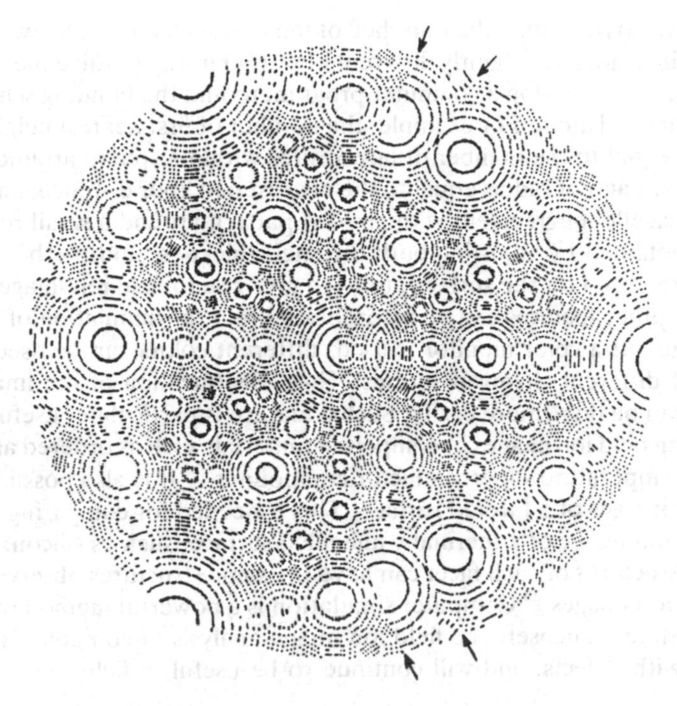

Fig. 3.8 (*b*) Computer simulation image of same Ir surface. In this simple simulated image, no image brightness variation is considered. Thus the boundary is less clearly seen. Brightness variations can be incorporated into the simulation. For alloys, different colors can be used for different chemical species.

In the thin shell model, the only criterion for imaging a surface atom is that it must lie within a thin shell of thickness p_0 from the spherical surface, or only the geometrical aspect of the surface is considered. As the field ion emitter surface is processed by field evaporation, there are other factors which can affect the field ionization rate above a surface atom. A bond model has been proposed which imposes an additional imaging criterion based on the bonding scheme of the surface atom.[21] In field ionization, the rate should depend on the electric field as well as on the density of the unoccupied electronic states inside the surface. It is therefore natural that two atoms with the same local electric field strength but with different bonding schemes can have different field ionization rates. The effect of atomic bonds should be very small beyond the first nearest neighbors. Of course the tip end form will also depend on the field evaporation process, which should be affected by the binding energies or the bonding schemes of surface atoms. Perry & Brandon[21] used a mathematical technique developed by Mackenzie *et al.*[22] to find the bonding schemes of surface atoms in different regions of a field emitter surface which lies within the thin shell of the thin shell model. Sanwald & Hren,[23] instead, use a slightly different but simpler way to solve the same problem. They develop a computer program to find the bonding scheme of each imaged atom. For example, the number of first nearest neighbor bonds is equal to the number of lattice sites within a sphere around the image point and of radius equal to the nearest neighbor distance, and so on. An excellent agreement with an FIM pattern around a small region can be obtained by adjusting both the shell thickness and the bonding scheme required for the surface atoms to give rise to field ion images.

Many other features can be incorporated into the simulation of field ion images. If a color monitor is used, different colors can be used for atoms of different chemical species. Thus identification of the imaged species can be done more easily. Such a feature is particularly useful for simulating field ion images of compound materials such as ordered alloys or high temperature superconductor compounds.[24] It is also possible to simulate images of a faceted surface. This is needed for comparing with the field ion image of a thermally annealed surface, such as silicon. The atomic structures of each facet can have the same structures observed in the field ion images.[25] Computer simulation is a powerful method which can facilitate immensely the field ion image analysis of compounds and lattices with defects, and will continue to be useful in field ion microscopy.

3.1.5 Field strength

In many field emission and field ionization experiments, field strength is a basic parameter which has to be known accurately before a lot of experimental data can be interpreted properly. Determination of field strength at the field emitter surface and field distribution above the field emitter surface in field electron and field ion emission, however, is not an easy task because of the complicated geometry of the tip. In field emission, the validity of the Fowler–Norheim theory has been established experimentally to within about ±15%, and the current density as a function of the field has been tabulated.[26] Thus it is possible to determine the field strength simply from the field emission current density. The field strength so determined cannot of course be more accurate than ±15%.

It is not always convenient in a field ion emission experiment to switch to the field emission mode to measure the field emission current density. Thus a much more convenient method for estimating the field strength has been proposed by Müller & Young.[27] They find that there is a field strength where the field ion microscope gives the 'best field ion image', or the field ion image appears to be the sharpest, or to have the best spatial resolution and image contrast. They find that this best image field can be agreed consistently to about ±1% by different observers. For helium, the best image field is determined from the field emission current density to be about 4.4 V/Å. This author and his co-workers, in general, find that the best image fields determined by different observers are consistent to within about ±5% when the tip temperature is around 20 K. Using this method, the best image fields for Ne, Ar and H_2 are found to be about 3.8 V/Å, 2.2 V/Å and 2.2 V/Å, respectively. On the basis of the calibration method of Sakurai & Müller,[28] the accuracy of the best image fields is now known to be much better than the ±15% originally expected from the field emission method. They determined the field strength from a measurement of the field ion energy distribution.

As discussed in Section 2.1.4, although the critical ion energy deficits in field ion energy distributions do not change with the field strength, the peak positions as well as the distribution widths do change with the field strength. Sakurai & Müller assume that the field strength at the location where the field ionization has the maximum rate, or the peak position of the field ion energy distribution, is independent of the field strength, or is a constant. On the basis of this assumption, they are able to determine the field strength to within ±1% accuracy. However, there is really no sound physical basis for such an assumption and in fact de Castilho & Kingham[29] point out, on the basis of a theoretical calculation, that the assumption is not rigorously correct, and a small correction has to be made. Judging from the relatively small correction needed, the field strength determined from this method is probably accurate to within ±5%, much better than the method based on field emission current density measurement. The best image field for helium field ion images determined from this new method is again about 4.4 V/Å, consistent with the result of Müller & Young. For routine determination of field strength ion emission experiments, one invariably uses the method of the best image field. The accuracy of this method, of course, has now been ascertained, by field ion energy distribution measurement, as about ±5%.

3.1.6 Field distribution

In field ion microscopy, one would also like to know how the field distributes itself above an emitter surface. This information is important in the quantitative interpretation of many field ion emission phenomena and experiments. It is also important in calculating the ion trajectory to enable a proper aiming in an atom-probe analysis. Unfortunately, not only does each tip have its own particular shape, but the presence of lattice steps also complicates the situation immensely. There are so far no reliable calculations for the field distribution above an emitter surface, nor for predictng the ion trajectory, nor yet for where the probe-hole should be aimed in single atom chemical analysis. Fortunately a method of proper atom-probe aiming can be very easily found with an imaging atom-probe experiment. As will be discussed later in greater detail in the section describing imaging atom-probes, proper aiming positions of different crystal planes can be found by comparing a field ion image with a field desorption image of the same surface. Thus, at least for the purpose of atom-probe aiming alone, there is no real need for detailed calculations of the ion trajectory.

While a proper aiming of the atom-probe can be experimentally determined, information on field lines and on equipotential lines is difficult to derive with an experimental method because of the small size of the tip. Yet this information is needed for interpreting quantitatively many experiments in field emission and in field ion emission. We describe here a highly idealized tip–counter electrode configuration which may be useful for describing field lines at a short distance away from the tip surface but far enough removed from the lattice steps of the surface. The electrode is assumed to consist of a hyperboloidal tip and a planar counter-electrode.[30] In the prolate spheroidal coordinates, the boundary surfaces correspond to coordinate surfaces and Laplace's equation is separable, so that the boundary conditions can be easily satisfied.

The prolate spherical coordinates (α, β, ϕ) are related to the rectangular coordinates by

$$\begin{aligned} x &= a \sinh \alpha \sin \beta \cos \phi, \\ y &= a \sinh \alpha \sin \beta \sin \phi, \\ z &= a \cosh \alpha \cos \beta, \end{aligned} \tag{3.5}$$

where $0 < \alpha < \infty$, $0 < \beta < \pi$, $-\pi < \phi < \pi$, and a is a scale factor. The triply orthogonal coordinate surfaces are: (1) prolate spheroids with foci at $(0, 0, \pm a)$ for α = constant, (2) double-sheeted hyperboloids of revolution confocal with the spheroids for β = constant, and (3) planes passing through the z-axis for ϕ = constant. The quantity $d = a|\cos \beta|$ is the distance between the apex of a hyperboloid and the origin. In this coordinate system, Laplace's equation is

$$\frac{1}{a^2(\sinh^2 \alpha + \sin^2 \beta)} \left[\frac{1}{\sinh \alpha} \frac{\partial}{\partial \alpha} \left(\sinh \alpha \frac{\partial u}{\partial \alpha} \right) + \frac{1}{\sin \beta} \frac{\partial}{\partial \beta} \left(\sin \beta \frac{\partial u}{\partial \beta} \right) + \left(\frac{1}{\sinh^2 \alpha} + \frac{1}{\sin^2 \beta} \right) \frac{\partial^2 u}{\partial \phi^2} \right] = 0. \tag{3.6}$$

If we set the boundary conditions so that $u = V$ on the hyperboloid denoted by $\beta = \beta_0$ $(\beta_0 > \pi/2)$ and $u = 0$ on the plane $(\beta = \pi/2)$, the solution to Laplace's equation which satisfies these boundary conditions is

$$u = V_0 \ln\left(\tan\frac{\beta}{2}\right)\Big/\ln\left(\tan\frac{\beta_0}{2}\right). \tag{3.7}$$

The electric field has only a β-component and is given by

$$E = E_\beta = -\frac{V}{a}\left\{\ln\left(\tan\frac{\beta_0}{2}\right)\sin\beta\,(\sinh^2\alpha + \sin^2\beta)^{1/2}\right\}^{-1}. \tag{3.8}$$

In this case the surfaces β = constant correspond to the equipotentials and the surfaces α = constant correspond to the surfaces on which the electric field lines are fixed. The distance from the origin to the focus, a, is related to the distance between the apex and the origin, d, and the radius of curvature of the tip, r_t, by $a = [d(d + r_t)]^{1/2}$.

This solution can also be expressed in cylindrical coordinates (ρ, ϕ, z) as

$$E = -V\left[a\ln\left(\tan\frac{\beta_0}{2}\right)\right]^{-1} 2^{1/2}[(\rho^2 + z^2)^2 + 2\rho^2 - 2z^2 + 1]^{-1/4}$$
$$\times \{[(\rho^2 + z^2)^2 + 4\rho^2 - 2z^2 + 1]^{1/2} - \rho^2 - z^2 + 1\}^{-1/2}, \tag{3.9}$$

where ρ and z are in units of a. The relation between the cylindrical coordinates and prolate spheroidal coordinates can be expressed in compact form as $z + i\rho = \cosh(\alpha + i\beta)$. Along the axis ($\alpha = 0$ or $\rho = 0$),

$$E = E_z = -V\left\{a\ln\left(\frac{\tan\beta_0}{2}\right)\right\}^{-1}(1 - z^2)^{-1}. \tag{3.10}$$

An example for the field cones and equipotential surface is shown in Fig. 3.9 for $d = 1.2$ mm and $r_t = 420$ Å. The vertical line represents a position of $5r_t$ away from the tip. The field lines are drawn so that their density is proportional to the field strength. Field distributions and equipotential surfaces of other tip shapes have also been investigated, particularly as regards the field emission current density distribution,[24,31] but will not be discussed here.

3.2 Atom-probe field ion microscope

The atom-probe field ion microscope is a device which combines an FIM, a probe-hole, and a mass spectrometer of single ion detection sensitivity. With this device, not only can the atomic structure of a surface be imaged with the same atomic resolution as with an FIM, but the chemical species of surface atoms of one's choice, chosen from the field ion image and the probe-hole, can also be identified one by one by mass spectrometry. In principle, any type of mass analyzer can be used as long as the overall detection efficiency of the mass analyzer, which includes the detection efficiency of the ion detector used and the transmission coefficient of the system, has to be close to unity.

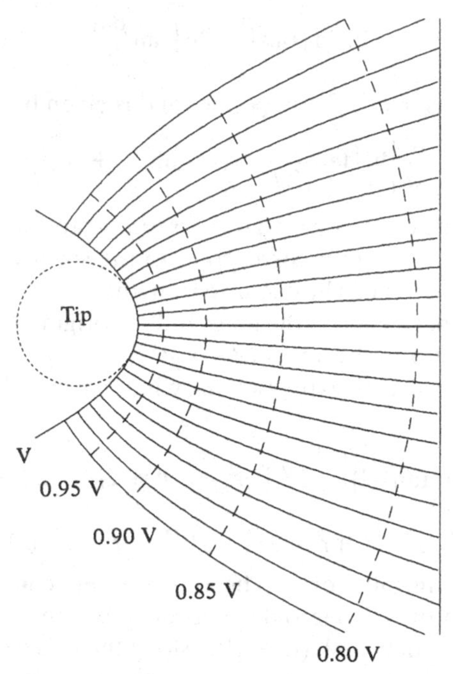

Fig. 3.9 Field distribution and equipotential surfaces for a hyperboloidal tip with a planar counter-electrode. The tip radius is 420 Å, and the tip to counter-electrode distance is 1.2 mm. The vertical line represents a position $5r_t$ away from the tip. The field lines are drawn so that their density is proportional to the field strength.

Two types of mass spectrometers have been used in the past. These are the time-of-flight (ToF) mass spectrometer[32] and the magnetic-sector mass spectrometer.[33,34] The time-of-flight system has the advantage of being able to detect all the desorbed ions covered by the probe-hole all at once with a reasonable mass resolution. The sensitivity is therefore limited only by the ion detector used and the transmission coefficient of the system, which can easily be made to be 100 per cent. With the rapid progress in the development of very fast electronics, time-of-flight techniques are finding wider and wider applications, and almost all the existing atom-probes in the world today are of this type. The magnetic sector type atom-probe has the advantage of achieving good mass resolution very easily. It is no longer uncommon for magnetic sector mass spectrometers to have a resolution better than 100 000, which cannot yet be easily achieved with a time-of-flight mass spectrometer despite the rapid progress in very fast electronic time measuring devices. Unfortunately, this type of mass spectrometer, when designed to have high resolution, can only detect ions within a very small mass range at one time. Although one can scan the entire mass range by varying the magnetic field strength, mass scanning defeats the sensitivity of the atom-probe. Using a Chevron double channel plate as the ion detector, mass lines covering about 10 to 20 a.m.u. can be displayed at one time, and a mass line containing only a few ions can be unmistakenly identified.[34] Unfortunately, the dark current of the Chevron double channel plate is not negligible, and there is the severe problem of signal to noise ratio in this system. So far the magnetic sector atom-probe does not even have the sensitivity to analyze the composition of one layer of atoms on a field ion emitter surface, not to mention a few atoms. For some special applications, such as the formation of metal hydride and metal helides[35] and ion mass and kinetic energy analysis in field ionization and field evaporation, and in liquid metal ion sources,[36] the magnetic sector atom-probe is very useful as a highly sensitive spectrometer. It is, however, of limited applicability as the atom-probe of a single atom chemical analysis mass spectrometer. Such an instrument requires a true single ion detection sensitivity over the entire mass range all at once. Therefore, so far the magnetic sector atom-probe has not found applications as a true 'atom-probe'. Our discussion will therefore emphasize the time-of-flight atom-probes.

The original function of the atom-probe is for the chemical analysis of the atoms of one's choice. It is however possible to extend the function of the atom-probe to ion kinetic energy analysis[37] and ion reaction rate measurement,[38] or general spectroscopy, with the same sensitivity, as has already been described in Chapter 2.

3.2.1 Linear time-of-flight atom-probe

The first time-of-flight atom-probe developed by Müller *et al.*[32] and the subsequent systems developed by various investigators[39] are the linear-type, high voltage pulse atom-probes. The principle of the atom-probe has already been shown in Fig. 1.4, but is repeated here in Fig. 3.10 with more details. Basically, the sample tip is mounted on either an external or an internal gimbal system, with the provision of cooling the tip down to near 20 K. At the screen assembly there is a small probe-hole 1 to a few millimeters in diameter. In some cases the probe-hole is placed behind the channel plate screen assembly, which can be swung aside after the tip orientation has been adjusted to have the atoms of one's choice properly aimed by the probe-hole. Behind the probe-hole is a time-of-flight tube, several tens to a few meters in length and terminated with an ion detector. For the ion detector, a Chevron double channel plate or a channeltron may be used. The former has an ion detection efficiency of about 60%, and can be improved by having a negatively biased grid placed in front of

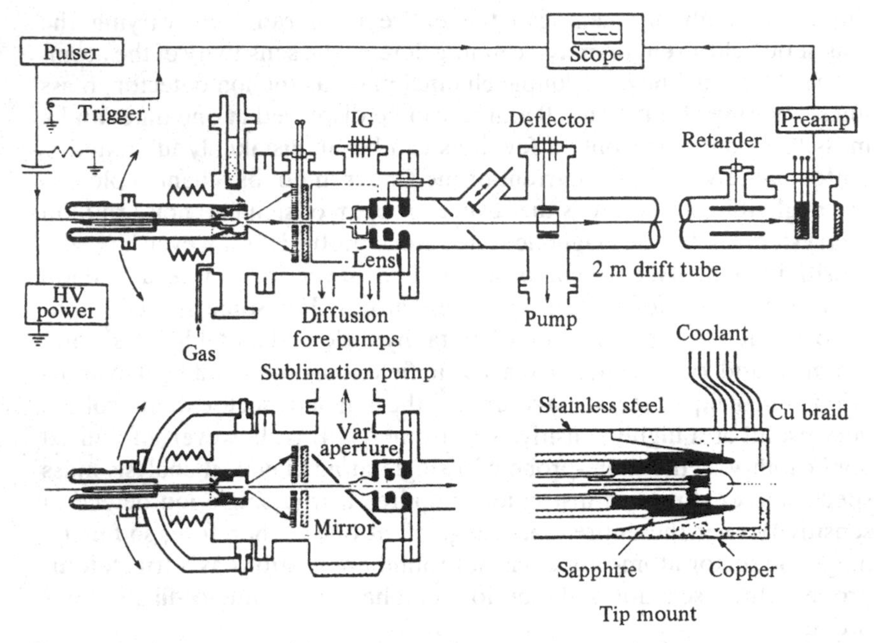

Fig. 3.10 Linear type high voltage pulse time-of-flight atom-probe with an external gimbal.

the channel plate face to suppress the escape of the secondary electrons.[40] Channeltron has the advantage of having a near-100% detection efficiency, although the response time is much slower and can thus deteriorate slightly the mass resolution of the atom-probe.[41] For cooling the tip, either a cold finger or a Displex helium refrigerator can be used. In the latter case, direct mounting of the tip to the refrigerator is possible if one uses an external gimbal system, but then a provision for vacuum lock tip changing is most difficult. Otherwise heat conduction through a copper braid can also be used to cool down the tip, as shown in Fig. 3.10.

To operate the atom-probe, the tip orientation is adjusted until the images of those surface atoms intended for chemical identification fall into the probe-hole. The tip is kept at a d.c. voltage slightly below the d.c. evaporation voltage of the sample. The surface is then pulse field evaporated slowly with ns-width high voltage pulses. Although field evaporation occurs over the entire emitter surface, only ions of the intended surface atoms can go through the probe-hole, enter the flight tube and reach the ion detector. The flight times of the ions can be measured with either an oscilloscope, a waveform digitizer or an electronic timer. For long flight tube, linear type ToF atom-probes, electronic timers such as the LeCroy model 4208 time-digital-converter (TDC), with 1 ns time resolution, or model 4204 TDC, with 156 ps time resolution, are most convenient. The model 4208 can count 8 hits in one trigger, so losing count of ions will not occur unless the field evaporation is done at much too high a rate. For a proper compositional analysis of the sample, the field evaporation rate has to be slow, as will be discussed in Section 3.2.7. The pulsed-field evaporation can also be induced by laser pulses. Pulsed-laser time-of-flight atom-probes have many advantages, as will be described later. Here we will discuss only the high voltage pulse operated linear type atom-probe.

The mass-to-charge ratios of ions are related to the d.c. tip voltage and the pulsed voltage according to K.E. $= \frac{1}{2}Mv^2 = ne(V_{dc} + \alpha V_p)$, where n is the charge state, V_{dc} is the d.c. voltage of the tip, V_p is the height of the voltage pulse and α is a pulse enhancement factor. The pulse enhancement factor arises from the problem of properly terminating the transmission line of the very fast high voltage pulse right at the tip, and thus superposition by the reflected wave of the high voltage pulse at the tip cannot be completely avoided. The pulse factor has a value somewhere between 1 and 2, and depends on how well and how close to the tip the high voltage pulses are terminated. To avoid excess reflection and waveform deterioration, the 50 Ω resistor termination of the transmission line should be done as close to the tip as possible. The pulse enhancement factor can be determined from the flight times of ions of known mass-to-charge ratios. As the velocity v is related to the measured flight time by $v = l/(t + \delta)$, where δ is a time-delay constant which arises from the fact that electric signals take time to generate and to travel across the connecting cables etc., the mass-to-charge ratios of ions are given by

$$\frac{M}{n} = C(V_{dc} + \alpha V_p)(t + \delta)^2, \qquad C = \frac{2e}{l^2}. \tag{3.11}$$

The linear type high voltage pulse time-of-flight atom-probe, while simpler in design, is now being gradually replaced with other types of atom-probes because its mass resolution is very limited. In ns-width high voltage pulse field evaporation, the total voltage needed is at least 15% higher than the voltage needed for slow d.c field evaporation because of the much higher field evaporation rate: of the order of 10^7 layers/s compared with $\sim 10^{-2}$ layer/s in d.c. field evaporation.[42] Thus unless the pulse voltage is at least about ~18% of the d.c. tip voltage, one will not be able to pulse field evaporate the surface without also simultaneously slowly d.c. field evaporating the surface atoms. When the composition of a compound or an alloy is sought, it is possible that one of the species may be preferentially field evaporated from the plane edges by the d.c. holding field, and the composition of the sample obtained with the atom-probe will not then be correct. It is well recognized through actual measurements with alloys that to derive the correct composition of an alloy, the high voltage pulse must be at least 15% of the total voltage, or at least ~18 per cent of the d.c. voltage.[43,44] This high value of the pulse voltage needed for a proper compositional analysis with the atom-probe will produce a very large kinetic energy spread of the field desorbed ions, which in turn will deteriorate the mass resolution of the atom-probe. The cause of the large ion energy spread can easily be understood without going into detailed calculations. It arises from the fact that an ion can be field evaporated at any time within the duration of a pulse. If an ion is field evaporated at the very beginning of the pulse, it will gain almost the full energy of the total applied voltage, or $ne(V_{dc} + \alpha V_p)$, as within a few ns the ion will have already traveled a large distance from the tip surface and will have gained nearly the full energy of the accelerating voltage. On the other hand, if an ion is field evaporated at the very end of the high voltage pulse, it will gain only an energy of neV_{dc}, since the ion will not be able to see the voltage pulse. Thus these two ions will have an energy difference of $ne\alpha V_p$. This is the maximum possible energy spread in high voltage pulse field evaporation. It is a very large energy spread and easily exceeds several hundred eV in most cases. Although the real energy spread will be somewhat smaller, it is of the order of a few hundred eV. Because of this ion energy spread, the mass resolution of the linear type high voltage atom-probe is limited to no better than 200 at the half peak height, and usually it is much worse. The resolution of the system cannot be improved either by using a longer flight tube or by using an electronic timer of better time resolution: as will be discussed in detail in Section 3.2.5.

It is now well recognized that chemical identification of even those materials composed of the most commonly used elements requires a much better mass resolution, since most elements have more than one isotope, and isotope overlapping is often a serious problem for a reliable compositional analysis of samples. This is the reason that the linear type high voltage pulse ToF atom-probe is now gradually being replaced by other types.

3.2.2 Flight-time-focused time-of-flight atom-probe

There are at least two different ways to improve the mass resolution of the atom-probe. The most straightforward and the easiest is to use a different kind of pulse field evaporation which does not produce a large ion energy spread as in pulsed-laser stimulated field evaporation.[45] This will be discussed in Section 3.2.4. The other method tries to have ions of greater kinetic energy travel a larger distance (ions of smaller kinetic energy travel a shorter distance) to reach the ion detector. This can be done by passing field evaporated ions through a 90° electrostatic lens and tilting the Chevron double channel plate at an angle, as done by Müller & Krishnaswamy.[46] An ion with a slightly larger kinetic energy will travel in a circular path of greater radius, and the detector is also tilted further away from this ion path. An ion of smaller kinetic energy will travel in a circular path of smaller radius, and the detector is also tilted closer to this ion path. Thus the two ions can be made to arrive at the detector at nearly the same time. Unfortunately, ions with an energy dispersion cannot be made to all arrive at the detector simultaneously with this simple 90° configuration, and the mass resolution is not greatly improved by this flight-time compensation device. A detailed first order theory of Poschenrieder suggests several flight-time-focusing configurations which combine straight drift paths with toroidal sector fields.[47] With these configurations, ions with the same mass-to-charge ratio but with a few per cent kinetic energy spread arrive at the detector isochronously and are also focused to the same spot. One of the configurations, using a 163° spherical electrostatic deflector as shown in Fig. 3.11, was first adapted to an atom-probe by Müller and Krishnaswamy[46] at The Pennsylvania State University with a greatly improved mass resolution. They used oscilloscope for measuring the flight times and recorded with a photographic method. A mass resolution, defined by $(\Delta M/M)_{50\%}$, as high as $\pm\frac{1}{4834}$ was achieved for Rh.

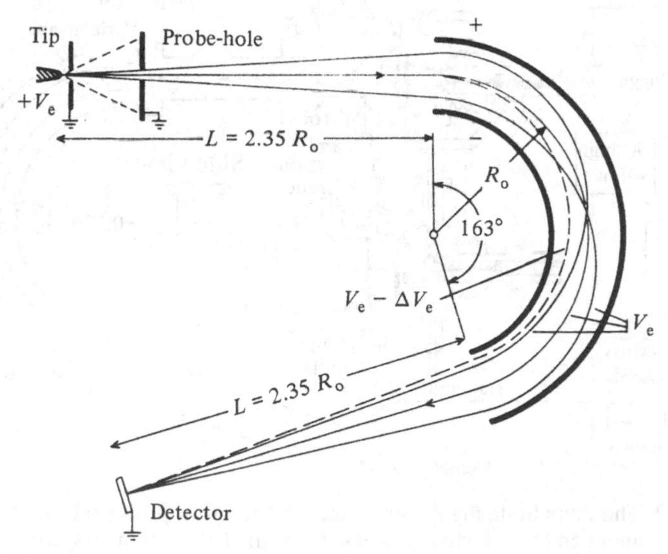

Fig. 3.11 Poschenrieder ion flight time focusing scheme which uses a 163° spherical electrostatic lens.

This Penn State system has been continuously developed by Tsong and co-workers over the years to incorporate fast electronic timers for measuring ion flight times and to automate data collection with a data processor. A schematic of this system in its early form is shown in Fig. 3.12. A mass spectrum of Pt–5% Au sample taken with this system is already shown in Fig. 1.5. The majority of atom-probes now operating in the world are of this type. It appears that the earliest systems of this type developed outside Penn State are those constructed by A. R. Waugh and M. J. Southon[48] at Cambridge University and by O. Nishikawa and co-workers[49] at Tokyo Institute of Technology. There is now a commercial system, the VG-F100, co-developed by G. D. W. Smith of Oxford University and Vacuum Generator Ltd in the UK. Nishikawa's system is the first one to incorporate a provision for a vacuum lock tip changing without losing the capability of heating the tip resistively. The VG-F100 has a separate chamber for tip processing. Most of the existing atom-probes, including the Penn State atom-probe, now use an internal gimbal system for tip orientation adjustment and a Displex helium refrigerator for the tip cooling, and can also use laser pulses for pulsed-field evaporation of the tip sample. For the time measuring device, LeCroy model

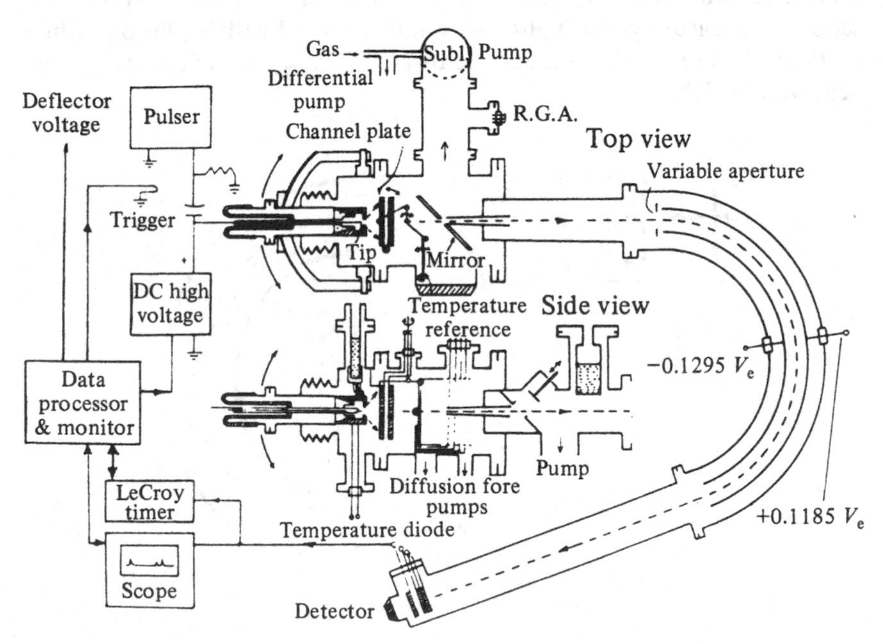

Fig. 3.12 The Penn State flight-time-focused ToF atom-probe in its middle stage of development and the electronic block diagram. This system now uses a LeCroy 4208 TDC for measuring flight times. The gimbal system has also been changed to an internal one as shown in Fig. 3.4c.

4208 TDC with 1 ns time resolution is the most popular one. With a flight path length of 2 to 3 m, atom-probes of this type can easily achieve a mass resolution of 1000 to 2000 at the 50% peak height depending on the fine adjustment of the focal points, the size of the probe-hole and the electronics of the system used.[50] The high voltage pulse operated flight-time-focused atom-probe, once properly adjusted and fine tuned, is very easy to use and an excellent mass resolution is always achieved without the need for fine tuning the operational parameters of the system each time, as has to be done in pulsed-laser atom-probes of the linear-type. This type of atom-probe is therefore the most convenient for material analysis, and is now the most popular. The number of systems operating around the world seems to increase steadily with time. There are several systems in Japan and in the United States. Additional systems can be found in Britain, Sweden, Germany, etc. The disadvantages are the elaborate mechanical designs needed for the Poschenrieder lens, and the need for a careful high voltage pulse termination. The use of an artificial flight-time focusing scheme also prevents the instrument from being used as an ion energy and ion reaction rate, or reaction time, analyzer. Pulsed-field-evaporation by high voltage pulses can be done only for samples of good electrical conductivity, and thus the material applicability is also severely limited. This difficulty can be overcome by using laser pulses, as will be discussed in detail in Section 3.2.4. Most of the flight-time-focused atom-probes now also have the capability of using laser pulses for pulsed-field evaporation, so that poor conductor samples can be analyzed.

3.2.3 Imaging atom-probe

In ordinary time-of-flight atom-probes, only surface atoms from a very small area, a circular area of a few atomic diameters covered by the probe-hole, are analyzed. If one enlarges the probe-hole to cover the entire emitter surface, it becomes the imaging atom-probe. An imaging atom-probe was first developed by Panitz and was called the 10 cm atom-probe.[51] It is an adaptation of an idea discussed by Walko & Müller,[52] who found that a field desorption image could be formed in field evaporation of a tip surface if the gain of the channel plate was cranked up sufficiently. Of course, the electric signals generated can be used for time-of-flight mass spectrometry. With the availability of large Chevron channel plates, imaging of single ions is no longer a problem and field desorption images can be routinely obtained together with ion signals. Thus an imaging atom-probe is basically a field ion microscope equipped with a more sensitive Chevron double channel plate assembly capable of detecting single ions and of measuring their flight times, and can also give rise to a greatly intensified image of the desorbed ions. It is a time-of-flight mass spectrometer as well as an ion microscope. A schematic diagram of an imaging atom-probe is shown in Fig. 3.13. Of course, the pulsed-field evaporation needed for a time-of-flight mass analysis can be induced by either high voltage pulses or by laser pulses. In Fig. 3.14, a ToF mass spectrum of silicon, obtained with a pulsed-laser imaging atom-probe, is shown, while Fig. 3.15 depicts a flight-time gated helium field desorption image of a Rh tip together with a He field ion image of the same surface.

The flight path of a typical imaging atom-probe is only about 10 to 15 cm, which is mainly limited by the size of the Chevron double channel plate available. With such a short flight path and a very large reception angle of the ion detector, and thus a possible large difference between the ion flight path lengths, the mass resolution is severely limited. Even if a spherical Chevron double channel plate is used, it is difficult to place the tip right at the exact center of the spherical surface, and a path difference of about 1% of ions is about the optimum adjustment one can reasonably achieve. In addition, the flight time is usually less than 1 μs unless the tip voltage is exceptionally low. The mass resolution is usually no better than 30 to 60 at half peak height. A resolution as high as 200 has been claimed, but it is not easily achievable. Thus only very carefully selected systems can be studied without uncertainties in the mass identification. It is not unusual to make mistakes in identification of chemical species and ion species. An example is the detection of Si_2^+ and Si_3^+ in pulsed-laser stimulated field evaporation of silicon. The correct ion species are now known to be Si_4^{2+} and Si_6^{2+}. Thus some of the conclusions drawn from imaging atom-probe studies are not without uncertainties. One therefore

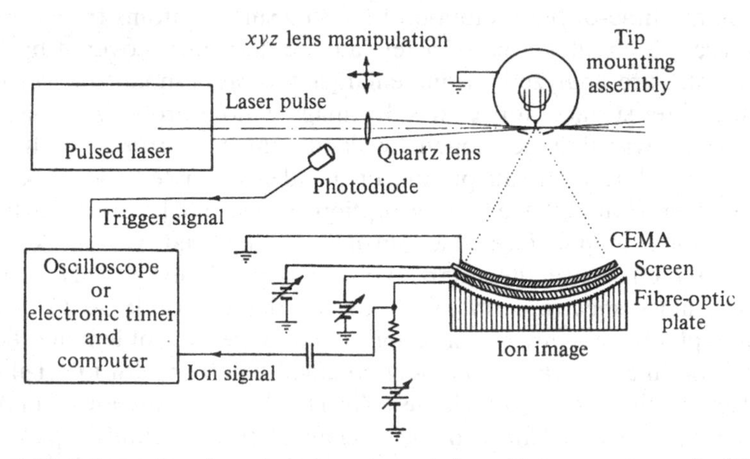

Fig. 3.13 Schematic of an imaging atom-probe which uses the pulsed-laser technique.

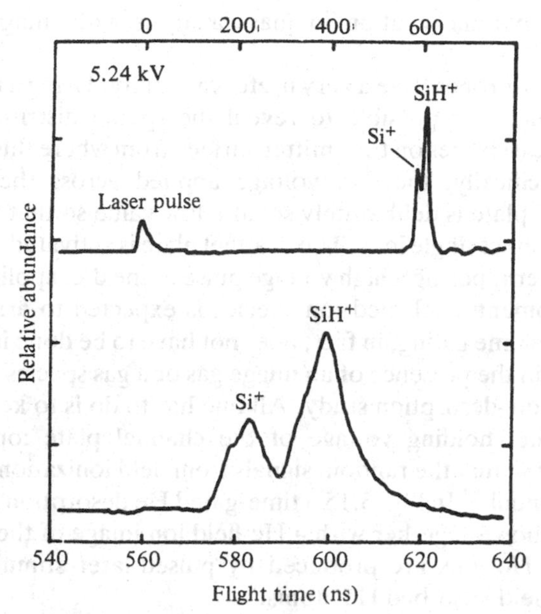

Fig. 3.14 A ToF spectrum of a high purity silicon whisker in hydrogen image gas, obtained with a pulsed-laser imaging atom-probe. Below 80 K, silicon is an excellent insulator. However, the tip surface can be easily field evaporated with the stimulation of laser pulses.

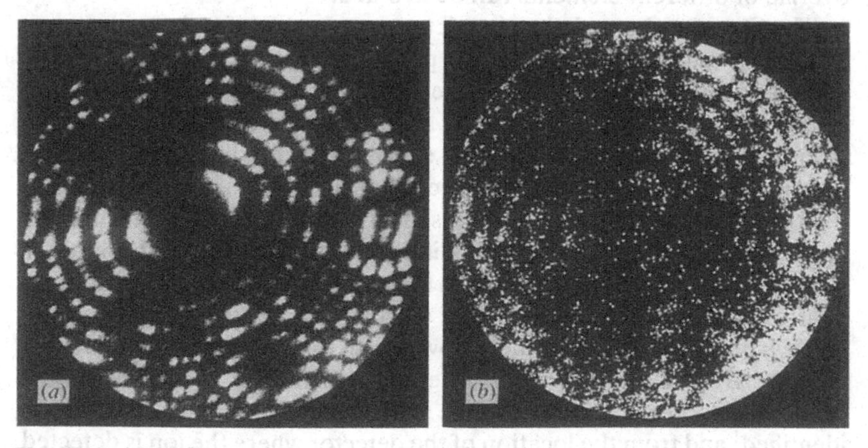

Fig. 3.15 (*a*) Helium field ion image of a Rh tip.
(*b*) Time-gated, pulsed-laser field desorbed helium ion image of the same Rh tip surface.

must exercise extreme caution in mass analysis with imaging atom-probes.

Imaging atom-probes have a very useful capability. Using a flight-time gating technique, it is possible to reveal the spatial distribution of a selected chemical species on the emitter surface from where this species is desorbed. Specifically, the d.c. voltage applied across the Chevron double channel plate is deliberately set at a low value so that the gain is insufficient to detect single ions. The channel plate is activated for only 40 to 60 ns, by superimposing a high voltage pulse to the d.c. applied voltage only at the moment a selected ion species is expected to arrive at the detector.[53] This time gating, in fact, does not have to be done in vacuum. It can be done in the presence of an image gas or a gas species to be used for an adsorption–desorption study. All one has to do is to keep the gas ion and the d.c. holding voltage of the channel plate ion detector sufficiently low so that the random signals from field ionization of the gas are negligibly small.[54] In Fig. 3.15 a time-gated He desorption image of a Rh surface is shown together with a He field ion image of the same tip. The desorbed He ions are produced by pulsed-laser stimulated field desorption of field adsorbed He atoms.

Imaging atom-probes, although severely limited in mass resolution, are very useful where one seeks information about the spatial distribution of chemical species on the emitter surface as well as in the bulk. They find many applications in studies of metallurgical problems,[55] in studies of chemisorptions and surface reactions[56] and oxidation of metals,[57] etc., as will be discussed in later chapters. In using imaging atom-probes, it is important to select the system carefully and intelligently so that mass overlap of different elements can be avoided.

The Chevron channel plate ion detector assembly of an imaging atom-probe can also be replaced by a position sensitive particle detector combined with a data processor, as reported by Cerezo *et al.*[58] (A position sensitive detector was used earlier for the purpose of field ion image recording and processing.[59]) With such a detector both the chemical identity and the spatial origin on the emitter surface can be found for each field evaporated ion. This 'position sensitive atom-probe' can be used to study the spatial distribution of different ion species on the emitter surface as well as inside the bulk of the emitter with a spatial resolution nearly comparable to the FIM. For such a purpose, one carries out the field evaporation at an extremely slow rate so that no more than one ion is detected from the entire field ion emitter surface in each pulsed field evaporation. From the flight time of the ion its chemical species is identified, and from the location of the detector where the ion is detected the spatial origin of the ion is located. With a fast data processor, a two-dimensional distribution of chemical species on the tip surface can be mapped out and displayed in a monitor in different colors. By a proper calibration of the number of ions detected for each surface layer, a three-dimensional distribution of chemical species in the tip sample can also be mapped out. The mass resolution of this imaging atom-probe can also be greatly improved by correcting for the difference between the flight path of an ion traveling at an angle from the tip axis and that of one traveling along the tip axis. Although the mass resolution will never be as good as the probe-hole type atom-probes because of its very short flight path, the capability of mapping out the two-dimensional and three-dimensional distributions of chemical species on the tip surface and in the bulk of the tip is indeed a very attractive one. This instrument can also be operated like an ordinary imaging atom-probe, i.e. using a time gating technique. The depth resolution and accuracy of the chemical analysis will be slightly compromised, but data collection can be accelerated. Its detector can also be divided into many sections, with each section timed by a separate electronic timer. In doing so, the field evaporation rate can be increased but can still avoid having two ions arriving at the same timer at the same time (or within the time resolution of the timing device), thus a two-dimensional chemical analysis of a sample can be accelerated. Undoubtedly this new instrument will be further developed and will become a powerful tool for the microanalysis of surfaces and interfaces of materials in material science applications.

Let us divert a little of our attention to the problem of aiming in the regular, or the probe-hole type, time-of-flight atom-probes, or atom-probes with a very small probe-hole. An ideal atom-probe is of course the one having an excellent mass resolution, 100% detecting efficiency, great versatility, ease of operation, and with 100% accuracy in the aiming of atoms of one's choice. To be 100% accurate in atom-probe aiming, one would like to have the trajectory of a field evaporating ion to follow exactly the same trajectory of the image gas ions. In pulsed-laser field desorption of field adsorbed gas species, the desorbed gas ions follow very closely the same paths as field ions; thus the field desorption image is very similar to the field ion image as shown in Fig. 3.15. Obviously this is not the case in field evaporation of tip atoms, as can be seen by comparing a field desorption image of Ir and a helium field ion image of the same Ir surface as shown in Fig. 3.16. In fact, the field desorption image exhibits dark zones of a width as great as several atomic diameters. These zones are of course where field evaporated ions can never be detected if the probe-hole is aimed there. There are also bright zones where field evaporated ions tend to arrive with an excessively high number. There are at least two possible reasons for forming such dark and bright zones. One is the focusing and defocusing effect of the electric field line above the emitter surface. A field evaporation end form is not a perfectly

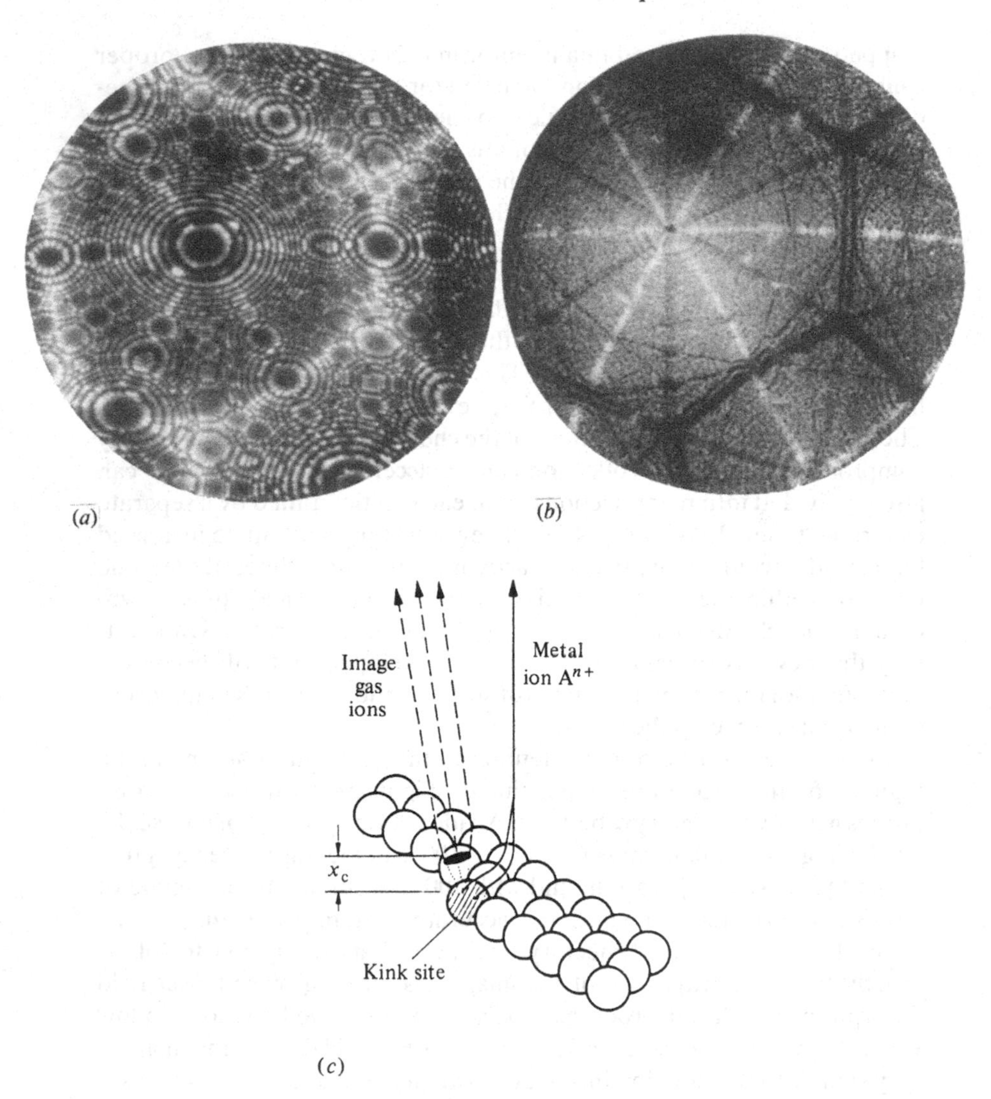

Fig. 3.16 (*a*) Helium field ion image of an Ir tip.

(*b*) Vacuum field desorption image of the same tip surface showing dark and bright bands. In the probe-hole type ToF atom-probes, if the probe-hole is aimed at a dark band area, no ions can be detected. On the other hand, if the probe-hole is aimed at a bright band area, the ion yield will be unusually high. These bands are produced by the focusing and defocusing effect of ions by the field lines, and also by interaction between field evaporating atoms and their neighbor atoms.

(*c*) Schematics explaining why interactions among surface atoms will change the ion trajectory of a field evaporated ion from those of image gas ions. The dark circle is the ionization disk for the shaded surface atom.

smooth spherically shaped surface. Instead, it has many edges of the lattice steps and bumps and flat facets. Field evaporated ions near protruding zone lines tend to diverge out of the zones, thus these zones will appear dark in the field desorption image. Field ionization, on the other hand, occurs more readily in these zones because of the higher local electric field. Thus field ion image intensity is higher in these zones. These features can be clearly seen in the images of Fig. 3.16. Another possible explanation, proposed by Waugh *et al.*,[60] is that during field evaporation the attractive and repulsive forces existing between a field evaporating atom and its surrounding lattice atoms may be sufficiently large to change the trajectory of the field evaporated ion from those of the image gas ions. It should be most interesting to investigate this effect theoretically and experimentally, as field desorption images may shed some light on the directional dependence of the interatomic forces acting among surface atoms.

In doing single atom atom-probe analysis, or even in compositional analysis of surface layers, it is important to recognize the possibility of large aiming errors. Usually for a large facet one detects few ions if the probe-hole is aimed at the very center of the facet. If instead, the probe-hole is aimed at a spot slightly inside the plane edge of the top surface layer, the detection yield is quite good. In some cases, aiming error is not a problem. An example is the analysis of adsorbed atoms of very low desorption field on a very large facet. These atoms can be desorbed without field evaporating surface atoms. For detecting these atoms, the probe-hole should be large enough to cover the entire surface layer without the risk of detecting substrate atoms. Unfortunately it is not possible to predict the aiming error of surface atoms from theoretical calculations. As has just been discussed, it is possible that by carefully comparing field desorption images and field ion images taken with an imaging atom-probe, detailed aiming errors and even the interactions of surface atoms on various planes can be investigated.

3.2.4 Pulsed-laser time-of-flight atom-probe

The fast pulsed-field evaporation needed for a time-of-flight spectroscopy can also be done with laser pulses. Irradiation of field ion tips with a laser was investigated in the middle 1960s by Müller *et al.* for the purpose of creating lattice defects by rapid heating and quenching of crystal lattices.[61] As the laser unit used, a ruby laser, was much too powerful and the stability of laser was not very good in that early stage of laser development, the experiment was not successful. Even though laser heating induced lattice defects and field evaporation of metal tips could be observed, the lack of control of the experiment forced Müller eventually to abandon the attempt. The possibility of a photon-enhanced field ionization and selective field evaporation had already been discussed in the book by Müller & Tsong.[62] In 1975–6, Tsong *et al.*[63] investigated photon-enhanced field ionization using a tunable pulsed-laser unit. With Rhodamine 6G dye operated at 300 nm (4.16 eV) and an average power of 6 mW and pulse width of 2 ms, they were able to detect an enhancement by a factor of 4 in the field ionization signals from an aluminum oxide tip. During the course of this investigation, Tsong & Block were also able to field evaporate tip surfaces, in a perfectly controlled manner, by pulsed laser stimulation for both W and aluminum oxide tips, or tips of both a metal and an insulator.

During 1976 and 1977, many investigators were interested in analyzing semiconductors such as Si with the atom-probe, at that time still operated only with high voltage pulses. Unfortunately it was found that the ns HV pulse needed for atom-probe operation could not be effectively transmitted across the tip because of the tip's high resistivity, and no pulsed field evaporation could be induced.[64] In the summer of 1977, at the Symposium on Application of Field Ion Microscopy to Materials Science, organized by David Seidman at Cornell University, the difficulty of analyzing silicon with the atom-probe was discussed. During the discussion, Tsong suggested that this problem could easily be overcome by using laser pulses since he and Block had already observed earlier well-controlled pulsed-field evaporation of both metal and insulator tip surfaces with laser pulses. Several possible advantages were pointed out in the discussion and were later published in an article.[45] Unfortunately, in part because of the early unsuccessful attempt of a laser experiment by Müller *et al.*[61], the suggestion was met with great skepticism. Many thought that it would be impossible to focus a laser beam onto the tiny little tip with good reproducibility, and no controlled field evaporation could be done. Melmed *et al.* therefore pushed in a different direction by using HV pulses of greater width, up to a few hundred ns, for the pulsed-field evaporation of doped silicon.[65] Because of the wide skepticisms, progress in the development of a pulsed-laser time-of-flight atom-probe was delayed. Only in 1980 did Kellogg & Tsong[66] finally demonstrate that laser pulses could indeed be used for time-of-flight atom-probe operation with good control; mass spectra of silicon were also shown. Drachsel, Nishigaki & Block also developed a similar system called a photon-enhanced field ionization mass spectrometer.[67] When a high resolution pulsed-laser time-of-flight atom-probe was developed by Tsong *et al.*[68] the potential and advantages of this technique were fully recognized.

Let us discuss some of the advantages and disadvantages of the pulsed-laser time-of-flight atom-probe as compared with the HV pulse atom-probe. Some of the advantages are as follows:

1. Laser pulses induce field evaporation and desorption mainly by a heating effect. The energy spread of the ions should be of the order of kT, or less than 0.1 eV when laser power is not excessively high. This is negligible compared with that produced by high voltage pulses. Thus excellent mass resolution can be achieved without the need of having an elaborate Poschenrieder lens. Even if photo-excitation effects occur, the energy spread will still be only a few eV (unless of course the laser power is excessively high, where multiple photon-excitation can occur as already discussed in Section 2.2.6), which is still smaller by a factor of about 100 compared with that of HV pulse field evaporation.

2. One of the most serious shortcomings of the HV pulse atom-probe is the inability to pulse field evaporate materials of very low electrical conductivity such as a high purity silicon. This is due to the difficulty of transmitting ns HV pulses across the tip. This limitation is overcome by the use of laser pulses which can be focused right to the tip apex. Thus the material applicability of the atom-probe is greatly expanded by the use of laser pulses for the pulsed-field evaporation.

3. Although the mass resolution of HV pulse atom-probe can be greatly improved by the use of a Poschenrieder flight-time focusing lens, this artificial lens excludes the atom-probe to be used as an ion energy spectrometer. A pulsed-laser atom-probe is not only a high resolution mass spectrometer, but also an ion energy analyzer, which in turn can be exploited to measure ion reaction times and rates with excellent time resolution by using an ion reaction time amplification scheme.[69]

4. A serious problem in applying the HV pulse ToF atom-probe to study gas surface interactions is the excessively high electric field needed to pulse field desorb adsorbed species. The desorbed species are often field dissociated into small fragments bearing little resemblance to the states of the adsorbed species. In fact some of the adsorbed species cannot be field desorbed without field evaporating the surface. In pulsed-laser stimulated field desorption, adsorbed species are either thermally desorbed (sub-ns flush thermal desorption), or desorbed by thermally enhanced field desorption. The field strength needed can be less than a third of that of ordinary field desorption, or less than a tenth in the effect of the applied field. The artifacts of the electric field can be greatly reduced and a more meaningful study of gas–surface interactions can be made.[70] It will eventually be possible to desorb a selected adsorption species by resonance absorption of photons, although so far this has not yet been achieved in pulsed-laser atom-probe studies.

5. The mechanical construction of the pulsed-laser atom-probe is greatly simplified without the need of properly terminating the ns HV pulses very close to the tip. There is also no need of a flight-time focusing lens. Very short duration laser pulses can be much more easily produced

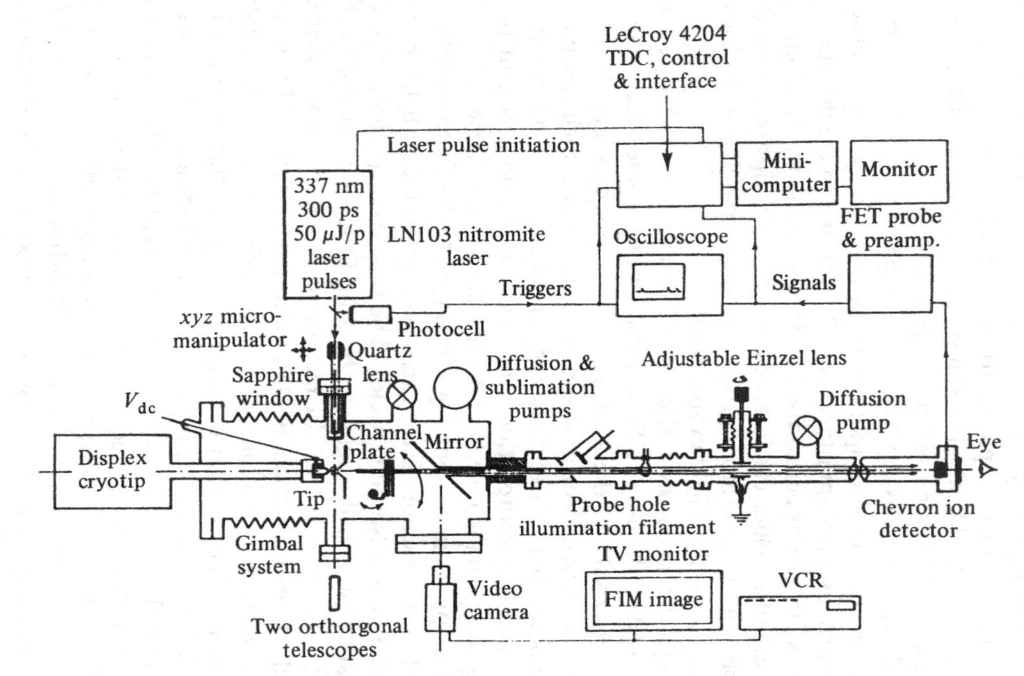

Fig. 3.17 Schematic of the Penn State high resolution pulsed-laser ToF atom-probe. The flight path length of this system is now ~778 cm. It uses two LeCroy 4204 TDCs of 156 ps time resolution for flight time measurement.

than high voltage pulses. Femtosecond width laser pulses are now available from commercial laser units which will be very useful for studying dynamics and kinetics of surface reactions, gas–surface interactions and various surface excitations using the pulsed-laser ToF atom-probe.

There are of course some drawbacks of pulsed-laser atom-probes as compared with HV pulse atom-probes. Focusing of the laser beam onto the tip needs some experience and also takes some time, although once one learns how to do it, it usually takes no more than a few minutes to achieve the degree of focusing desired. The mass and energy resolution of the system are very sensitive to the laser power and other experimental parameters, as each mass line is really an ion energy distribution. It is also found that in compositional analysis of PtRh alloys the pulsed-laser atom-probe does not give correct values even though this may simply indicate that a proper operating condition of the measurement has not yet been found.[44] The pulsed-laser atom-probe is in fact found to give correct compositions of compound semiconductor samples, whereas the HV pulse atom-probe does not.[71] The author believes that as long as the operation conditions are properly chosen, both HV pulse atom-probes and pulsed-laser atom-probes all should be able to give correct values in the compositional analyses of all materials. Unfortunately, there is no universal condition for this purpose.

The first pulsed-laser atom-probes, an imaging atom-probe developed by Kellogg & Tsong[66] and a photon-enhanced field ionization mass spectrometer developed by Drachsel, Nishigaki & Block,[67] are not high resolution systems, although many of the expected advantages of the pulsed-laser atom-probe are confirmed. In 1982, a high resolution system was developed in the author's laboratory at Penn State.[68] This system has been under continual renovation. Figure 3.17 shows the system at the time of writing. The system has a flight path length of 778 cm. It now uses two LeCroy model 4204 TDCs, with a time resolution of 156 ps, in series, a tip voltage power supply of 2 parts in 10^6 per hour stability, and a six and a half digit high precision and high stability DVM for the tip voltage measurement. But even when an oscilloscope was used for recording the ion flight times and the flight path was only about 2 m in length, the mass resolution was sufficient to see clearly the formation of HeW^{3+} and He_2W^{3+} ions as shown in Fig. 3.18. With two orthogonal cross-haired telescopes, the position of the tip can be reproduced to better than 0.1 mm. Care is also exercized to maintain a constant room temperature. A GHz response time preamplifier is used for the ion signals. With these precautions, a precision of 1 to 2 parts in 10^5 can be routinely achieved in ion mass and ion kinetic energy analysis. With the use of an elaborate calibration procedure, to be discussed in Section 3.2.6, an accuracy of

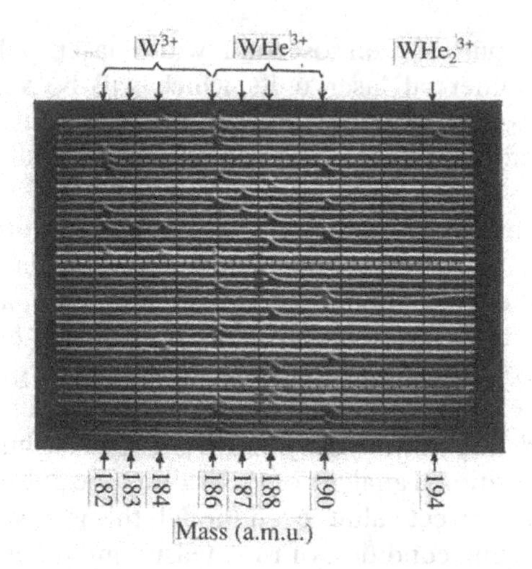

Fig. 3.18 To show more vividly how ions are counted one by one, we show a small time window of an oscillograph containing atom-probe signals of W^{3+} and HeW^{3+} and He_2W^{3+} ions, taken with the Penn State pulsed-laser ToF atom-probe in its early stage of development when the flight path length was ~200 cm and the laser pulse width was still 5 ns.

better than 1 to 2 parts in 10^5 can also be achieved in ion mass and ion kinetic energy measurements. This corresonds to 0.000 05 to 0.001 u accuracy in ion mass measurement for light to heavy elements, and 0.1 eV accuracy out of a total energy of 5 to 10 keV in ion kinetic energy analysis. In ion reaction time analysis, a resolution better than 20 femtoseconds[69] has been achieved even before some of these new improvements of the system are made. Many of the mass spectra and ion energy distributions shown in Chapter 2, such as Figs 2.5, 2.14, 2.15, 2.17, 2.18, 2.24 and 2.26, are obtained with this system at various stages of its development. The system can easily separate isotopes of the same mass numbers, such as N_2^+, Fe^{2+} and Si^+, D_2^+ and $^4He^+$, H_3^+ and $^3He^+$, etc. Figure 3.19 gives an example where mass separated ion energy distributions of He^+ and D_2^+ are shown. The data were derived when the system was 7.78 m in length. Their onset flight times differ by 34 ns, exactly that calculated from the system constants at 5.50 kV.

The power of the laser unit needed for the atom-probe operation is not really well defined. It depends on the pulse width, the stability and the reproducibility of the degree of focusing of the laser beam, the cone angle of the tip, the materials to be studied, and the wavelength and reflectivity of the surface of the tip material at that wavelength, etc. A numerical calculation by Liu & Tsong,[72] shown in Fig. 3.20, indicates that to produce a temperature rise of a few hundred K, the energy flux should be a few 10^6 J cm^{-2} s^{-1}. Temperature pulses of a few hundred K suffices to

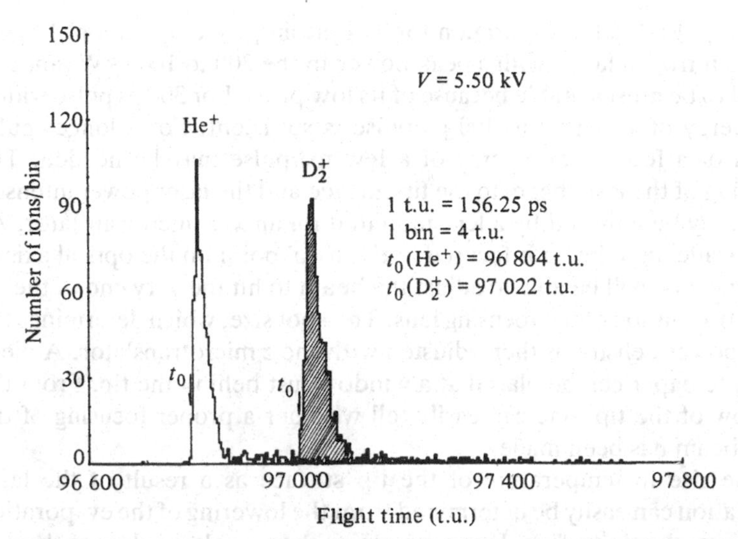

Fig. 3.19 Mass separated energy distributions of pulsed-laser field desorbed $^4He^+$ and D_2^+ ions obtained with the Penn State pulsed-laser ToF atom-probe when the flight path length was 778 cm. Their onset flight times are separated by 34 ns, exactly that calculated from the system constants and their critical ion energy deficits at 5.5 kV.

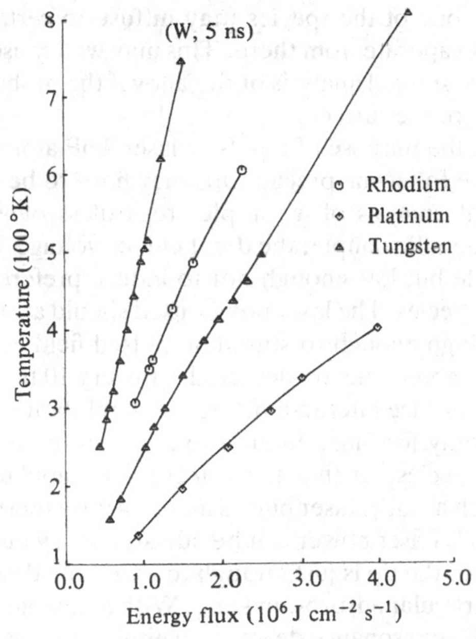

Fig. 3.20 The peak temperature reached by pulsed laser irradiation of a tip as a function of the energy flux of the laser pulse for Rh, Pt and W tips. The pulse width is 300 ps. Data are also given for 5 ns pulse width for W.

induce pulsed-field evaporation for ToF atom-probe operation. In general, a nitrogen laser, with a peak power in the 200 to 1000 kW range, is found to be most suitable because of its low price. For 300 ps pulse width, an energy of about 50 to 100 μJ/pulse is sufficient. For a longer pulse width of a few ns, an energy of a few mJ/pulse may be needed. The focusing of the laser beam to the tip surface and the laser power intensity can easily be adjusted by a lens mounted on an *xyz* microtranslator. As the parallel light beam is focused to the focal point on the optical axis of the lens, one will be able to adjust the beam to hit the very end of the tip by *xy* translation of the focusing lens. The spot size, which determines the laser power density, is then adjusted with the *z* microtranslator. A piece of white paper can be placed at a window just behind the tip. From the shadow of the tip, one can easily tell whether a proper focusing of the laser beam has been made.

The rise in temperature of the tip surface as a result of the laser irradiation can easily be determined from the lowering of the evaporation voltage of the tip surface. If one maintains the tip voltage close to the d.c. field evaporation voltage, the pulsed-field-evaporation needed for time-of-flight atom-probe operation can be induced by laser heating of only less than 100 K. If one uses too high a power of the laser, then temperature and field induced surface diffusion of surface atoms can occur. For alloys one of the species may diffuse to certain sites of the surface and field evaporate from there. This may well cause an erroneous result in the compositional analysis of the alloy if the probe-hole is aimed at the wrong spot of the surface.

Depending on the purpose of a pulsed-laser ToF atom-probe experiment, a few of the following precautions may have to be exercized. For the compositional analysis of a sample, regardless of whether it is a metallic or non-metallic sample, the d.c. field, or voltage, should be kept as high as possible but low enough not to induce preferential d.c. field evaporation of a species. The laser power used should also be kept as low as possible, just high enough to stimulate pulsed-field evaporation at a very low rate, i.e. about one ion detected for every 10 to 20 laser pulses. In a study of gas–surface interaction, the d.c. field should be kept as low as possible, i.e. only just high enough to field ionize pulsed-laser thermally desorbed species, so that the effect of the applied field can be minimized. In such a study, laser pulses induce a flush thermal desorption in a ns time period. Laser power can be adjusted to a value such that the temperature rise of the tip is just enough to thermally desorb a chemical species from a particular adsorption state. With a tunable laser, it should be possible to induce resonance desorption either by exciting a particular vibrational mode, or to an electronic excited state.[73] With this resonance photo-desorption it should be possible to desorb selected species from the surface. Another precaution one should take is the elimination of the effect of the earth's magnetic field. It is found that as the flight tube reaches a few meters in length, light ions such as H^+, H^{2+} and He^+ etc. can be lost simply by bending of the flight path out of the detector by the earth's magnetic field if the tip voltage is very low. This error can be corrected by making a magnetic shield around the flight tube. The shielding can be easily done by wrapping the flight tube with a high-μ metal sheet of ~0.5 mm in thickness.

3.2.5 Resolution of time-of-flight atom-probe

A time-of-flight spectrometer can be used as a mass analyzer, an ion kinetic energy analyzer, and an ion reaction time analyzer. We will consider here only what factors affect the resolution of the system in mass analysis.[74] The same consideration can easily be extended to find the resolution in other analyses. There are at least two kinds of mass resolution. One refers to the ability of the system to separate two ion species of nearly equal masses in the same mass spectrum. This is related to the sharpness of the mass lines, or the full width at half maximum (FWHM) of the mass lines. The other refers to the ability of the system to distinguish two ion species of nearly identical masses, but not necessarily in the same mass spectrum. This latter mass resolution is related to the sharpness of reference points in the mass lines such as the onset flight times of the ion species, and the overall long-term stability of the system. This latter resolution determined also how accurately the instrument can measure the mass of ion species. Although this latter resolution is more closely related to ion kinetic energy analysis and is as important as the former one, we will consider here only the former kind, or the conventional kind, of mass resolution.

The uncertainty in the mass measurement can be derived from eq. (3.11) to be

$$\frac{\Delta M}{M} = \left[\left(\frac{2\Delta l}{l}\right)^2 + \left(\frac{2\,\Delta t}{t}\right)^2 + \left(\frac{\Delta V}{V}\right)^2\right]^{1/2}, \qquad (3.12)$$

where Δl is the spread in the flight path lengths of ions, Δt is the uncertainty in the flight time measurement which includes resolution of the timer and the duration of the pulsed-field evaporation and also the response time of electronics and detectors, etc., and ΔV includes the voltage fluctuations of the tip voltage power supply, the inaccuracy in the voltage measurement and the intrinsic ion kinetic energy spread in the pulsed-field-evaporation. Let us take Δl to be the maximum difference in the flight path lengths of ions taking the shortest and longest possible flight paths. If the tip is placed at the focal point of the einzel lens, the ions will travel along the axial direction after passing through the einzel lens and Δl can thus be minimized. Assuming that both the probe-hole and the Chevron channel plate face are perpendicular to the axial direction, it can then be shown that

$$\Delta l \simeq \tfrac{1}{8}\frac{fd^2}{r_p^2}, \tag{3.13}$$

where d is the diameter of the probe-hole, f is the focal length of the einzel lens and r_p is the distance from the probe-hole to the tip. To reduce Δl, a short focal length einzel lens is preferred, but this will make the alignment extremely difficult. Using a probe-hole of about 2 mm diameter, place the einzel lens at about $\frac{1}{4}$ of the length of the flight tube, and r_p of about 160 mm, the factor, $2\Delta l/l$, is about 10^{-4} to 5×10^{-6} for a flight tube length of 1 to 7.78 m. This value can be reduced by using a smaller probe-hole and a longer flight tube. This factor does not depend on whether HV pulses or laser pulses are used for the atom-probe operation. For obtaining accurate data using an ultra-high-resolution system, flight path changes produced by room temperature changes have to be avoided also. The coefficient of linear expansion of stainless steel is about 1×10^{-5} °C^{-1}. It is also important in this respect to have the tip position reproducible to within 0.05 to 0.1 mm every time a new tip is installed. By using two cross-haired orthogonal telescopes, the tip position can be made reproducible to within about 0.1 mm or better.

The time uncertainty comes from several sources. These include the width of the desorption pulses, the time resolution of the time measuring devices, the response time of the ion detector, the timer triggering pulses and the electronics used in the signal amplification and recording, etc. Assuming that best available electronics are used for the system, then the time uncertainty comes mainly from the width of the pulsed field evaporation and the time resolution of the electronic timer. In HV pulse atom-probes, the pulse width is typically a few ns. One may try to use pulses of shorter width, but then the pulse height will have to be increased. Increasing the pulse height will have a large deterioration effect on the mass resolution, as will be clear shortly. Because of this limitation, timers with a better resolution than 1 ns are not really needed. For a HV pulse atom-probe of length 200 cm, $2\Delta t/t$ is about 1×10^{-3} for a typical ion mass around 50 u. For pulsed-laser atom-probes the width of laser pulses can be as short as a few tens of femtoseconds, and the time uncertainty in field desorption can be of the order of 1 ps (adsorbed atoms can be field desorbed within the duration of the laser pulse), but it is about 1 ns in field evaporating tip atoms (here the cooling time of the tip is as important as the width of the laser pulse). For a pulsed-laser ToF atom-probe it is beneficial to use the best timer available. We find the Lecroy model 4208 TDC, with a time resolution of 1 ns and which can record up to 8 signals in one trigger, and the model 4204 TDC, with a time resolution of 156.25 ps, but which can record only 1 signal in one trigger, are very satisfactory. For pulsed-laser ToF atom-probes, the time uncertainty is much less than one part in 10^5 and is of much less concern than other uncertainties such as the flight path length.

The most severe limitation in the mass resolution of the linear type HV pulse atom-probe comes from the $\Delta V/V$ term, which accounts for both the uncertainty and fluctuation of the applied tip voltage and the ion kinetic energy spread in pulsed-field evaporation. As already explained in connection with the pulsed-laser ToF atom-probe, the maximum ion energy spread in HV pulse field evaporation is $n\alpha eV_p$, where n is the charge state, α is the pulse enhancement factor, and V_p is the height of the HV pulse. Real measurements[75] show that a $\Delta V/V$ as large as 0.03 is common for HV pulse atom-probes although it can be reduced by using smaller high voltage pulses, but it will not then give a correct composition in the analysis of materials. This factor cannot be reduced either by lengthening the flight tube or by using a better timer or more stable HV power supplies. This factor is detrimental to the further improvement of the linear type HV pulse ToF atom-probe, and is the reason that the much more elaborate design of the flight-time-focused HV pulse ToF atom-probe becomes necessary. In pulsed-laser ToF atom-probes, $\Delta V/V$ is mainly limited by two factors, namely the fluctuations and stability of the d.c. HV tip power supply used and the intrinsic energy spread of ions in field evaporation and desorption. Very stable and low-noise HV power supplies are now available commercially, and the energy spread due to the d.c. tip power supply is not a major limitation. Pulsed-laser stimulated field desorption at low laser power is induced by a heating effect as discussed in Section 2.2.6, and thus the ion energy spread due to laser pulses is of the order of a few kT, or about a few tens of meV. This is negligible compared with the several to ~30 keV kinetic energy of field evaporated ions. Even if a quantum effect in photo-excitation is important in photon stimulated field evaporation of some materials such as semiconductors and insulators, the ion energy spread will still be only a few eV, the photon energy, and is still smaller than that in HV pulse field evaporation by a factor of 100. The main limitation comes from the natural width of the ion energy distribution of field evaporated ions, which is 1 to 2 eV, depending on the field and the charge state of the ions. This term is intrinsic to the ion formation process and cannot be improved by better electronics or a longer flight tube, but can be improved again by a flight-time-focusing technique such as a Poschenrieder lens. Even without a flight-time-focusing lens, a mass resolution of better than 1000 is very easy to achieve with a linear type pulsed-laser ToF atom-probe. As has been mentioned at the beginning of this section, there are other kinds of resolution of time-of-flight spectrometers. For ion energy measurements, the precision is not limited by the intrinsic width of the ion energy distribution, and the energy uncertainty due to this term, $\Delta V/V$, comes only from the voltage fluctuation and instability of the tip d.c. HV power supply, which can be less than about 2 parts in 10^6 at the present time.

In general, to improve the mass resolution of the ToF atom-probe, or for that matter any ToF spectrometer or any other instrument, one should try to identify the main sources limiting the resolution and improve from there. It is also important to consider what are the principle functions of the instrument, i.e. whether it will be used mainly for mass analysis or for ion kinetic energy analysis. For the latter purpose, an artificial flight-time-focusing scheme cannot be used.

3.2.6 Accurate calibration of the time-of-flight atom-probe

In a ToF atom-probe analysis, the mass-to-charge-ratios of ion species can be calculated from their flight times according to eq. (3.11), provided that the length of the flight tube, the time delay constant and the pulse enhancement factor are all accurately known. In fact none of these factors can be separately measured with great accuracy. A better method is to find values of these factors using the flight times of known ionic species. Ionic masses are of course known from a standard table to an accuracy of better than 7 or 8 significant figures, thanks to years of mass spectrometries. Once these factors are calibrated, the mass-to-charge ratios of any ion species can then be calculated from their flight times using eq. (3.11). We will discuss here high precision methods originally devised for calibrating the linear type pulsed-laser ToF atom-probe,[76] but later also adapted to calibrate the flight-time-focused atom-probe.[77] For simplicity, we will discuss first the method for the linear type ToF atom-probe.

Each mass line in the linear type pulsed-laser ToF atom-probe is really an ion energy distribution. The onset flight time of an ion species represents the flight time of those ions with the maximum kinetic energy, or ions with a known kinetic energy of $(neV_0 - \Delta E_c)$, where ΔE_c is the critical ion energy deficit. Since ionic masses are known accurately from an isotope table, from the onset flight times of known ion species, accurate values of the instrument constants can be found. For this purpose, eq. (3.11) is rearranged in

$$t_0 = \frac{1}{C^{1/2}}\left[\frac{M}{n\left[V_0 - \dfrac{\Delta E_c}{ne}\right]}\right]^{1/2} - \delta, \qquad (3.14)$$

where $C = 2e/l^2$ is called the flight path constant of the system. For pulsed-laser atom-probes, no high voltage pulses are used and $\alpha = 0$. The calibration method involves with measuring the onset flight times of ion species with their masses and critical ion energy deficits already accurately known, and then plotting these flight times t_0 against $\{M/[n(V_0 - \Delta E_c/ne)]\}^{1/2}$ using a linear regression method. The slope of the straight lines gives a value of C, and the intercept of the line gives a value of δ. It is important that a double-precision numerical method rather than a graphic method or a single-precision numerical method is used, as the latter methods lack the precision needed for an accurate calibration of the system. Liou & Tsong used 25 flight times of inert gas ions, He, Ne and Ar, and N_2 ions taken at tip voltages of 2–9 kV, for a precision calibration of the Penn State pulsed-laser ToF atom-probe. Data are listed in Table 3.2. For field desorbed gas ions, the critical ion energy deficit is simply given by $I - \phi_{av}$, where ϕ_{av} is the average work function of the flight tube wall which is assumed to be 4.5 eV. Values of these system constants obtained are: $\delta = 20.1$ ns, $C = 0.003\,187\,805$ u μs^{-2} kV^{-1}. The excellent linearity of the plot is demonstrated by the value of the coefficient of determination obtained, which is 0.999 999 996. It differs from a perfect linearity, 1, by only 4×10^{-9}. Using these two system constants, the flight path constant and the time delay constant, the ionic mass or critical energy deficit of any ion species can be accurately calculated from its onset flight time. Table 3.3 lists ionic masses of gas species and solid materials measured with this atom-probe. They differ from standard table values by only 0.000 01–0.001 u for light to heavy elements. Conversely, one may use the ionic masses from the standard table for calculating the critical energy deficits of ion species. As is clear from Table 3.3, values of critical energy deficit obtained differ from those calculated from eq. (2.37) by only 0.1 to 0.2 eV if one takes Q_n, the activation energy of field desorption and evaporation, to be 0. It should be noted that values of the flight-path constant and the time delay constant will change every time changes are made with the instrument. A new calibration has to be done. It is also found that the accuracy of the calibration does not depend critically on the value of ϕ_{av} used as long as a value within about 4.3–4.8 eV is used. The accuracy of this calibration depends only on the precision and the linearity, not the accuracy, of flight-time and voltage measuring devices, and the accuracy of the ionic masses listed in the standard isotope table. Ionic masses of most isotopes are of course known to the accuracy of seven or eight significant figures.

The same method can be modified slightly for the calibration of the flight-time-focused atom-probe. To simplify the method of calibration, it is best to keep the ratio V_p/V_{dc} constant at 0.2 for all the measurements. This ratio is chosen for the purpose of not having preferential d.c. field evaporation from plane edges of a low evaporation field species in the

Table 3.2 *Data for various gas species*

Ion species	Mass u	$\Delta E_c/n$ keV	V kV	$[M/(V-\Delta E_c/e)]^{1/2}$ $(u/kV)^{1/2}$	t_0 μs
He^+	4.002 054	0.020 1	2	1.421 738 894	25.163 0
			2.5	1.270 352 913	22.481 0
			3	1.158 885 913	20.506 0
			3.5	1.072 403 173	18.973 0
			4	1.002 779 373	17.740 0
			4.5	0.945 164 306	16.720 0
			5	0.896 460 505	15.856 0
			5.5	0.854 584 753	15.115 0
			6	0.818 077 619	14.469 0
			6.5	0.785 881 961	13.899 0
			7	0.757 210 925	13.391 5
			7.5	0.731 465 043	12.935 5
			8	0.708 178 516	12.523 0
			8.5	0.686 983 184	12.147 5
			9	0.667 583 696	11.804 0
Ne^+	19.991 890	0.017	4	2.240 380 446	39.659 0
			5	2.003 002 462	35.453 0
			6	1.827 963 103	32.356 0
			6.5	1.756 058 243	31.082 0
			7	1.692 021 611	29.949 0
			7.5	1.634 515 541	28.930 5
N_2^+	28.005 599	0.011 1	3.5	2.833 206 613	50.158 0
			4	2.649 694 831	46.912 0
Ar^+	39.961 834	0.011 3	3.5	3.384 471 748	59.922 0
			4	3.165 242 724	56.044 0

Best fit data: $\delta = 20.1$ ns, $C = 0.003\,187\,805$. Linearity: $\gamma^2 = 0.999\,999\,996$

atom-probe analysis of an alloy, or a compound, so that the composition of the alloy, or the compound, can be analyzed correctly. With this fixed value, the flight time of an ion species should be given by

$$t = \frac{1}{C^{1/2}}\left[\frac{M}{n(1+0.2\alpha)V_{dc}}\right]^{1/2} - \delta. \tag{3.15}$$

The flight time here refers to the peak position of a mass line rather than the onset flight time of a mass line. The calibration method consists of the following steps:

1. Presume a value of α.
2. Measure the flight times of different ion species of known M/n as a function of V_{dc} over a wide mass range and a wide voltage range. V_p

Table 3.3 *Ionic masses and critical energy deficits of gas ions*

Ion species	Ion mass u	Measured ion mass u	ΔM u	$\Delta E_c/n$ eV	Measured $\Delta E_c/n$ eV
He^+	4.002 054	4.002 061 ± 0.000 34	0.000 001 ± 0.000 34	20.1	20.09 ± 0.38
Ne^+	19.991 890	19.991 540 ± 0.002 3	0.000 350 ± 0.002 3	17.0	16.97 ± 0.45
Ar^+	39.961 834	39.962 814 ± 0.003 4	0.000 980 ± 0.003 4	11.3	11.41 ± 0.33
N^{2+}	28.005 599	28.005 723 ± 0.002	0.000 124 ± 0.002 5	11.1	11.14 ± 0.53
Rh^2	51.452 201	51.452 736 ± 0.000 3	0.000 530 ± 0.000 61	11.15	11.23 ± 0.04
W^{3+}					
182	60.648 9	60.649 70 ± 0.001 1			15.14 ± 0.16
183	60.982 8	60.984 55 ± 0.003 1			15.37 ± 0.50
184	61.316 6	61.316 58 ± 0.001 2			15.10 ± 0.18
186	61.984 2	61.982 67 ± 0.002 2	0.001 2 ± 0.000 68	15.1	14.88 ± 0.47
					15.12 ± 0.17

is always adjusted to the same proportion, and the deflector of the Poschenrieder lens, V_{def}, is always adjusted to be equal to $(1 + 0.2\alpha)V_{dc}$. For the Penn State atom-probe, approximately $-0.13V_{dc}$ is applied to the inner electrode and $+0.12V_{dc}$ is applied to the outer electrode.

3. Make a best linear plot of t vs. $[M/nV_{def}]^{1/2}$ by using the method of linear regression. From the slope and intercept of this plot, accurate values of C and δ are obtained for this particular value of α.
4. The same measurement is repeated by assuming different values of α. The coefficient of determination or correlation factor r^2 is calculated for each value of α. The best value of α is the one with its value of r^2 closest to one. In a calibration done by Ren *et al.*[77] for the Penn State flight-time-focused atom-probe, the following values were obtained:

$$\alpha = 1.05, \qquad C = 0.030\,392\,9\ \mathrm{u\,\mu s^{-2}\,kV^{-1}},$$
$$\delta = 0.035\,3\ \mu\mathrm{s}, \qquad r^2 = 0.999\,998\,8.$$

It is found that with this calibration method ionic masses can be derived with an accuracy of 0.01 to 0.05 u for light to heavy elements. With a proper calibration, the shape of the mass lines also appears much more symmetric and a better mass resolution is achieved. But again, the system cannot be used for an ion energy analysis because of the flight-time-focusing scheme.

3.2.7 Methods of counting single ions and compositional analysis

Atom-probe compositional analyses differ from most macroscopic techniques where compositions are usually derived from the signal intensities measured for various chemical species. In atom-probes, compositions are calculated from the number of individual ion signals detected for different chemical species. When an electric pulse is registered by the timer it is counted as the detection of one ion. No provision is made to find the number of ions detected from the signal pulse height. Any instrument which counts single ions or particles directly from the number of ion signal pulses registered in the detector has to consider the possibility of signal pulse overlap due to the limited time resolution of the detector assembly. This problem is especially severe if the resolution of the system is better than the minimum time width the ion detector can resolve. If two ions of the same mass-to-charge ratio arrive at the detector within Δt, which is smaller than the resolution of the pulse detection device and flight time measuring device, then the two ions will be registered as one, and one ion count is lost. Thus majority species of a sample will be undercounted since the probability of signal overlap is greater for the majority species than for the minority species. This problem can easily be alleviated simply by operating the atom-probe at a very low counting rate, say one ion detected for every twenty HV pulses or laser pulses. The probability of signal overlap becomes negligibly small, and the composition derived should then be accurate, provided that there is no other source of error, such as preferential d.c. field evaporation of a species, etc.

Errors due to signal overlap can also be corrected with a statistical method. Statistical methods for such purposes have been developed in nuclear physics where particle countings are very common. Based on a method commonly used for particle counting with a low time resolution Geiger–Müller counter, a statistical method has been developed for analyzing atom-probe data.[78] We will discuss this method in some detail for two purposes. The first is to illustrate the statistical nature of atom-probe data, so that when other circumstances arise similar analyses can be made based on similar statistical methods. The method to be presented is also directly applicable to other surface analytical techniques such as secondary ion mass spectroscopy (SIMS) when they are operated in a single ion counting mode. Second, even though correct composition can be derived by simply operating the atom-probe in a very low counting rate, such operation is not practical if the minority species has a very low concentration, as with impurities in a sample. The statistical method should still be useful for such an analysis.

Consider a binary alloy. Because of the statistical nature of single atom counting, the statistical uncertainty in the compositional analysis is given by

$$\delta = 100\% \times \frac{(N_A N_B)^{1/2}}{(N_A + N_B)^{3/2}}, \tag{3.16}$$

where N_A and N_B represent, respectively, the number of A and B atoms in the data set. It is only a few steps to show that in order to determine the concentration of species A, f_A, to within an accuracy of $\pm\delta$, the least number of ions, N, collected should be

$$N = f_A(1 - f_A)/\delta^2. \tag{3.17}$$

Assuming that f_A is 0.01, and we would like to determine this value with the atom-probe to an accuracy of $\pm 5\%$ of f_A, then the number of ions collected should be at least 39 600. This is already near an impractically large number in the atom-probe analysis. It quickly becomes impractical to determine accurately the concentration of a minority species with the atom-probe at a low counting rate if its concentration is smaller than ~0.5%. The following statistical analysis points to a method with which this difficulty can be overcome.

It is a well recognized fact that in field ion microscopy field evaporation does not occur at a constant rate because of the atomic step structures of the tip surface. For the sole purpose of a compositional analysis of a sample, one should try to aim the probe hole at a high index plane where the step height is small and field evaporation occurs more uniformly. But even so, the number of atoms field evaporated per HV pulse or laser pulse within the area covered by the probe-hole will not be the same every time. It is reasonable to assume that the field evaporation events are nearly random even though there has been no systematic study of the nature of such field evaporation events. Let the average number of atoms field evaporated per pulse within the area covered by the probe-hole area be n. The probability that n atoms are field evaporated by a pulse is then given by the Poisson distribution

$$P_n(\bar{n}) = \frac{\bar{n}^n}{n!}\exp(-\bar{n}). \tag{3.18}$$

$\bar{n}$ is the average number of atoms evaporated per pulse. Consider a binary system with the true fractional abundances represented by f_{A} and f_{B}. Field evaporation of a surface layer normally takes place at the plane edge. Assuming that the atom-probe is operated under the condition that no preferential d.c. field evaporation of a species occurs, then the two components should field evaporate at a rate solely determined by the receding rate of the layer edge. The probability that exactly n_{A} of the n atoms field evaporated are A atoms is given by

$$p_{n_{\mathrm{A}}}(n) = \frac{n!}{n_{\mathrm{A}}!(n-n_{\mathrm{A}})!} f_{\mathrm{A}}^{n_{\mathrm{A}}} f_{\mathrm{B}}^{(n-n_{\mathrm{A}})}. \tag{3.19}$$

A field evaporated ion is detected if it reaches the detector and successfully registers a count. This probability is the product of the quantum efficiency of the detector and the transmission coefficient of the system; it is denoted by e. The probability that all of the n_{A} ions of A-atoms fail to produce an A signal is $(1-e)^{n_{\mathrm{A}}}$, and the probability that an A signal is successfully registered is therefore given by $[1-(1-e)^{n_{\mathrm{A}}}]$. The average number of A signals registered per pulse is therefore given by

$$\begin{aligned}\bar{n}_{\mathrm{A}} &= \sum_{n=0}^{m}\sum_{n_{\mathrm{A}}=0}^{n}[1-(1-e)^{n_{\mathrm{A}}}]p_{n_{\mathrm{A}}}(n)P_n(\bar{n}) \\ &\approx \sum_{n=0}^{\infty}\sum_{n_{\mathrm{A}}=0}^{n}[1-(1-e)^{n_{\mathrm{A}}}]p_{n_{\mathrm{A}}}(n)P_n(\bar{n}) \\ &= 1-\exp(-ef_{\mathrm{A}}\bar{n}),\end{aligned} \tag{3.20}$$

when each pulse removes only a fraction of an atomic layer. In the equation, m is the number of atoms covered by the probe-hole. The same equation can also be derived from a slightly different point of view.

$$\bar{n}_A \approx \sum_{n_A=0}^{\infty} P_{n_A}(f_A \bar{n})[1-(1-e)^{n_A}] = 1 - \exp(-ef_A\bar{n}). \quad (3.21)$$

Similarly, for B-atoms,

$$\bar{n}_B = 1 - \exp(-ef_B\bar{n}). \quad (3.22)$$

The fractional abundances for A- and B-type atoms, as registered in the atom-probe signals, are given by

$$F_A = \frac{\bar{n}_A}{(\bar{n}_A + \bar{n}_B)}, \quad (3.23)$$

$$F_B = \frac{\bar{n}_B}{(\bar{n}_A + \bar{n}_B)}, \quad (3.24)$$

The quantities F_A and F_B are the 'apparent abundances' of the sample for A- and B-type atoms, respectively. These are what an atom-probe analysis gives directly. To find the true abundances, f_A and f_B, one may solve eqs (3.21) to (3.24) for them using a numerical method. For this purpose, it is necessary that the value of $e\bar{n}$ is known. This quantity can be derived by comparing the measured relative abundance of two isotopes of the same element with that listed in the isotope table. There are a few asymptotic behaviors of interest:

1. When $e\bar{n} \to 0$, $F_A \to f_A$ and $F_B \to f_B$. The atom-probe data give directly the true composition of a sample if and only if the pulsed-field evaporation is done at a very slow rate, a procedure now practiced by most investigators.
2. An atom-probe with a low overall detection efficiency e gives a more accurate composition of a sample directly. Such a system, however, has a lower sensitivity and thus a much more limited capability.
3. When $e\bar{n} \gg 1$, both F_A and F_B approach 0.5 as a limit for all combinations of f_A and f_B.

This statistical analysis can be easily extended to a multiple-component system. For an N-component system with fraction abundances $f_1, f_2, \ldots, f_N$, one has

$$\bar{n}_j = 1 - \exp(-ef_j\bar{n}) \quad (3.25)$$

and

$$F_j = \frac{\bar{n}_j}{\sum_i \bar{n}_i}, \quad \text{for } j = 1,2\ldots, N. \quad (3.26)$$

It is important to recognize the statistical nature of atom-probe data even if the detection efficiency of the system is 100%. While the need of a statistical correction can easily be avoided just by collecting data at a very slow field evaporation rate, the method presented can be used to analyze impurities of very low concentrations. A slightly modified numerical method and an almost identical method have later been proposed, but the basic idea is the same. The effect of the overlap of ion signals can also be alleviated by a new detector design, and by measuring the pulse height of each signal. The former method divides the ion detector into several sections and counts the number of ions registered in each section.[79] The chances of having more than one ion of the same mass-to-charge ratio hitting the same detector are then much smaller. Obviously one may also use a position sensitive ion detector combined with a fast data processor to avoid almost completely the problem of signal overlap, provided that one can afford the large number of electronic timers needed. In the later method one assumes that on average the height of a signal pulse should be proportional to the number of ions detected.[80] Unfortunately there is a need to suppress ion signals of very low pulse height so that noise signals will not be counted. This need will complicate considerably the calibration of pulse height to the number of ions detected. This method has the advantage that there is no need to field evaporate at a very low rate, so that impurities of very low concentration can be more easily analyzed.

3.2.8 A method for ion reaction time amplification

In general, ion reaction rates can be observed directly in a time-of-flight spectrometer with a time resolution comparable to that of the system, which is about 10^{-9} to 10^{-10} s. The rate measurement can achieve a much better time resolution by using an ion reaction time amplification method. With this method, very fast ion reactions can be measured with a time resolution much better than the time resolution of the system. It is with this method[69] that the field dissociation reaction of $^4HeRh^{2+}$ was measured with a time resolution of about 20 femtoseconds when the time resolution of the system was still only ~1 ns.

An ion reaction time amplification means that those ion reactions taking place in a very short time period $\delta\tau$ will have their detection stretched over a much longer period δt so that the ion detector and the system electronics can adequately respond. The amplification factor can be defined as

$$A \equiv \delta t/\delta\tau. \tag{3.27}$$

The basic idea is to build the time-of-flight mass spectrometer into two electrically well shielded sections, the acceleration–reaction section of very short length l and a field free flight section of very large length L, or

$l \lll L$, as shown in Fig. 3.21. Thus the total flight time of an ion is determined almost entirely by the kinetic energy of the ion when it leaves the acceleration–reaction section and the length of the field free region. This energy, however, depends entirely on the location in the acceleration–reaction section where the ion is formed. The location, of course, depends on the rate of the ion reaction, or the ion reaction time. As this ion reaction time amplification method is not really confined to atom-probe studies, the basic principle will be illustrated by a generalized ion reaction as represented by

$$(M+m)^{n+} \xrightarrow{\text{ion reaction}} M^{n+} + m^{+} + e^{-} \quad \text{(into metal).} \tag{3.28}$$

The ion reaction can be a photo-dissociation reaction or a further field ionization reaction followed by Coulomb repulsive dissociation. For simplicity, the compound ions, $(M+m)^{n+}$, are assumed to be formed right at the surface. Let us assume that an $(M+m)^{n+}$ ion is dissociated by the ion reaction at x where $x < l$. The total flight time of the M^{n+} ion formed at x is given by

$$t \approx \frac{L}{v_{\mathrm{f}}(x)} = L\left\{\frac{2neV_0}{M}\left[1-\left(\frac{m}{m+M}\right)\frac{x}{l}\right]\right\}^{-1/2}, \tag{3.29}$$

where $v_{\mathrm{f}}(x)$ is the final velocity of the M^{n+} ion when it enters into the field free region. This velocity depends on x where it is formed. A difference in the flight times of δt for two ions will correspond to a difference of δx in the locations where they are formed, which in turn will correspond to a difference of $\delta\tau$ in the times of their formation. They are related to each other by

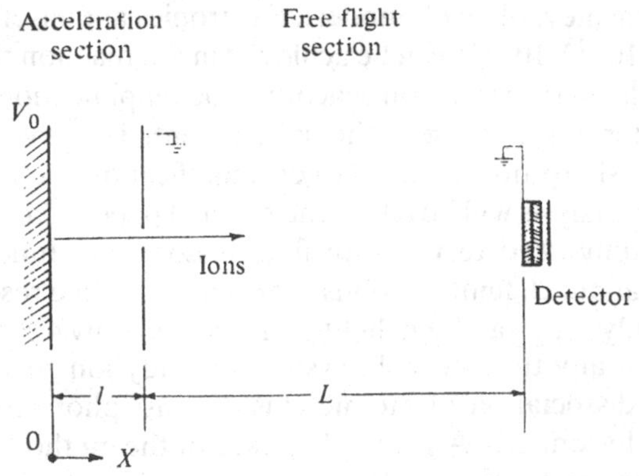

Fig. 3.21 Geometrical scheme for ion reaction time amplification as discussed in the text.

$$\delta t = \frac{\frac{neV_0}{M}\left(\frac{m}{m+M}\right)\frac{L}{l}\delta x}{\left\{\frac{2neV_0}{M}\left[1-\left(\frac{m}{m+M}\right)\frac{x}{l}\right]\right\}^{3/2}}, \tag{3.30}$$

and

$$\delta\tau(x) = \delta x/v(x), \tag{3.31}$$

where

$$v(x) = \left[\frac{2neV_0}{(M+m)}\frac{x}{l}\right]^{1/2} \tag{3.32}$$

is the velocity of the compound ion $(M+m)^{n+}$ just before its dissociation. Combining the last four equations, one finds

$$A(x) = \frac{\left(\frac{Mx}{ml}\right)^{1/2}}{2\left(\frac{m+M}{m}-\frac{x}{l}\right)^{3/2}}\frac{L}{2l}. \tag{3.33}$$

This amplification factor is proportional to L/l, and is also dependent on the location where the ion reaction occurs. Maximum amplification occurs at $x = l$, $A(l) = (m/2M)(L/l)$, or if the reaction occurs right after the parent ion has gained the full energy of the acceleration voltage. Equation (3.33) suggests that whenever possible one should measure the flight time distribution, or the energy distribution, of the ion species with the smaller mass.

It is quite practical to have an ion reaction time amplification factor of a few thousands. For example, with an l of say 1.5 mm and L of 7.5 m, an amplification of 5000 is achieved. Thus with a system time resolution of 1–0.1 ns, easily achievable with modern electronics and laser units, a time resolution of 10^{-13}–10^{-14} s can be achieved in ion reaction time measurements using the very simple ion reaction time amplification scheme. For field ion emitter tips, because of the rapid potential drop very close to the surface of the sharp tip, a much larger amplification factor is achieved. Details of the analysis will be left to the original paper.[69] This scheme has been successfully used to measure field dissociation time with a time resolution of about 20 femtoseconds even when the time resolution of the system was only ~1 ns and the flight path was still only 4.2 m in length. It can be used in any time-of-flight system to study ion reactions such as spontaneous dissociation of atomic cluster ions, photo-fragmentation, Coulomb explosion, etc. A great advantage of the method is its extreme simplicity, since only a long flight tube and a small probe-hole separating the reaction–acceleration region and the free flight region are needed.

3.2.9 Magnetic sector atom-probe

The single atom chemical analysis capability required of an atom-probe, can in principle also be attained with another type of atom-probe which combines a field ion microscope, a probe-hole and a single ion detection sensitivity mass spectrometer of some type. In the magnetic sector atom-probe a sensitive magnetic sector mass spectrometer is used.[33,34] Figure 3.22 shows a magnetic sector atom-probe which uses a Chevron channel plate for detecting single ions over a mass range of about 10 a.m.u.[34] As mentioned earlier, this type of atom-probe is severely hampered by its limited range of masses which can be detected at a given magnetic field strength used. Although different mass ranges can be scanned by changing the magnetic field strength, this procedure defeats the whole purpose

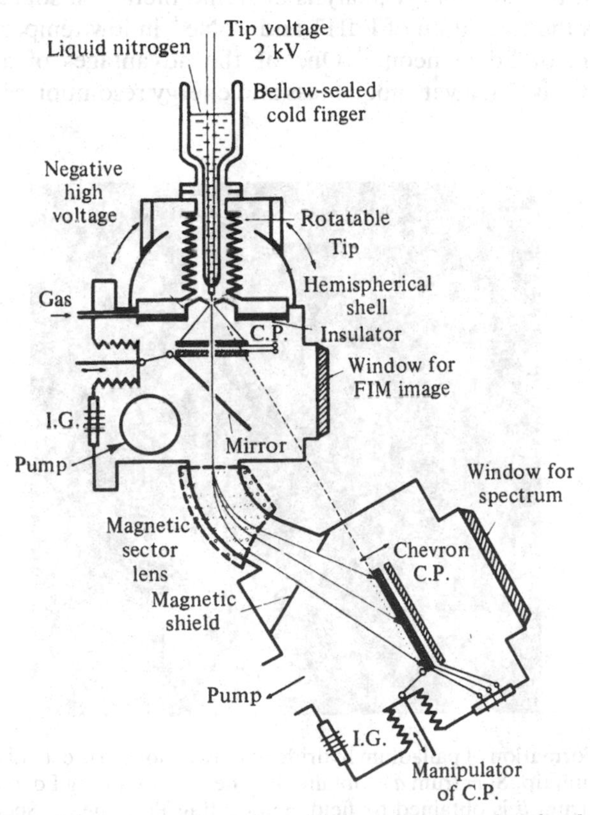

Fig. 3.22 Magnetic sector atom-probe which uses a second order focusing 60° sector field. The line spectrum is displayed on a 7.5 cm diameter Chevron channel plate and a proximity focused screen, which can show the impact of single ions.

of single atom chemical analysis of a sample of unknown chemical species. With the rather large noise level of the Chevron channel plate in d.c. mode operation, it is impossible to analyze the composition of one atomic layer even if all the atoms have their masses falling into the same mass window range. The author believes that it is possible to use a position sensitive ion detector to remedy this limitation if the detector has a very low noise level, and the magnetic sector atom-probe may be very useful for some special applications where the mass range needed is not large. The mass window can of course be widened by using a large strip of detector, or several detectors covering different mass ranges of interest.

Up to the time of this writing, magnetic sector atom-probe has been used mainly as a sensitive field ion and field desorption mass spectrometer and ion momentum analyzer. It has been successfully used to study helide ion, neide ion and hydride ion formation in field evaporation and field desorption, and ion energy analysis of liquid metal ion sources. Figure 3.23 shows the formation of PdH^+ and $PdNe^+$ in low temperature field evaporation of Pd in neon.[34] One of the advantages of a magnetic spectrometer is that a very good mass and energy resolution can be easily

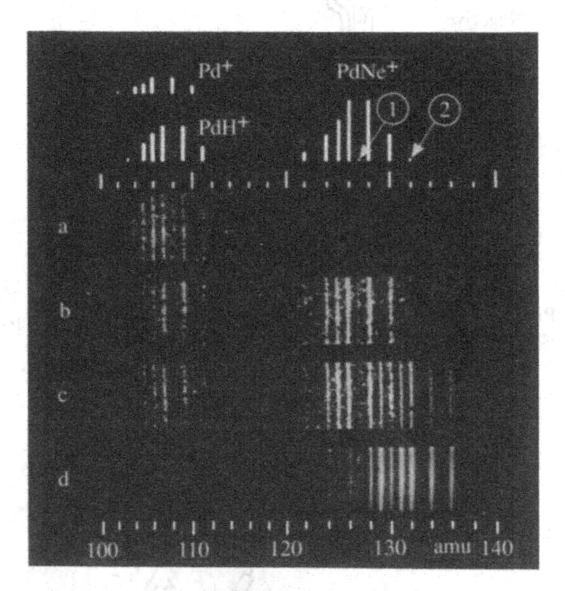

Fig. 3.23 Formation of palladium hydride and neide ions in d.c. field evaporation of a palladium tip. Spectrum *a* is obtained by field evaporating Pd in hydrogen at 78 K. Spectrum *b* is obtained by field evaporating Pd in neon. Spectrum *c* is a double exposure of two spectra, first as in *b*, second replacing Ne with Xe at a 22% reduced field. Spectrum *d* is obtained in field ionization of Xe. There are 9 isotopes of Xe.

achieved. Up to the moment of this writing, however, it has not been successfully used for compositional analysis of samples with a similar sensitivity as a time-of-flight type atom-probe. With the availability of position sensitive particle detectors, it may be possible to enlarge the mass window considerably and the magnetic sector atom-probe may one day find wider applications.

References

1 See for examples: Direct imaging of atoms in crystals and molecules, *Proc. Nobel Symp.*, **47**, ed. L. Kihlborg, Royal Swedish Academy of Science, 1979; also D. J. Smith in *Chemistry and Physics of Solid Surfaces VI*, 413, Springer, Berlin, 1986.

2 See for example P. Hansma & J. Tersoff, *J. appl. Phys.*, **61**, R1 (1987) and references therein.

3 E. W. Müller, *Z. Physik*, **131**, 136 (1951).

4 E. W. Müller & T. T. Tsong, *Field Ion Microscopy, Principles and Applications*, Elsevier, New York, 1969.

5 E. W. Müller, *Z. Physik*, **106**, 541 (1937).

6 T. T. Tsong & R. J. Walko, *physica status solidi*, **a12**, 111 (1972).

7 See for examples: W. J. Tegart, *The Electrolytic and Chemical Polishing of Metals*, Pergamon Press, Oxford, 1959; P. A. Jacquet, *Metallurgical Rev.*, **1**, 157 (1961); I. S. Brammer & M. A. P. Dewey, *Specimen Preparation for Electron Metallography*, American Elsevier, New York, 1966; J. W. Edington, *Practical Electron Microscopy in Materials Science*, Macmillan Philips Technical Library, Eindhoven, 1974.

8 This table is an incomplete compilation of data available in the literature, mostly from the original book of E. W. Müller & T. T. Tsong.

9 C. F. Douds, MS Thesis, Pennsylvania State University, 1957.

10 B. Ralph & D. G. Brandon, *Phil. Mag.*, **8**, 919 (1963); K. M. Bowkett, J. J. Hren & B. Ralph, in *Third European Reg. Conf. Electron Microscopy, Prague*, 1964, 191, Publishing House of Czech. Acad. Sci.; S. Ranganathan, *ibid.*, 265.

11 A. J. Melmed & R. J. Stein, *Surface Sci.*, **49**, 645 (1975).

12 H. Morikawa, T. Terao & Y. Yashiro, private communication.

13 C. L. Chen & T. T. Tsong, unpublished data.

14 H. W. Fink, *IBM J. Res. Develop.*, **30**, 460 (1986).

15 O. Nishikawa, K. Hattori, F. Katsuki & M. Tomitori, *J. de Physique, Coll.*, **49**, C6-55 (1989).

16 D. G. Brandon, *J. Sci. Instrum.*, **41**, 373 (1964).

17 A. J. W. Moore, *J. Phys. Chem. Solids*, **23**, 907 (1962).

18 E. W. Müller & K. Bahadur, *Phys. Rev.*, **102**, 624 (1956).

19 M. Drechsler & P. Wolf, in *Proc. IV Intern. Congr. Electron Microscopy, Berlin*, 1958, Springer, Berlin, 1960, Vol. 1, 835.

20 A. J. W. Moore & S. Ranganathan, *Phil. Mag.*, **16**, 723 (1967).

21 A. J. Perry & D. G. Brandon, *Surface Sci.*, **7**, 422 (1967); *Phil. Mag.*, **16**, 119 (1967).

22 J. K. Mackenzie, A. J. W. Moore & J. F. Nickolas, *J. Phys. Chem. Solids*, **23**, 185 (1962).

23 R. C. Sanwald & J. J. Hren, *Surface Sci.*, **7**, 197 (1967).
24 B. Young, *Science*, **239**, cover page (1988).
25 See T. T. Tsong, D. L. Feng & F. M. Liu, *Surface Sci.*, **199**, 421 (1988).
26 R. H. Good, Jr & E. W. Müller, *Handbuch der Physik*, 2nd edn, Springer, Berlin, **XXI**, 176 (1956).
27 E. W. Müller & R. D. Young, *J. Appl. Phys.*, **32**, 2425 (1961).
28 T. Sakurai & E. W. Müller, *Phys. Rev. Lett.*, **30**, 532 (1973).
29 C. M. C. De Castihlo & D. R. Kingham, *Surface Sci.*, **173**, 75 (1986); *J. de Physique*, **45**, C9-77 (1984).
30 M. Chung, N. M. Miskovsky, P. H. Cutler, T. E. Feuchtwang & E. Kazes, *J. Vac. Sci. Technol.*, **B5**, 1628 (1987).
31 R. Haefer, *Z. Physik*, **116**, 604 (1951); J. A. Becker, *Bell System Techn. J.*, **30**, 907 (1951); W. P. Dyke, J. K. Trolan, W. W. Dolan & G. Barnes, *J. Appl. Phys.*, **24**, 570 (1953); M. Drechsler & E. Henkel, *Z. angew. Phys.*, **6**, 341 (1954).
32 E. W. Müller, J. A. Panitz & S. B. McLane, *Rev. Sci. Instrum.*, **39**, 83 (1968).
33 D. F. Barofsky & E. W. Müller, *Surface Sci.*, **10**, 177 (1968).
34 E. W. Müller & T. Sakurai, *J. Vac. Sci. Technol.*, **11**, 878 (1974).
35 T. Sakurai, Ph.D. Thesis, Pennsylvania State University, 1974; S. Kapur, Ph.D. Thesis, Pennsylvania State University, 1979.
36 T. Sakurai, R. J. Culbertson & G. H. Robertson, *Appl. Phys. Lett.*, **34**, 11 (1979).
37 T. T. Tsong & T. J. Kinkus, *Phys. Rev.*, **B29**, 529 (1984).
38 T. T. Tsong & Y. Liou, *Phys. Rev. Lett.*, **55**, 2180 (1985); T. T. Tsong, *Phys. Rev. Lett.*, **55**, 2826 (1985).
39 S. S. Brenner & J. T. McKinney, *Surface Sci.*, **23**, 88 (1970); P. J. Turner, B. J. Regan & M. J. Southon, *Vacuum*, **22**, 443 (1972).
40 J. A. Panitz, *Crit. Rev. Solid State Sci.*, **5**, 153 (1975).
41 Channeltron was used in the early atom-probe of S. S. Brenner & J. T. McKinney. 100% detection efficiency was emphasized by T. Sakurai, T. Hashizume & A. Jimbo, in *Appl. Phys. Lett.*, **44**, 38 (1984).
42 T. T. Tsong, *J. Chem. Phys.*, **54**, 4205 (1971); *J. Phys. F: Metal Phys.*, **8**, 1349 (1978).
43 M. Yamamoto & D. N. Seidman, *Surface Sci.*, **118**, 535 (1982).
44 G. D. W. Smith and co-workers have compared atom-probe compositional analysis of various materials using a HV pulse atom-probe and a pulsed-laser atom-probe, and have found proper conditions for analysis of alloys and semiconductors. See A. Cerezo, C. R. M. Grovenor & G. D. W. Smith, *Appl. Phys. Lett.*, **46**, 567 (1985).
45 T. T. Tsong, *Surface Sci.*, **70**, 228 (1978).
46 E. W. Müller & S. V. Krishnaswamy, *Rev. Sci. Instrum.*, **45**, 1053 (1974).
47 W. P. Poschenrieder, *Int. J. Mass Spectrom. Ion Phys.*, **6**, 413 (1971); 9, 357 (1972).
48 A. R. Waugh & M. J. Southon, *Surface Sci.*, **89**, 718 (1979).
49 O. Nishikawa, K. Kurihara, M. Nachi, M. Konishi & M. Wada, *Rev. Sci. Instrum.*, **52**, 810 (1981).
50 A very carefully fine tuned system can be found at the Oak Ridge National Laboratory under M. K. Miller.
51 J. A. Panitz, *Rev. Sci. Instrum.*, **44**, 1034 (1973); *J. Vac. Sci. Technol.*, **11**, 206 (1974).
52 R. J. Walko & E. W. Müller, *Physica Status Solidi*, **a9**, K9 (1972).

53 J. A. Panitz, *Prog. Surface Sci.*, **8**, 219 (1978).
54 G. L. Kellogg & T. T. Tsong, *Surface Sci.*, **110**, L559 (1981).
55 See for example B. Ralph, S. A. Hill, M. J. Southon, M. P. Thomas & A. R. Waugh, *Ultramicroscopy*, **8**, 361 (1982).
56 C. F. Ai & T. T. Tsong, *Surface Sci.*, **138**, 339 (1984); *J. Chem. Phys.*, **81**, 2845 (1984).
57 G. L. Kellogg, *Phys. Rev. Lett.*, **54**, 82 (1985).
58 A. Cerezo, T. J. Godfrey & G. D. W. Smith, *Rev. Sci. Instrum.*, **59**, 862 (1988).
59 Th. Schiller, U. Weigmann, S. Jaenicke & J. H. Block, *J. de Physique*, Collog., **47**, C2-479 (1986); J. Witt & K. Müller, *Phys. Rev. Lett.*, **57**, 1153 (1986).
60 A. R. Waugh, E. D. Boyes & M. J. Southon, *Surface Sci.*, **61**, 109 (1976).
61 E. W. Müller, S. B. McLane & O. Nishikawa, *Proc. 6th Intern. Congr. Electron Microscopy, Kyoto*, Vol. 1, 235 (1966).
62 E. W. Müller & T. T. Tsong, *Field Ion Microscopy, Principles and Applications*, Elsevier, New York, 1969, 276.
63 T. T. Tsong, J. H. Block, M. Nagasaka & B. Viswanathan, *J. Chem. Phys.*, **65**, 2469 (1976).
64 Attempts were made by A. J. Melmed and J. A. Panitz, and also by T. T. Tsong and Y. S. Ng, and others.
65 A. J. Melmed, M. Martinka, T. Sakurai & Y. Kuk, *Proc. 27th Intern. Field Emission Symp.*, Tokyo, eds Y. Yashiro & N. Igata, 1980, 151.
66 G. L. Kellogg & T. T. Tsong, *J. Appl. Phys.*, **51**, 1184 (1980).
67 W. Drachsel, S. Nishigaki & J. H. Block, *Int. J. Mass Spectrom. Ion Phys.*, **32**, 333 (1980); S. Nishigaki, W. Drachsel & J. H. Block, *Surface Sci.*, **87**, 389 (1979).
68 T. T. Tsong, S. B. McLane & T. J. Kinkus, *Rev. Sci. Instrum.*, **53**, 1442 (1982).
69 T. T. Tsong, *J. Appl. Phys.*, **58**, 2404 (1985).
70 C. F. Ai & T. T. Tsong, *Surface Sci.*, **138**, 339 (1984); *J. Chem. Phys.*, **81**, 2845 (1984).
71 O. Nishikawa, O. Kaneda, M. Shibata & E. Nomura, *Phys. Rev. Lett.*, **53**, 1252 (1984).
72 H. F. Liu & T. T. Tsong, *Rev. Sci. Instrum.*, **55**, 1779 (1984); H. F. Liu, H. M. Liu & T. T. Tsong, *J. Appl. Phys.*, **59**, 1334 (1986).
73 T. J. Chuang, *Surface Sci. Rep.*, **3**, 1 (1983).
74 The analysis presented here follows essentially that presented in ref. 68.
75 S. V. Krishnaswamy & E. W. Müller, *Rev. Sci. Instrum.*, **45**, 1049 (1974).
76 T. T. Tsong, Y. Liou & S. B. McLane, *Rev. Sci. Instrum.*, **55**, 1246 (1984). Y. Liou & T. Tsong, to be published in *J. de Physique*.
77 D. M. Ren, T. T. Tsong & S. B. McLane, *Rev. Sci. Instrum.*, **57**, 2543 (1986).
78 T. T. Tsong, Y. S. Ng & S. V. Krishnaswamy, *Appl. Phys. Lett.*, **32**, 778 (1978).
79 D. Blavett, A. Bostel & J. M. Sarrau, J. de Physique, **47**, C2-473 (1986).
80 O. Nishikawa, K. Oida & M. Tomitori, *J. de Physique*, **47**, C7-515 (1987).

4

Applications to surface science

4.1 Atomic structure of solid surfaces

The field ion microscope, when operated at low temperatures, has a resolution of better than 3 Å if the tip radius is less than a few hundred Å, which is sufficient to resolve the atomic structures of most solid surfaces. For closely packed surfaces, atomic spacings are usually less than 2.8 Å. But even such surfaces can be resolved if the tip temperature is below 20 K, the tip radius is less than 150 Å and the plane size is also very small so as to contain no more than, say, a few tens of atoms, as can be seen from images shown in Fig. 2.32. For a material to be able to be imaged in the field ion microscope, however, the low temperature evaporation field of the material has to be comparable to or higher than the best image field of the image gas used for the field ion imaging. In field ion microscopy, to eliminate the problem of contaminating a field evaporated clean surface and of destroying the surface by a field induced chemical etching from chemically reactive gases, one almost always uses an inert gas for the field ion imaging. The ionization energies of inert gases are higher than most chemically reactive gases, thus requiring a higher field ionization field, or an image field. This image field can serve to protect the clean, field evaporated emitter surface from contamination and from a field induced chemical etching by chemically active residual gases inside the vacuum chamber. At the present time, the lowest ionization energy inert gas often used in field ion microscopy is Ar, which has a best image field of about 2.2 V/Å. In principle, any material with its low temperature evaporation field comparable to or higher than this value can be imaged in the field ion microscope. These materials include most metals, with the exception of the alkali metals, their alloys, semiconductors, and many compound materials such as LaB_6 and high temperature oxide superconductors, as will be discussed in the following sections. With the availability of convenient image intensification devices such as the microchannel plate and image intensifier, the much more easily achievable UHV, and better method of purifying gases, it is possible that the range of materials which can be studied in the field ion microscope will continue to widen by using lower ionization energy inert gases such as Kr and Xe as these technologies continue to advance.

4.1.1 Atomic structures of field evaporated solid surfaces

In the past, most field ion microscopists focused on studying the atomic structure of field evaporated surfaces. Many pretty field ion micrographs have been published in the literature for most metals and many alloys and some compounds also. We show here only a micrograph of W of very small tip radius, a Rh tip of intermediate radius, and a Pt of very large tip radius, in Figs 4.1, 4.2 and 4.3. When the tip radius is very small, the field ion image shows very good spatial resolution, but the number of atoms in a net plane is rather small, and only a very small number of low index facets are developed. The image intensity is fairly uniform over the entire emitter surface. When the tip radius is very large, the spatial resolution becomes less good with only a few low index surfaces atomically resolved, but many high index planes are now well developed. The brightness variations of different crystallographic regions also become very pronounced and the micrograph forms a very well ordered and yet rather artistic pattern, as can be seen in Fig. 4.3 for Pt. If the tip radius is small enough and the cooling temperature is below ~20 K, then the atomic structure of even the most closely packed surface can be resolved. For

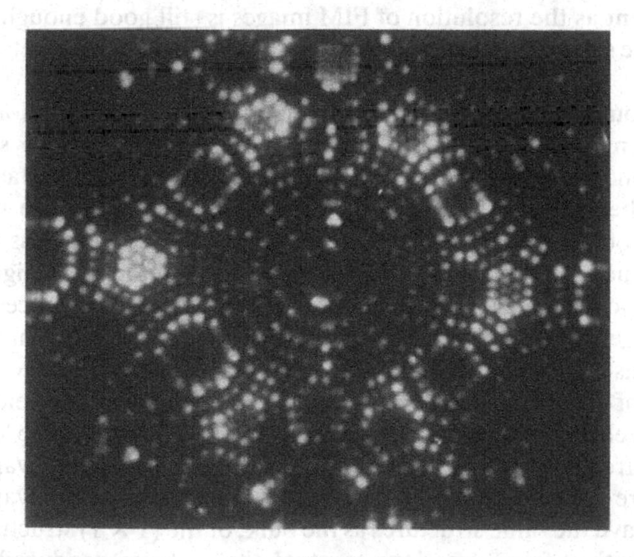

Fig. 4.1 Helium field ion image of a (110) oriented bcc tungsten tip of radius ~140 Å where atoms in the {001} and {111} are resolved. The image covers a ~100° extension angle of the emitter surface.

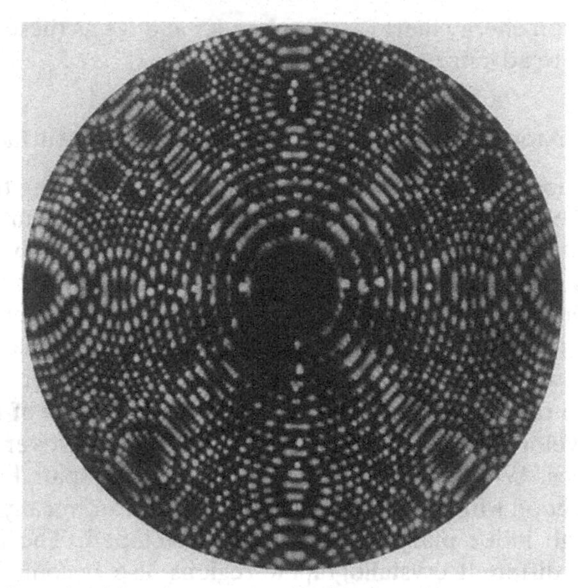

Fig. 4.2 Helium field ion image of a (001) oriented fcc rhodium tip with an extension angle of ~50° where most atoms in high index planes are resolved.

most purposes, however, tip radii in 150 to 600 Å range are most convenient as the resolution of FIM images is still good enough and the size of the surfaces is also large enough to contain fair sized facets for FIM studies.

It is found that for metals, low temperature field evaporation almost always produces surfaces with the (1 × 1) structure, or the structure corresponding to the truncation of a solid. A few such surfaces have already been shown in Fig. 2.32. That this should be so can be easily understood. For metals, field penetration depth is usually less than 0.5 Å,[1] or much smaller than both the atomic size and the step height of the closely packed planes. Low temperature field evaporation proceeds from plane edges of these closely packed planes where the step height is largest and atoms are also much more exposed to the applied field. Atoms in the middle of the planes are well shielded from the applied field by the itinerant electronic charges which will form a smooth surface to lower the surface free energy, and these atoms will not be field evaporated. Therefore the surfaces produced by low temperature field evaporation should have the same structures as the bulk, or the (1 × 1) structures, and indeed with a few exceptions most of the surfaces produced by low temperature field evaporation exhibit the (1 × 1) structures.

If field evaporation is done at high temperatures where jumping of an

Fig. 4.3 Helium field ion image of a (001) oriented platinum tip of very large tip radius, taken by Müller in the 1960s, which shows a regionally varying image brightness pattern.

atom to neighbor sites becomes possible, the surface produced may not have the (1 × 1) structure of the crystal. Under such conditions, surface atoms in a small plane may rearrange to assume an atomic structure of lowest surface free energy in the presence of the applied field, which may not be the same as the atomic structure of lowest free energy in the absence of the applied field. The effect of the applied field in changing the atomic structure of a small surface has been observed for the Rh (001) when the surface is field evaporated at a field below 3.5 V/Å with pulsed-laser heating,[2] and of a small atomic cluster has been observed for a W adatom-cluster on the W (110) surface.[3]

For materials of large field penetration depth,[4] mostly covalent and ionic bond materials, field evaporation can occur almost randomly from plane edges as well as from the middle of a plane. Surfaces prepared by low temperature field evaporation are usually not very well ordered. This is of course the reason that well ordered field ion images of silicon have not been obtained by field evaporation despite the considerable efforts by numerous investigators over many years.[5,6,7] Sometimes this random nature of field evaporation can be alleviated by using field induced chemical etching of the surface with a chemically reactive gas such as hydrogen.[8] Hydrogen reduces the evaporation field of silicon by almost 50% by forming silicon hydride ions. Formation of hydride apparently is more effective for kink site atoms where there are many more dangling bonds available for reacting with gas molecules. Thus field evaporation proceeds much more regularly from the layer edges, and a better surface can be developed. Silicon surfaces prepared by field evaporation in hydrogen usually show very well the atomic ring structures of the surface layers of low index surfaces such as the (111). Unfortunately, no atomic arrangement of a surface can be revealed by such a method. Even if an atomic structure can be revealed by this method, it will still represent only a structure produced under the influence of chemisorption of the gas species, not the thermodynamically equilibrated structure of the clean surface.

It is important to distinguish a field evaporated surface from a thermally equilibrated surface.[9] Let us repeat the differences. The atomic structure of a surface prepared by low temperature field evaporation reflects the nature of field evaporation as a result of the particular field penetration depth of the material. The atomic structure of a surface prepared by field evaporation at a temperature where an atom can jump to neighbor sites should reflect the atomic structure of lowest surface free energy in the presence of the applied field, or a term corresponding to the polarization energy has to be included in the surface free energy. When a surface is prepared by field induced chemical etching the structure of the surface should reflect the etching process. All these atomic structures are not necessarily the atomic structure of the thermally equilibrated, clean surface.

4.1.2 Atomic reconstructions of metal surfaces

It is now well established that the structure of a thermally equilibrated surface may be quite different from that obtained by simply truncating a solid.[10] Surface atoms may rearrange to form a structure according to the lowest surface free energy of that surface. The determination of the atomic structure of solid surfaces is one of the important endeavors in surface science, as the atomic structure will decide both the physical and the chemical properties of the surfaces. Surface reconstruction will also play an important role in the crystal growth and epitaxy of thin films.

Many surface structure analytical techniques have been used to investigate the atomic structure of solid surfaces. Basically, they can be categorized into reciprocal space techniques and direct space techniques. In a reciprocal space technique, diffraction patterns are obtained with a surface sensitive particle beam such as low energy electrons. In real space techniques, either macroscopic particle channeling methods[11] are used to derive information about the atomic structures or microscopic atomic resolution microscopes[12,13,14,15] are used to image directly the atomic structure of the surfaces. Microscopic techniques have recently attracted a lot of attention from many surface scientists for the following reasons. First, it is now well recognized that even surfaces prepared by the most elaborate procedures are not atomically perfect and flat on a macroscopic scale like many surface scientists used to think they were. They usually contain lattice steps, patches of very irregular atomic structures, and many lattice defects. Many surface properties are to a very large extent affected by these irregularities.[16] Second, it is now well recognized that islands of different atomic structures may coexist on a thermodynamically very well equilibrated surface.[17] If a macroscopic technique is used to derive the atomic structure from a sample area of macroscopic size, the atomic structure information obtained is really that due to superposition of these different atomic structures and it may be difficult to derive all these different atomic structures from that information. Third, atomic resolution microscopies have now reached a resolution where not only the atomic structure of a flat plane can be observed, but also the atomic structures of 2-dimensional lattice defects on the surface. Already some of the outstanding questions in the atomic structure studies have been resolved by atomic resolution microscopies. For example, from scanning tunneling microscope (STM) investigations[12] it is now generally agreed that a model proposed by Takayanagi[18] is the correct atomic structure for the (7×7) reconstructed silicon surface. The (1×1) to (1×2) and the

(1 × 1) to (1 × 5) surface reconstructions of the (011) and (001) surfaces of fcc metals such as Au, Pt and Ir have also been successfully studied with the STM,[19,20] with the transmission electron microscope (TEM)[13,21] and with the field ion microscope.[14,15,22] Some of the FIM surface atomic reconstruction studies will be described below.

(*a*) *Bcc* {001} *surfaces*

In field ion microscopy of surface atomic reconstruction, the first surface studied appears to be the W (001). A low energy electron diffraction (LEED) study by Yonehara & Schmidt finds a temperature dependence of the atomic structure of the W (001): above 300K the LEED pattern has (1 × 1) symmetry and below 300 K the pattern changes to a c(2 × 2) symmetry.[23] This structure change was originally suspected to be due to adsorption of hydrogen or other impurities. The effect was later shown by Felter *et al.*[24] and Debe & King[25] to be a property of the clean W (001) surface, or a temperature induced structure phase transition. At first a vertical shift model was proposed by Felter *et al.*[24] to explain the c(2 × 2) LEED pattern. Debe & King[26] proposed a lateral shift model. The LEED pattern arises from a W (001) surface with anti-phase domains having *p2mg* symmetry and a $(\sqrt{2} \times \sqrt{2})$R45° configuration, in a (1 × 1)-structured sea, with all atoms in the reconstructed domains being displaced (1) laterally and periodically along ⟨11⟩ directions, and (2) vertically away from the second layer atoms by equal amounts. They further interpreted their LEED data to show that no reconstruction occurred within 20 Å of a surface step.[26] This lateral shift model is now generally accepted, although there are some exceptions.

In field ion microscopy, Tsong & Sweeny[27] find that the structure of the W (001) plane can be fully resolved if the tip temperature is low enough and both the tip radius and the plane size are also very small. Some of the images have already been shown in Fig. 2.32*c*. The image structure obtained is (1 × 1), and neither a c(2 × 2) image structure expected from the vertical shift model nor an image structure expected from the lateral shift model is observed. In view of the very small size of the FIM W (001) surfaces, no more than 40 Å in diameter, this FIM study appears to be in good agreement with the result of Debe & King; they find no reconstruction within 20 Å of a lattice step. However, an FIM study by Melmed *et al.*[28] finds that field evaporation of a W (001) surface in the temperature range between 15 K and 400 K all develops into a c(2 × 2) pattern. They interpret this result as supporting the vertical shift model. A more detailed study by Tung *et al.*[29] finds that during field evaporation at high temperature, above 400 K, some atoms (shown as A in Fig. 4.4) field evaporate at lower voltages than do the other atoms. This results in an array of vacancies having p(2 × 2) symmetry. At a slightly higher

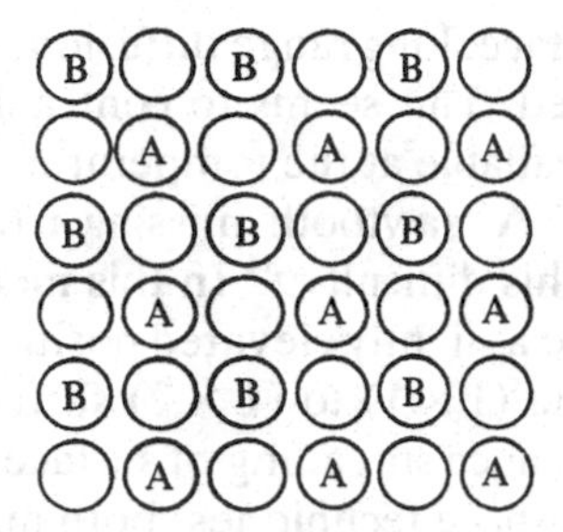

Fig. 4.4 According to Tung *et al.*, when a W (001) surface is field evaporated above 400 K, atoms labelled A are field evaporated at lower voltages, resulting in an array of vacancies having p(2 × 2) symmetry. At a slightly higher voltage, a second set of atoms (B) field evaporates. These leave a second set of vacancies. The remaining atoms form a layer of the $(\sqrt{2} \times \sqrt{2})R45°$ configuration.

voltage, a second set of atoms (B) field evaporates; these leave a second p(2 × 2) array of vacancies. The remaining atoms after field evaporation of A and B atoms have the $(\sqrt{2} \times \sqrt{2})R45°$ configuration. The two different stages of field evaporation of A and B atoms are interpreted to show two different vertical shifts of atoms in the reconstructed surface. While this FIM result is most interesting, the interpretation has not in general been accepted by other investigators. In this author's opinion, one must distinguish the structure of a field evaporated surface from that of an equilibrated surface in the absence of the applied field. There is no question that the peculiar field evaporation behavior of the W (001) surface is a result of the same particular energetics and atomic interactions as are responsible for the surface atomic structure phase transformation. The field evaporation behavior, however, does not necessarily indicate the occurrence of vertical shifts of surface atoms, but may just reflect the different binding energies of surface atoms in different sites. It is this same binding energy difference which produces the surface atomic structure phase transformation. The important thing is that the structure phase transformation of the W (001), found by LEED studies, manifests itself in the field evaporation behavior of the surface in the FIM which can be clearly observed.

(*b*) *Fcc* {110} *surfaces*

Macroscopic experiments, both reciprocal-space techniques and direct-space techniques, conclude that the (110) surfaces of Pt, Ir and Au reconstruct to a (1 × 2) structure.[30] Several atomic models have been proposed,[31] of which the 'simple missing row model' agrees better with various experimental data. In this model every other [110] atomic row is missing from the reconstructed (1 × 1) surface. To transform a (1 × 1) surface into a (1 × 2) surface, long range diffusion, of the order of the size of the surface, is required. This seems to contradict the mass-transport surface diffusion data available at the temperature where surface reconstruction can occur.[32,33] A 'sawtooth missing row model' is therefore proposed to overcome this difficulty.[34] In this model, every other [110] atomic row is displaced to a slightly elevated position by a jump of ~2.8 Å of atoms in the row. The (1 × 1) to (1 × 2) surface reconstruction will then require no large-distance spreading of surface atoms or [110] atomic rows. While most direct-space techniques, both macroscopic and microscopic, agree that the simple missing row model is the correct atomic model for the (1 × 2) reconstructed surfaces of the Pt, Ir and Au (110), no atomically resolved image has been obtained with a microscopy. In TEM experiments, a cross-sectional view consistent with the simple missing row model can be clearly seen, but the atomic spacing within an atomic row is not really seen. Neither is the question of atomic transport and surface diffusion being answered in these studies. FIM studies are able to address both of these two questions simultaneously.

Let us discuss how one can distinguish FIM images of a simple missing row (1 × 2) fcc (110) surface from those of a sawtooth missing row (1 × 2) fcc (110) surface. Figure 4.5 shows the differences. For the (1 × 1) structure, closely packed [110] atomic rows are separated from each other by about 4 Å. When the top layer is field evaporated, the second layer has exactly the same (1 × 1) atomic arrangement, as can be seen from the side view of the same figure. For the simple missing row (1 × 2) surface, the top layer consists of closely packed [110] atomic rows separated by about 8 Å, or twice the normal spacing. The second layer, on the other hand,

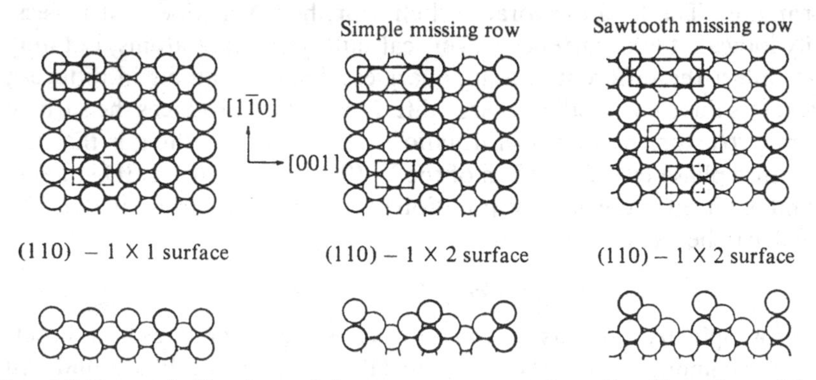

Fig. 4.5 Top and side views of the non-reconstructed, or (1 × 1), surface of fcc (110) surface and two models, simple missing row and sawtooth missing row models, of the (1 × 2) reconstructed surface. These two models can be distinguished from the side view; the first shows only one reconstructed layer whereas the other shows two reconstructed layers.

returns to the (1 × 1) structure. For the sawtooth missing row (1 × 2) surface, both the top and the second layer should both show closely packed [110] atomic rows separated by about 8 Å. Only from the third layer on will the structure return to the (1 × 1). Thus by field evaporation of a reconstructed surface these two models can be easily distinguished from the field ion images.

Field ion microscope studies of surface reconstruction of Pt and Ir (110) and other surfaces, carried out by Kellogg[14] and by Gao & Tsong,[15] observe the atomic structures of surfaces thermally annealed in the absence of the image field, not those of field evaporated surfaces as in FIM studies of the W (001) surface.[28,29] These studies represent the first FIM investigations of surface reconstruction performed under similar conditions as many other surface science studies. In Kellogg's experiment, a Pt tip is first field evaporated to produce a very small (110) surface of (1 × 1) structure. The image field is turned off and the tip is resistively heated to 310 K in UHV for a few tens of seconds. An FIM observation is made again. It is found that atoms rearrange to form [110] atomic rows of a spacing twice the original one. This rearrangement is found to occur for a very small surface containing as few as 5 atoms. Once the top layer is field evaporated, the layer underneath shows the (1 × 1) structure again. Thus the (1 × 2) reconstructed surface has a structure in agreement with the simple missing row model. It is also found that the edges of the second and third [110] layers, etc., all reconstruct to the simple missing row structure. In the investigation of Gao & Tsong,[15] two questions are addressed simultaneously, namely both the structure and the atom-transport process of the reconstruction. For this purpose, pulsed-laser heating is used. Again, a small to medium size Pt (110) surface of (1 × 1) structure is first prepared by low temperature field evaporation. The image voltage is then turned off, and the tip is subjected to irradiation of a laser pulse of 5 ns width. The power of the laser pulse is adjusted so that the temperature of the tip can be raised to about 350 to 400 K. The image voltage is turned on again, and changes in the atomic structure are carefully examined.This process is repeated as necessary to have the entire surface reconstructed. If the laser power is properly adjusted, small changes occur by irradiation of one laser pulse. Figure 4.6 gives an example of the pulsed-laser heating induced (1 × 1) to (1 × 2) atomic reconstruction of a Pt (110) surface. In Fig. 4.6(*a*) an image of a (1 × 1) surface of 5 atomic rows and 42 atoms, prepared by low temperature field evaporation, is shown. Although the image is slightly asymmetric because of the slightly non-symmetric tip shape, all the atoms in the top surface layer are fully resolved, and the structure is indeed (1 × 1). Between (*a*) and (*e*), heating by one laser pulse to ~400 K separates two successive images. As one can see, in four pulses the reconstruction is

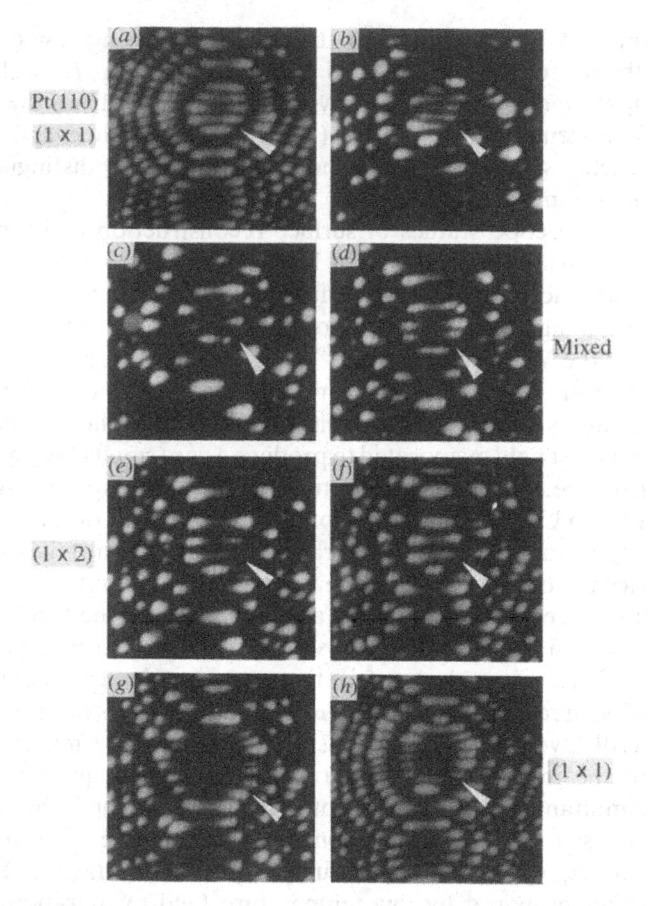

Fig. 4.6 (*a*) (1 × 1) Pt (110) surface prepared by low temperature field evaporation. From *a* to *e*, between each two photos is the application of one pulsed laser heating to ~400 K for 5 ns with the applied field turned off. Note the gradual reconstruction of the surface from (1 × 1) structure to (1 × 2) structure. At *e*, the reconstruction is complete. After *e*, the surface is gradually field evaporated to reveal the structure of the underneath layer, which still has the (1 × 1) structure. Thus the simple missing model is the correct model.

complete for the top surface layer, which shows a fully resolved atomic image of the (1 × 2) missing row structure. The surface is then gradually field evaporated at low temperature, from (*e*) to (*h*), which reveals that the underneath part of the second layer still has the (1 × 1) structure even though the edges of all the (110) layers are reconstructed. These atomically resolved micrographs thus clearly show that the simple missing row model is the correct one.

To be able to observe directly each atomic jump of the atomic reconstruction one would need a time resolution of about 10^{-13} second. This can never be done by a heating method as the tip, once heated, will cool down with a time constant of about one ns.[35] On the other hand, if one adjusts the laser power properly to observe one atomic jump in several laser pulses, the probabiliy of misinterpreting the atomic jumps from the images will be very very small. But even this is too difficult and time consuming if the size of the surface layer is large, especially if one considers the instability of the laser unit available. The second best method is to use surface layers of only a few atoms and to observe changes after irradiation of one laser pulse. This in fact is how Gao & Tsong did their experiment. It was found that atoms in a small [110] atomic row tended to jump together, as shown in Fig. 4.7, where each of the two atomic rows, one with two atoms and one with six atoms, is displaced by heating to ~350 K using one laser pulse of 5 ns width. One notices that occasionally atoms of the top surface layer can be lost to other surface layers. On the basis of observations of the reconstruction of over 20 small layers of 6 to 15 atoms and several larger layers of 40 to 50 atoms, and a statistical consideration,[36] the atom transport processes of the (1×1) to (1×2) atomic reconstruction are deduced to be as follows: (1) Atom transport tends to occur by cross-channel jumps of an entire 'plane edge' [110] atomic row. Thus a (1×1) layer is gradually spread in size by jumps or rolling of atomic rows in the [001] directions. (2) For a large surface where such spreading jumps of [110] atomic rows are less efficient, or are restricted by the available terrace spaces, the [110] atomic rows are broken up into small [110] row-fragments of two to several atoms by both lateral jumps (i.e., jumps in the [110] directions, or along the [110] surface channel) and cross-channel jumps (i.e., jumps in the [001] directions). Single-atom fragments are rarely observed. (3) From the consideration of

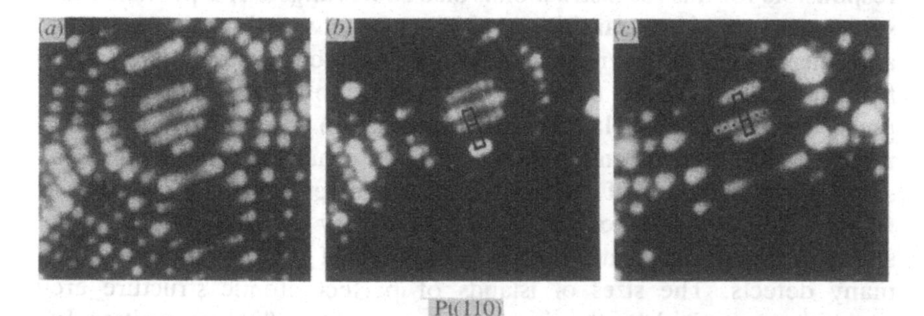

Fig. 4.7 (*a*) Atomically resolved (1×1) Pt surface prepared by low temperature field evaporation. (*b*) and (*c*) show how atoms jump by pulsed-laser heating of the surface to ~350 K for 5 ns. Atoms in a small [110] atomic row tend to jump together, or in a highly correlated way.

the probabilities and the great stability of the [110] atomic row, indicating a large nearest-neighbor bond strength of atoms on this surface, it is concluded that atoms in a small [110] atomic row jump simultaneously or in a highly correlated manner. This mode of surface diffusion, or mass transport, made the atom-transport required for the (1 × 1) to (1 × 2) surface reconstruction of fcc (110) surfaces much more efficient, and thus explains the discrepancy between the transition temperature of the atomic reconstruction and the mass-transport diffusion data available at that temperature of these metals. This atom-transport process is in great contrast to many other experimental observations[37] as well as computer simulations, where surface atoms are always concluded or assumed to move by jumps of individual atoms. Similar atomic structures and atom-transport processes have been found for the Ir (110) surface, and the Pt and Ir (131) surfaces. Details can be found in the original papers.[15,16]

One may argue that even the largest plane obtainable in an FIM experiment is very small compared with the size of planes existing on a macroscopic surface. The atomic reconstruction and atom transport process observed in FIM experiments may not represent those of macroscopic surfaces. This argument, in the opinion of this author, is not necessarily a correct one. As long as the size of a surface is larger than the range of atomic interaction responsible for the atomic reconstruction, the FIM result should be a valid one. Obviously for the (1 × 1) to (1 × 2) atomic reconstruction to occur, the atomic interaction must be attractive at the nearest neighbor//[110] bond and at the ~8 Å//[001] bond, but repulsive at the ~4 Å//[001] bond. Plane sizes in FIM experiments are much larger than 8 Å. For surface reconstruction with a large periodicity such as the (1 × 5) reconstruction of the Ir and Au (001) surface or the (5 × 20) reconstruction of the Pt (001) surface, atoms in fact move closer together than those in the (1 × 1) structure. Thus atomic interaction responsible for this reconstruction is also short ranged. It is possible that some structure phase transitions, which occur at very low temperatures and thus require only very small effects in the atomic interactions, arise from long range effects of the atomic interactions. For such structure phase transitions, FIM planes may be too small to give valid results. With regard to the atom transport process, one usually falsely idealizes a surface by a flat plane of macroscopic size. However, from STM experiments it is now well recognized that even a very carefully prepared flat surface contains many lattice steps and regions of great irregularities and many defects. The sizes of islands of perfect atomic structure are probably comparable to the size of a large facet of a field ion emitter. In the (1 × 1) to (1 × 2) atomic reconstuction of the fcc (110) surfaces, spreading of [110] atomic rows can easily occur by extending the lattice steps already existing on the macroscopic size surface. When a terrace is completely exhausted, atomic rows may jump over to the terrace of the next lower layer.

Another FIM study of fcc (110) surfaces is the chemisorption induced surface atomic reconstruction of the Ni (110). LEED studies find that the atomic structure of a surface may change as a result of chemisorption.[38] Obviously this effect is closely related to surface reactivity and therefore is of great interest to surface scientists. An example is the (1×1) to (1×2) atomic reconstruction of the Ni (110) surface around 200 K by chemisorption of hydrogen or deuterium. No reconstruction of this surface can be observed in UHV. Although the most popular structure model of the reconstructed surface is a paired model[39] in which the closely packed [110] atomic rows of the top surface layer displace laterally to form paired atomic rows, there are studies which favor a simple missing row model[40] as discussed earlier. Kellogg[41] finds that when small clusters of Ni atoms, having the (1×1) structure, on the Ni (110) plane are exposed to 10^{-8} Torr of deuterium at 165 to 175 K, the atoms rearrange to a simple missing row structure. This observation is inconsistent with the popular paired atomic row model deduced from LEED and ion scattering experiments. The number of atoms contained in these small clusters in the FIM study is less than 15, which may be much too small to form a paired-row structure. On the other hand, we know that even with only 5 atoms, a simple missing row structure can be formed in surface reconstruction of the Pt and Ir (110) in UHV.

(c) *Fcc (001) surfaces*

Surface reconstruction of an fcc (001) plane is a little more complicated. For the Ir (001), the reconstruction shows a (1×5) symmetry in LEED diffraction patterns.[42] For the Pt[43] and Au[44] (001) there are some uncertainties because of the very large periodicity of the superstructures, and the structures are also sensitive to the annealing temperature and time, etc. (5×20), $\begin{pmatrix} 14 & 1 \\ -1 & 5 \end{pmatrix}$, $\begin{pmatrix} 14 & 1 \\ 0 & 5 \end{pmatrix}$, etc., and $\begin{pmatrix} 14 & 1 \\ -1 & 5 \end{pmatrix}$ and $c(26 \times 68)$, etc., have been proposed for the Pt (001) and for the Au (001), respectively.[45] An FIM investigation of surface reconstructions of the fcc (001) surface was first reported by Witt & Müller for the Ir (001),[46] and then by Tsong & Gao[47] for the Ir and Pt (001). Surface atomic reconstruction of the Ir (001) surface occurs only at a temperature between 800 and 1300 K where chemisorbed species such as oxygen along the tip shank and impurity atoms inside the bulk can diffuse to the field evaporated part of the emitter surface. The result obtained by Witt & Müller, which seems to be inconsistent with results obtained from many other studies when the data are properly interpreted, is most likely produced by an impurity segregation effect. From a time-of-flight atom-probe analysis, Ahmad & Tsong[48] find that by annealing an Ir or Pt–Ir tip to a temperature of about

1000 to 1200 K, impurity atoms, mostly sulfur, segregate to the surface. Vanselow & Munschau[48] also find that impurity segregation can be observed in a similar temperature range even for Pt tips of five nine purity. It is no wonder that the FIM result obtained by Tsong & Gao[47] is very different from that obtained by Witt & Müller.

For inducing surface reconstruction, but avoiding the difficulty of impurity segregation at the high annealing temperature of the tip, Tsong & Gao use pulsed-laser heating with laser pulses of 5 ns width. Laser heating and annealing can be localized to a very small volume of the tip, very close to the tip apex. As the main part of the tip is still kept at low temperature, low concentration impurity atoms are going to be trapped at the tip shank or the unheated part of the tip rather than diffusing to the field evaporated part of the surface. Another advantage of the pulsed-laser heating technique is that atomic steps and atomic transport process leading to surface atomic reconstructions and other phase transformations can be studied by freezing the surface at various intermediate stages of the transformation for microscopic observations, as has been done for studying reconstructions of the Pt and Ir (110) surfaces. A drawback of this technique is that the heating temperature cannot be determined with good accuracy. A method based on lowering of field evaporation voltage at elevated temperature gives an accuracy of about ± 20 to ± 50 K.

The Ir (001) plane is known to be reconstructed to a (1×5) diffraction symmetry above ~ 800 K.[50] Below that temperature, no reconstruction occurs. Models proposed from LEED and other particle scattering studies are rather complicated, but for the purpose of comparing with field ion images the 'superstructure' can be described as squeezing six [110] atomic rows together to form a closely packed quasi-hexagonal structure to occupy five [110] atomic row widths of the underlying (1×1) layer. As the width is a little too small, these atomic rows are slightly buckled, by about 0.25 to 0.5 Å, as shown in Fig. 4.8. A similar reconstruction is known to occur for the Au (001) and the buckled structures of the [110] atomic rows have been observed by TEM and STM. However, no direct microscopic observation of the quasi-hexagonal structure had been reported prior to the FIM study of Tsong & Gao,[47] which proceeds as follows. First an Ir tip is carefully annealed in UHV to degas the tip shank. A (1×1) surface of the Ir (001) is then prepared by low temperature field evaporation in helium image gas. The field is turned off, and a laser pulse is applied to heat the tip apex part for 5 ns to a desired temperature. The voltage is raised again to see if any atomic reconstruction or other changes of the surface have occurred. If not, the field is turned off once more, a laser pulse of slightly higher power is applied and the surface is examined again. It is found that surface

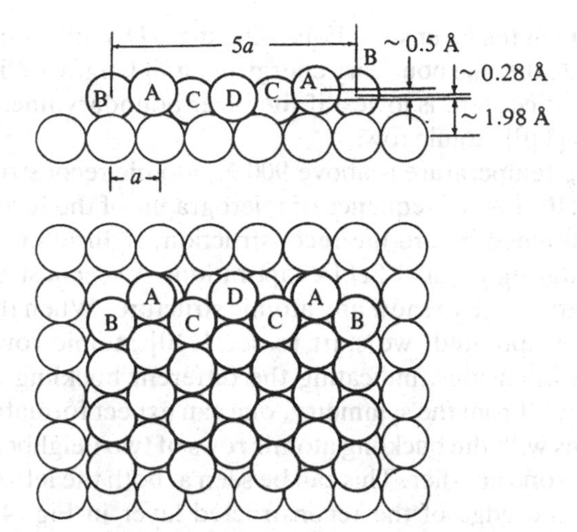

Fig. 4.8 Theoretical model of the (1 × 5) reconstructed surface of Ir (001). Six [110] atomic rows squeeze into 5 row-spacings of the (1 × 1) structure, resulting in a quasi-hexagonal, or pseudo-hexagonal, atomic arrangement with buckled [110] atomic rows.

reconstruction can be observed only if the heating temperature is above ~900 K. Below that temperature the originally nearly circular (001) layers around the (001) pole, produced by field evaporation, will gradually change to a much more square shape by diffusion of edge atoms along the (001) layer edges, as can be clearly seen in Fig. 4.9. The atomic structure remains unchanged. This can be seen as a shape change of a two-dimensional layer or a 'two-dimensional crystal' upon annealing to lower its 'surface' energy, or more precisely to lower its one-dimensional

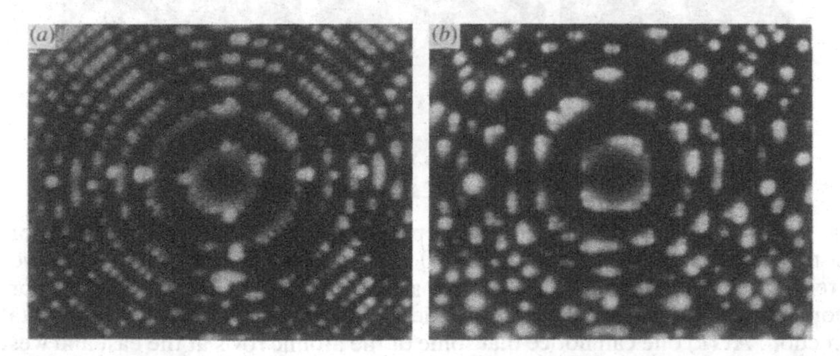

Fig. 4.9 The nearly circular shape of the (001) layers of an Ir tip, produced by low temperature field evaporation, gradually changes to a square shape. This two-dimensional crystal shape change occurs around 500 K, where no reconstruction of the surface occurs. The structure remains (1 × 1).

boundary energy of the layer step. Following the 2-D Wulff construction, we must conclude that the boundary energy per unit length, or line energy density of the lattice step, is lowest if the layer boundary lines up in the direction of the [110] atomic rows.

If the heating temperature is above 900 K, atomic reconstruction can occur. Figure 4.10 shows a sequence of micrographs of the reconstructed surface layer obtained before the reconstruction, right after the reconstruction, and during gradual field evaporation. As reconstructed, the surface layer is too large to show any atomic structure. When this layer is gradually field evaporated, we start to see [110] atomic rows of very different image intensities, indicating the different buckling heights of these atomic rows. From the symmetry, one can expect formation of 90°-rotation domains with the buckling atomic rows of two neighbor domains perpendicular to one another. This can be seen at both the left-hand edge and the right-hand edge of the reconstructed layer in Fig. 4.10*c*. The quasi-hexagonal arrangement of atoms in the reconstructed layer can sometimes be seen directly near the very edge of the layer when the layer is reduced in size. This structure, however, can be revealed most clearly by mapping the positions of atoms when the layer is gradually field

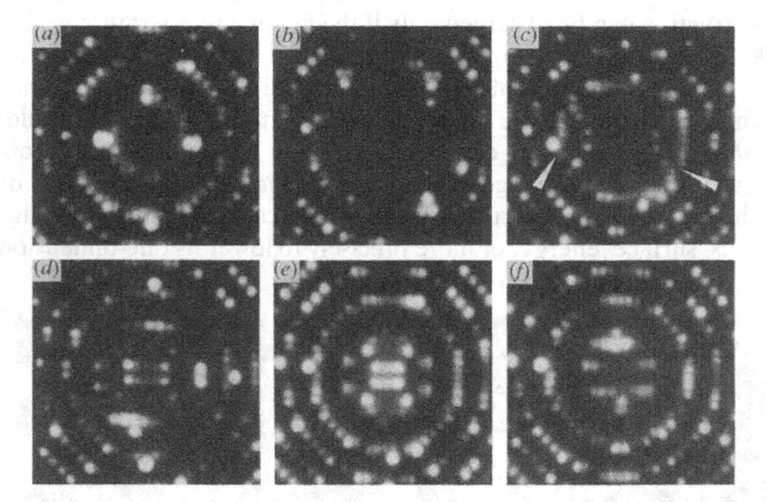

Fig. 4.10 (*a*) (1 × 1) Ir (001) surface prepared by low temperature field evaporation. When the surface is heated to ~900 K for 5 ns by a laser pulse, the top layer is reconstructed to the (1 × 5) structure as shown in (*b*). Gradual field evaporation, from (*c*) to (*f*), reveals the buckled [110] atomic rows in the horizontal direction. At (*c*) one can notice that some of the atomic rows at the east and west sides of the top layer are perpendicular to the other atomic rows, indicating formation of mutually perpendicular domains. The domain boundaries line up at 45° to the atomic row directions. If the laser power is properly adjusted, sometimes only a half of the top layer is found to be reconstructed.

evaporated. Figure 4.11 shows an example of such mapping. Indeed with these atoms, though they exhibit very different image intensity as a result of the different buckling height, the atomic arrangement is hexagonal. Thus the quasi-hexagonal atomic arrangement of the (1 × 5) reconstructed fcc (001) surface is established from an FIM study both from the image intensity variation of the (110) atomic rows and the hexagonal arrangement of these atoms. The atomic arrangement is established through a careful mapping of atomic positions when a reconstructed (001) layer is gradually field evaporated. It should be noted that using the rapid heating and quenching capability of the pulsed-laser heating, it is possible to have only a part of a (001) layer reconstructed while the rest of the layer remains unreconstructed. The structure of the intermediate states can be imaged.

Surface reconstruction of a Pt (001) surface is much more complicated. Heinz *et al.*[50] find that the reconstruction develops during an annealing process via a disordered phase. Above 700 K, two ordered phases, one similar to the (1 × 5) of Ir (001), 'Pt (001)-hex', and one, 'Pt (001)-hex-0.7°', are formed. The structures are quite complicated. Thus a few models, (5 × 20), $\begin{pmatrix} 14 & 1 \\ -1 & 5 \end{pmatrix}$ and $\begin{pmatrix} 14 & 1 \\ 0 & 5 \end{pmatrix}$ structures, etc., have been proposed by different authors. An FIM observation shows that when a nearly circular (1 × 1) structure top layer of a Pt (001) surface is heated by laser pulses to about 350 K, it starts to change to a 'skewed' square shape by diffusion of plane edge atoms along the plane edges. This two-dimensional crystal shape change is similar to that found for the top layer of the Ir (001) surface, but it is now accompanied by a small change in the shape from a square to a skewed square. Since the change in the degree of skewness is gradual, the atomic structure of this layer in general will not fit well to the underneath layer. Thus the surface can only show a disordered diffraction

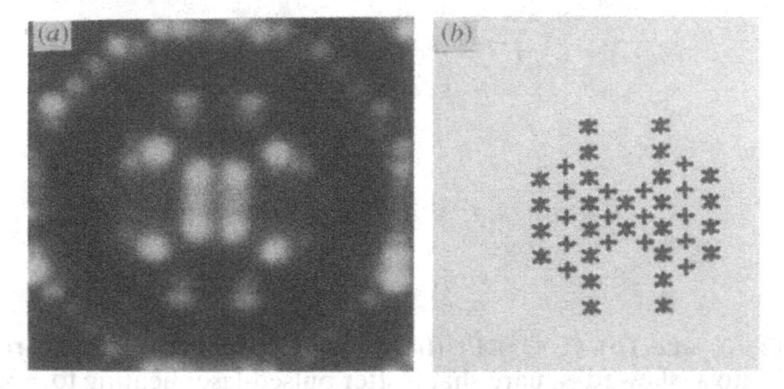

Fig. 4.11 The quasi-hexagonal atomic arrangement of a reconstructed layer can be found by mapping of the position of each of the atoms in the layer when the layer is slowly field evaporated.

symmetry even though the atomic arrangement is still very regular during the entire duration of the transformation. The skew angles reach about $11.0 \pm 1.5°$ in one direction and $4.0 \pm 1.5°$ in the other direction, as shown in Fig. 4.12. One notices that $\tan 4° \simeq \frac{1}{15}$, and $\tan 11° \simeq \frac{1}{5}$. If the heating temperature is above 700 K, then an image structure showing buckled [110] atomic rows is found. These atomic rows, however, do not line up exactly parallel to the [110] direction, but are tilted by about $4.0 \pm 1.5°$. The positions of atoms in the first layer therefore form a highly distorted hexagonal pattern, or surface atoms in the reconstructed layer form a distorted quasi-hexagonal structure. The FIM result is in better agreement with the $\left(\begin{smallmatrix}14 & 1\\ -1 & 5\end{smallmatrix}\right)$ model. FIM studies of atomic reconstructions, in general, not only confirm the diffraction symmetries derived with LEED, but also provide information on the atomic arrangements at different stages of the reconstructions. In addition, atomic steps and atom-transport processes can be studied with a time resolution of a few ns using a pulsed-laser heating technique.

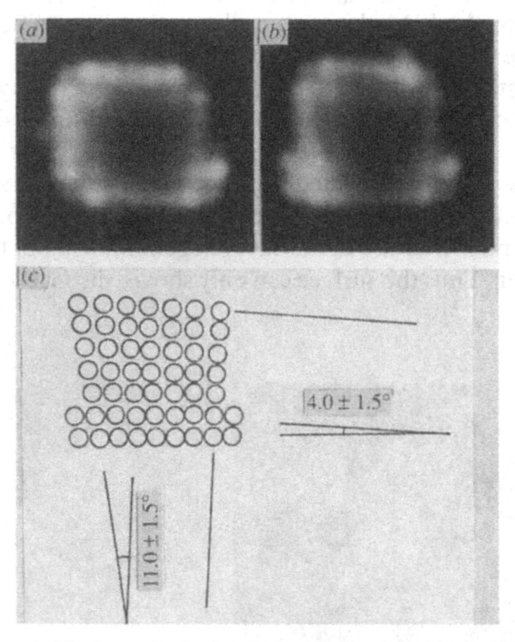

Fig. 4.12 (*a*) and (*b*) (1 × 1) Pt (001), which initially has a square structure, changes into a 'skewed' square shape after pulsed-laser heating to ~350 K. Since the change is gradual, they probably represent a disordered phase. (*c*) Map of (*b*) showing the skewness angles with respect to the atomic row directions of the underlying layer. Note that $\tan 4° \simeq \frac{1}{14}$ and $\tan 11° \simeq \frac{1}{5}$.

4.1.3 Image structure of alloy surfaces

For disordered alloys, field ion images show randomly distributed image spots. Two effects can give rise to this irregular image structure. First, there may be a preferential field evaporation of an alloy species; this will also destroy the regularity of the atomic arrangement of the alloy surface when it is developed by field evaporation. Preferential field evaporation is possible since not only the ionization energies of different alloy species can be different; so can their binding energies in the alloy surface be also. The evaporation fields calculated from eq. (2.35) should also be different for different species of the alloy.[51] Second, one of the alloy species may not be imaged as a result of preferential field ionization above an alloy species.[52] Studies of ordered alloys and formation of an alloy layer on a metal surface indicate that both processes can be important.[53,54] The random appearance of the FIM images of disordered alloys as well as metallic glasses makes a meaningful study of these surfaces by field evaporation almost impossible. A probable direction for future studies of alloy surfaces is to use thermal annealing or processing of these surfaces and image these surfaces without field evaporation after the annealing, as has been done in a successful FIM study of silicon surfaces which will be discussed in the next section. Field evaporation may destroy the surface because of preferential field evaporation of some alloy species.

The image structure of the field evaporated surface of an ordered alloy, however, is as regular as the surface of a pure metal.[54,55,56] In fact, because of the non-imaging of one of its species, and since only one of the sublattices is thus revealed, the field ion image often appears much better resolved overall. Figure 4.13(*a*) and (*b*) show field ion images of well ordered PtCo and Ni_4Mo alloys. In Ni_4Mo only Mo sublattice, which contains only 20% of atoms in the crystal is clearly imaged. Thus the field ion image appears much better resolved than pure metals. Also particularly well resolved are superlattice planes where every layer is mixed with two alloy species, but only one of them is imaged. For the fundamental planes where each layer contains only one of the alloy species, the image brightness as well as the magnification of the surface atoms can be very different for layers of different alloy species. Figure 4.14 gives a few examples of the image structures of ordered alloy surfaces, in this case the surfaces of WSi_2.

As one of the alloy species may not be imaged, either because it is completely field evaporated or from the lack of field ionization or both, it is possible to distinguish atoms of different alloy species from the field ion image alone without relying on atom-probe analysis. In fact this method can identify constituent atoms in an ordered alloy without the aiming error of the atom-probe. Misplaced atoms in an ordered alloy, i.e. atoms

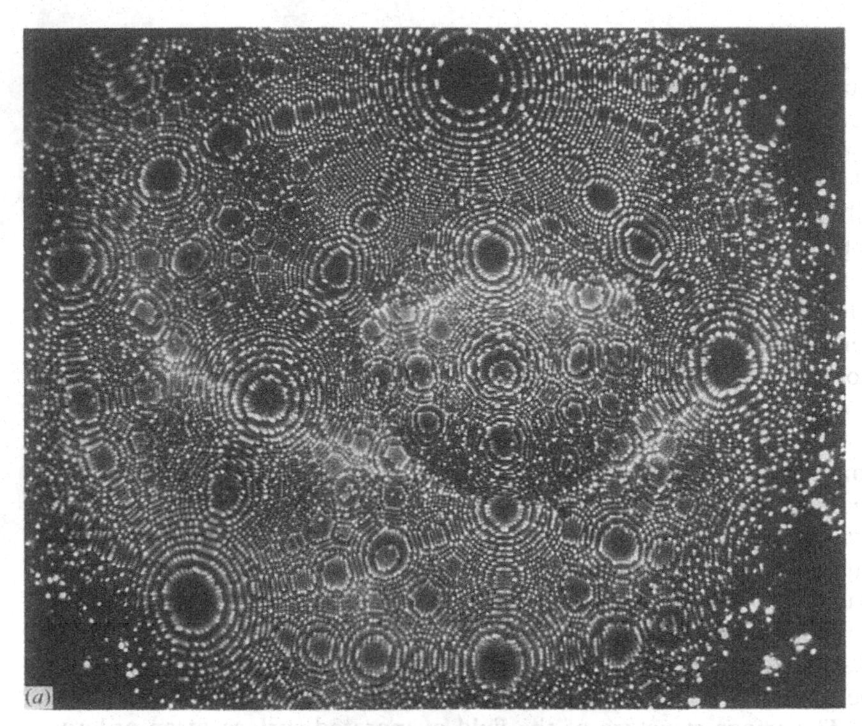
(a)

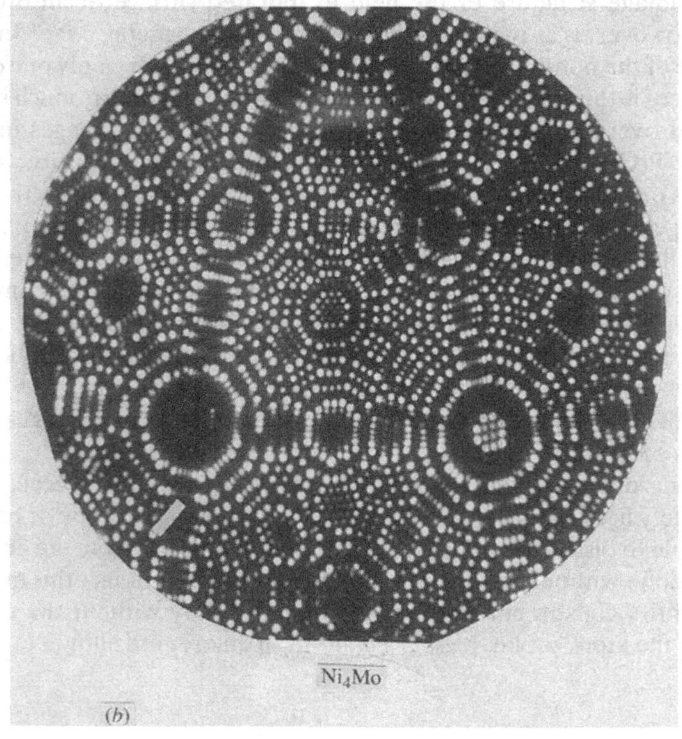
Ni_4Mo
(b)

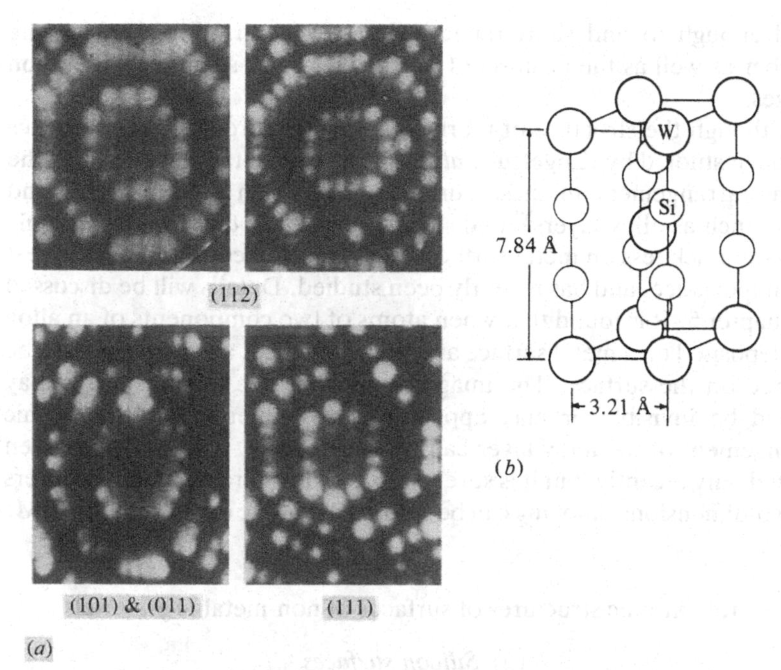

Fig. 4.14 (*a*) A few atomically resolved images of surfaces of WSi_2 where Si atoms are invisible. In tungsten silicide, only the W sublattice is seen in the field ion image. See Section 4.4.2 for discussions on silicide formation. (*b*) Atomic structure of WSi_2 crystal.

of one species which are misplaced from their own sublattice to the sublattice of the other alloy species, can be mapped out one by one from the field ion images. Using this procedure, the long range order parameter of a partially ordered Pt_3Co sample at different stages of ordering has been derived by direct counting of the fraction of misplaced atoms in the alloy.[57] The values obtained are found to be in good agreement with those found with an X-ray diffraction analysis. The short range order parameter is more difficult to derive, as inter-layer mapping of atomic positions is not reliable. With the availability of image digitizing techniques using computers, it is worthwhile to investigate once more what kind of accuracy one can achieve in the inter-layer mapping of atomic positions. With a proper choice of crystal planes, the accuracy may be

---

Fig. 4.13 (*a*) Field ion image of an ordered PtCo alloy. Co atoms do not give rise to image spots, thus only the Pt sublattice is imaged. The circular area near the center of the image is a thick layer of 90° orientation domain. (*b*) Field ion image of an ordered Ni_4Mo where Ni atoms either are invisible or appear much dimmer.

good enough to find short range order parameters directly from the number as well as the position of misplaced atoms seen in the field ion images.

Although the structure of thermally annealed ordered alloy surfaces has been studied by Kingetsu *et al.*,[58] there is no detailed analysis of the atomic arrangement in these surfaces. Formation of thin compound layers such as alloy layers[59] and silicide layers[60] of one to a few atomic layers in thickness on metal surfaces is a subject of considerable interest and importance, and has recently been studied. Details will be discussed in Chapter 5. It is found that when atoms of two components of an alloy are deposited on a metal surface and then annealed, an alloy layer can be formed on the surface. The image shows that one of the species may indeed be invisible or may appear much dimmer. Thus the atomic arrangement of the alloy layer can be mapped out. The work has been started only recently, but it is sure to attract the interest of many workers as two-dimensional alloying can be studied in atomic detail with the FIM.

4.1.4 Atomic structures of surfaces of non-metallic materials

(*a*) *Silicon surfaces*

Silicon is one of the most important materials in modern society because of its wide use in electronic devices. Silicon surfaces are also among the best studied of all surfaces.[61] During the 1960s, 1970s and 1980s, silicon surfaces and metal–silicon interfaces were extensively studied with a wide variety of techniques. A particularly well pursued subject is the (7×7) atomic reconstruction of the Si (111) surface. After two decades of intensive research and tens of proposed models, the problem is now at least temporarily settled, mainly from scanning tunneling microscope studies,[12] with the general acceptance of an atomic model proposed by Takayanagi[18] derived from studies by transmission high energy electron diffraction with the electron microscope. The Si (001) and (110) surfaces have also been very well studied. Field ion microscopy, though the first microscopy to achieve atomic resolution, has so far contributed little to the study of silicon surfaces despite continuous attempts by many investigators over the last 20 years.[5–8] Only in 1987 did Liu *et al.*[9,62] succeed in obtaining atomically resolved and well ordered images of silicon surfaces. Some of the FIM images of Si surfaces have a better quality than images of metal surfaces. As FIM studies of non-metallic materials are only preliminary, some of the interpretations and conclusions drawn may be subjected to later changes. Nevertheless, we will devote a little more space to how the atomic structures of silicon surfaces are studied with the FIM than to its use in the study of metal and alloy surfaces, since there has been little detailed discussion of field ion microscope studies of non-metallic materials in early books and reviews.

Before we discuss FIM studies of silicon surfaces, we will repeat here what we have already emphasized earlier. In field ion microscopy, a surface is almost always developed by field evaporation at low temperature, i.e. at a temperature where atomic jumps are impossible. For metals, the field penetration depth is less than 0.5 Å, much smaller than the atomic radius or step height of low index planes. Field evaporation proceeds in a very orderly manner from lattice steps of low index planes where the step heights are larger, and therefore plane edge atoms such as kink site atoms are in much more exposed positions, or shielding by electronic charges is much less complete. The surfaces produced should be the same as those produced by truncation of the solid, or the (1×1) structures. This procedure, which is very successful in producing well ordered (1×1) surfaces of metals in field ion microscopy, has in fact probably caused a delay in the progress of field ion microscope studies of surface atomic reconstructions for nearly thirty years! Only recently have careful studies of the atomic structures of thermally annealed and equilibrated surfaces been reported for metals, as discussed in the last section.

For covalent bond and ionic bond materials, the problem is even more serious. Field penetration depth of these materials under the high field of field ion microscopy is of the order of 10 Å for semiconductors.[4] Although the applied field penetrates into the semiconductor surface with a depth comparable to the Debye screening length, which is of the order of several times 10^6 Å, at a field of a few V/Å, a significant part of field penetration, or band bending, occurs within a depth of about 10 Å. Effective field penetration depths for silicon obtained from a calculation[4] are 9.4, 10.1, 11.2 and 13.3 Å respectively at 400, 300, 200 and 100 K. These effective field penetration depths are insensitive to the doping of silicon, and are much larger than the radius of the atoms and also the step heights of even the most closely packed surfaces. Since there is not much difference in the degree of protrusion of all surface atoms, as far as shielding from the applied field by electronic charges is concerned, surface atoms are more likely to be field evaporated almost randomly from all locations. Edge atoms of low index surfaces have a slightly smaller number of bonds, and therefore the binding energy is smaller and may still of course be field evaporated at a slightly lower field. Apparently this is what happens and silicon surfaces produced by low temperature field evaporation in ultra-high vacuum are not very well ordered, with only the ring structures of low index planes still visible, as shown in Fig. 4.15*a*.

Obviously the surface disordering effect of field evaporation can be

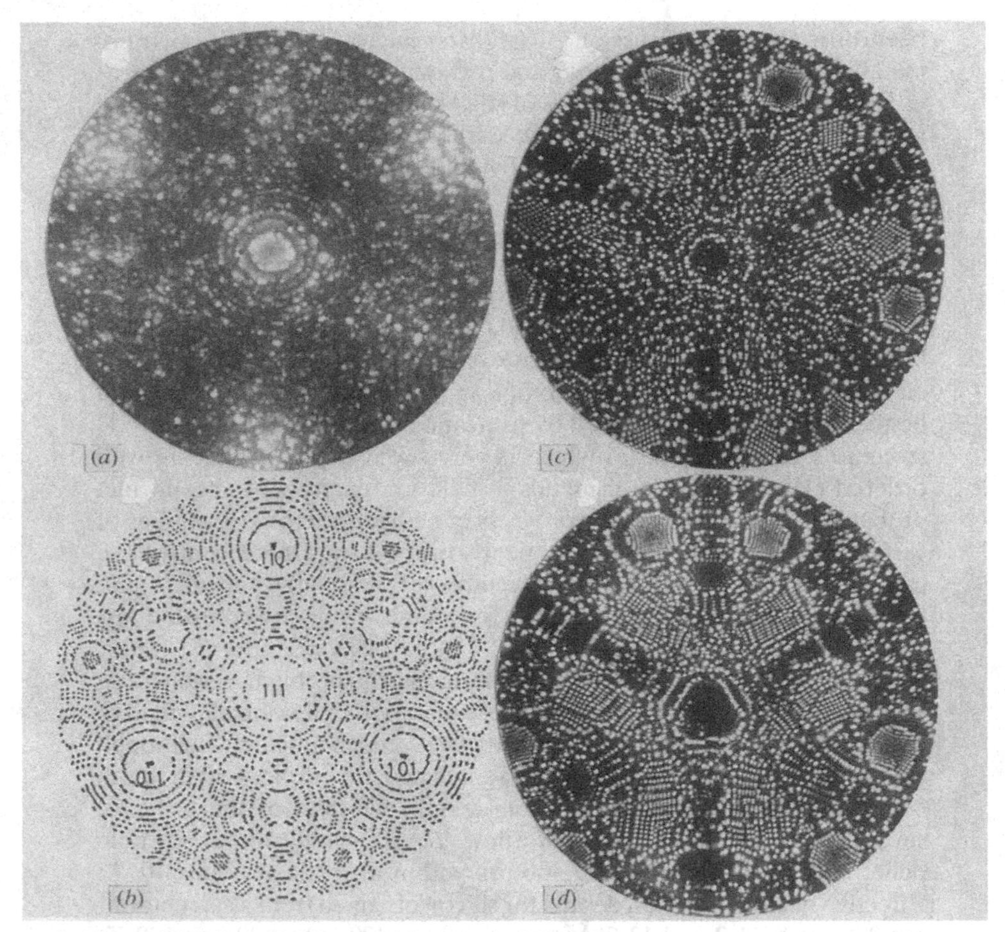

Fig. 4.15 (*a*) A 15 K He field ion image of a low temperature field evaporated silicon surface. (*b*) A computer simulation image of Si with (1 × 1) surfaces. (*c*) a 60 K Ne field ion image of a 720°C annealed silicon surface where well ordered atomic structures are developed at a few facets of the Si emitter surface. (*d*) When the Si tip is annealed at ~800°C, almost all the facets are well developed. The atomic structures of all these facets are completely reconstructed.

avoided if one prepares the emitter surface by a thermal annealing method. In fact, this is how reconstructions of metal surfaces have been observed and studied in most other techniques. There is, however, another trap here. In the case of surfaces which require relatively high temperatures to produce surface reconstructions, during the annealing impurity atoms inside the samples and along the tip shanks may diffuse to the surfaces, and may interfere with the real surface reconstruction. This problem can be overcome either by using pulsed-laser heating or by simply repeating the cycle of thermal annealing and low temperature field evaporation for many times. Each cycle will deplete a small fraction of impurities and eventually the surface will be clean enough to have true surface reconstruction. This later procedure is what Liu *et al.* used for studying atomic structures and surface reconstruction of silicon surfaces.

A silicon tip is first prepared by chemical polishing of a piece of thin, high purity silicon whisker. It is clamped in a Ni tube which is spot welded to a 5 mil Mo wire loop. A very small drop of silver paint fills the gap between the whisker and the Ni tube to insure a good electrical contact between the tip and the tip mounting Mo wire loop. The tip stem is about 3 to 5 mm in length. A freshly polished silicon tip is installed into the FIM chamber, and the chamber is pumped down to good vacuum as soon as possible to avoid excess oxidation of the silicon tip surface and the tip shank. When the vacuum reaches the low 10^{-8} Torr range the tip is first field evaporated in He or in hydrogen and then imaged in Ne for the purpose of examining the coarse structure of the tip, for instance the symmetry of the tip and the existence of extended defects at the tip surface, etc. As the resistance of a high purity silicon below 80 K is very high, it is often necessary to shine a beam of red light to the tip during FIM imaging. The resistivity is greatly reduced by the photo-conductivity effect.[5,6] Once a tip is established to have a good shape and there are no extended defects, it is then further processed. First the chamber is baked to about 150°C overnight to achieve vacuum in the 10^{-10} Torr range. The tip is annealed in vacuum to slightly over 800°C for from several minutes to over 1 hour. It is then imaged with high purity neon at around 60 K. Usually the image is quite disordered, most probably owing to impurity segregation.The surface is field evaporated and then annealed again. It is necessary to bake the system overnight again and then repeat the annealing and field evaporation many times. Only after numerous repetitions of this cleaning procedure may one see a well ordered 60 K neon image of the silicon emitter surface right after an annealing. It is found that well ordered silicon surfaces can be obtained only if the annealing temperature is in the range of 720–820 K. The surface is never very well ordered by annealing at a temperature outside this range. A well ordered surface structure, prepared by annealing, is destroyed if one field evaporates the surface again. Field evaporation is used only for shaping the emitter surface and also for removing surface impurities and contaminants. Repeated annealing and field evaporation are necessary to remove impurities and contaminants from both the tip part and the tip shank and perhaps some from the bulk. Figure 4.15(*c*) and 4.15(*d*) show images of silicon emitter surfaces annealed at 720 ± 30 and 800 ± 30°C. To be sure that the ordered images are not the image of adsorption layers of Ag and other impurities, the surface is analyzed in a high resolution pulsed laser ToF atom-probe right after an annealing. Beside a very small number of contaminant oxygen and carbon ions, which are found by atom-probe analyses to be present in most annealed emitter surfaces, only silicon atoms are detected. No Ag or Ni atoms are found on the surface. Good field ion images can be obtained with sharp tips, but the number of facets developed is too small. Good images with many facets are obtained for a tip radius around 600 to 650 Å, or at best image voltages of neon around 10 to 13 kV.

Identification of crystal facets turns out to be rather difficult for the reason that none of the surface structures observed seem to agree with the (1 × 1) structures of the expected planes, which are identified from their extension angles with the surrounding crystallographic poles. In other words, there are no atomic structures available for helping the identification of crystallographic planes. Thus a computer image simulation of faceted surfaces becomes essential for analyzing FIM images of silicon surfaces. For a thermally annealed field ion emitter surface, large facets develop along certain crystallographic poles. The thin shell model[63] is not a good representation and a faceted surface model has to be developed. The facet sizes are adjusted according to the field ion images, and the atomic structures are initially assumed to be (1 × 1), but can be adjusted later to fit with field ion image structures once the identification of the crystal plane is done. Figure 4.16(*a*), (*b*) and (*c*) show a simulated image, a map of crystallographic indices and a field ion image of a crystallographic triangle of silicon. It is found that only {023} poles and possibly also the {155} poles of the field ion image seem to match exactly the locations of the major facets of the simulated images. For the 720°C annealed surface, the {137} facets agree with simulated {137} poles almost exactly. There is one possible explanation of the slight discrepancy of the simulated pole positions and FIM facet positions. Facets developed on a thermally equilibrated field ion emitter surface may not have a spherical shape, and thus the facet of a given set of Miller indices may not be exactly at the pole position of the computer simulated image, which is based on stereographic projection of a spherical surface. The major facets developed on the emitter surface are indicated by dashed line contours, as shown in Fig. 4.16(*b*).

Atomic structures of thermally equilibrated high index surfaces of silicon have also been studied by Olshanetski & Mashanov using LEED.[64] They found that along the zone line connecting the {111} and {100} poles, {511), {311} and {211} facets are developed. The FIM study, on the other hand, found that the {411}, {311} and {522} are the best developed facets. The largest facets are the {311}, in agreement with the LEED study. But it is still relatively small, and also none of these facets show atomically resolved images. Along the zone line between {111} and {110), the LEED study found {221} and {331} to be the major

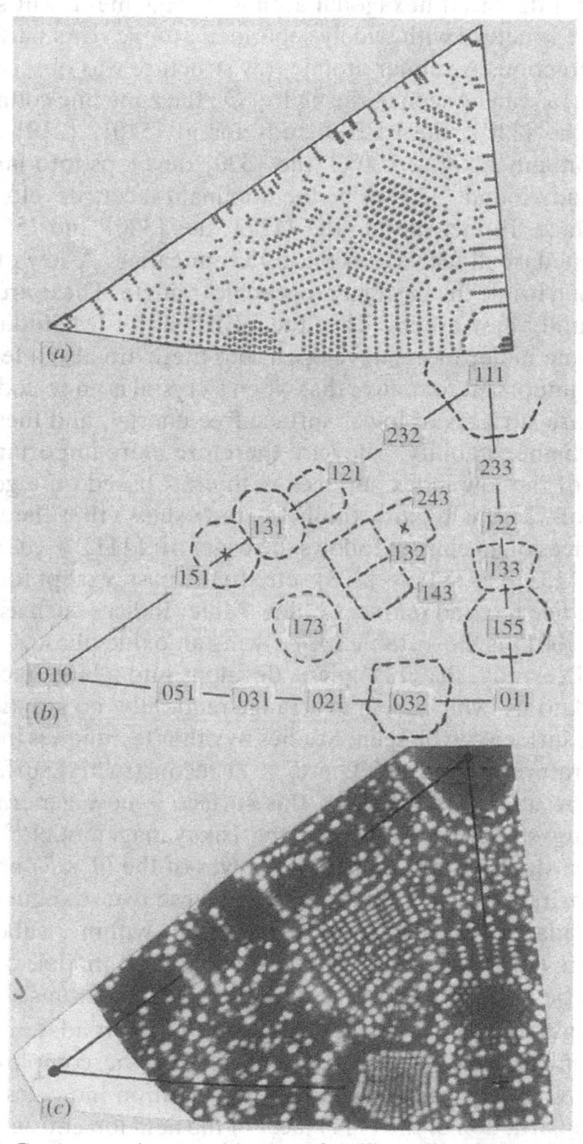

Fig. 4.16 (*a*) Computer simulated image of a silicon tip within a basic crystallographic triangle. In this image, no atomic reconstruction is assumed. (*b*) A map of the indices of the facets of the same surface. (*c*) A field ion image of the same triangular surface area. Note that all these surfaces are reconstructed.

facets. The FIM study, on the other hand, found the major facets developed to be the {551} and {331}. The {551} develops mostly into a structure of a distorted hexagonal atomic arrangement, but sometimes also shows a structure with widely separated atomic rows parallel to the zone line direction. A similar atomic row structure was observed for the {331} facets, as can be seen in the figure. On the zone line connecting the {100} and the {110}, the LEED study found {510}, {210} and {320} facets. Upon annealing to 800°C, the {320} develops into large facets. The FIM study found {320} to be the dominant facets developed on the emitter surface. In fact, beside the {111}, the {320} and {551} are the best developed of all facets at 800 ± 30°C annealing. A few other facets are developed inside the basic stereographic triangle. These are the {137} and {123} and a few others. Thus the LEED and FIM studies seem to agree on some major facets developed, but there are also a few discrepancies. It is important to realize that when a crystal is annealed the facets developed are surfaces of lower surface free energy, and these are also surfaces of higher stability. They are therefore more important surfaces than some of the low index surfaces of interest based on a geometrical consideration. In this regard, the FIM study shows that the stability of silicon surfaces of an emitter follows the order of: {111} > {023} > {155} > {141} > {133} > {131} > {123}, etc. In the past, except for the {111} surface, studies focused on faces of low Miller indices such as the {011} and {001}. {001} is most stable for growing an oxide phase.

In the FIM study, the (111) plane develops into a large facet. Except for some adatoms, which seem to arrange randomly, no atomic structure of the (111) surface can be seen. Studies by other techniques indicate that at a temperature around 800°C a (7 × 7) reconstructed surface should develop. The atomic structure of this surface is now generally agreed upon by many studies to be given by the Takayanagi model,[18] characterized by (1) a stacking fault in the two halves of the (7 × 7) unit cell, (2) nine atom pairs, or dimers, on the sides of these two triangular subcells, and (3) 12 adatoms of hexagonal arrangement within a subcell. While other features of the dimer–adatom–stacking fault model, or the DAS model, will be difficult to observe in the FIM, the adatoms should show up clearly in the field ion images. However no such adatoms have yet been seen in the field ion images. Considering the complexity of the model, one wonders whether the FIM observation indicates simply no (7 × 7) reconstruction of the (111) facet of the field ion emitter surface or whether there really are adatoms on the top layer of the reconstructed surface. It is more likely that the top layer of the (111) facet of a field ion emitter does not grow a (7 × 7) structure simply from the lack of a lattice step. Reflection LEED microscope[65] and reflection electron microscope[66] studies have found that (7 × 7) domain nucleates from lattice steps of the (1 × 1) surface when the sample temperature is gradually lowered to about 830°C. Without the existence of lattice steps, such a reconstructed surface may not grow. Terraces around the (111) pole of a field ion emitter surface may be too small to form the (7 × 7) structure, whereas the top (111) layer does not have a lattice step above it for nucleating the (7 × 7) structure.

Some of the best developed facets of the annealed emitter surface are the {023}, the {155} and the {123} at an annealing temperature of 800°C, and the {137} at an annealing temperature of 720°C. For the Si {023} surface, two image structures, shown in Fig. 4.17, are observed. They have the same unit cell size and are observed with almost equal probability at 800°C annealing. The difference in the surface free energy per atom of the two structures must be less than 0.02 eV. It is very difficult to determine reliably the magnification of the image. One possible method is to estimate the tip radius from the best image voltage. For neon, the best image field is approximately 3.7 V/Å, which is given by $V_{\text{biv}}/5r_t$, where V_{biv} is the best image tip voltage and r_t is the tip radius. This method gives the average magnification of the image from which the unit cell dimension can be estimated. This method is, however, very inaccurate in estimating the magnification of the atomic image of a small plane. A much better method is to use surfaces of known atomic structure to figure out the image magnification. Unfortunately, as far as one can tell, all the surfaces are reconstructed to unknown structures. There is really no known structure available on the annealed Si emitter surface which can be used for the calibration. Another method, which can also give an accurate calibration, is the mapping of the underlying substrate structure using surface diffusion of one adatom on the surface. This method

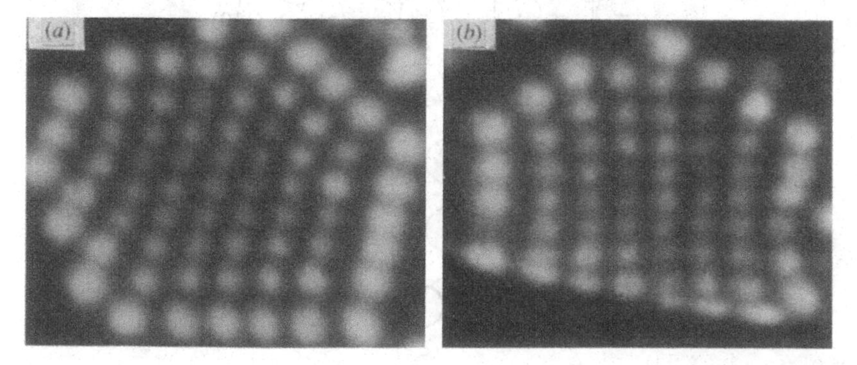

Fig. 4.17 (*a*) and (*b*) Field ion images of the two structures found for the Si {023} surfaces. These two structures are formed with nearly equal probabilities. They also have exactly the same unit cell size.

requires a long period of experiment with one adatom. The surface has to be very stable for such an experiment. At the present stage of technical development of studying silicon surfaces with FIM this is still impractical. An alternative method has been used by these authors. They assume that the paired atomic image structure observed for the (123) surface represents the Si–Si covalent bond distance of silicon lattice, or 2.35 Å. Magnifications based on the two above methods differ by a factor of about 2. The unit cell size for the image structure shown in Fig. 4.17(*a*) is either (11.4 ± 2.2 Å) × (8.8 ± 1.8 Å) with the rhombic angle of 73 ± 3°, or (5.7 ± 1.1 Å) × (4.4 ± 0.9 Å) with the same rhombic angle. For the rectangular structure unit cell, the size is either (10.6 ± 2.2 Å) × (8.0 ± 1.6 Å), or (5.3 ± 1.1 Å) × (4.0 ± 0.8 Å). The observed structures may therefore be either (1 × 2) or ($\frac{1}{2}$ × 1), as shown in Fig. 4.18. In the field ion images there is no information on vertical distances of surface atoms. Neither is there any information about the relative positions of atoms in the top layer and those in the underlying layers. Exact atomic arrangements in these reconstructed surfaces cannot be derived from field ion images, but must be modelled. The atomic arrangements shown in Fig. 4.18 are only suggested ones, not based on experimental data. The LEED study by Olshanetski & Mashanov favored a simple (1 × 2) structure, whereas a preliminary LEED study by Z. Q. Xue & Tsong found two diffraction patterns of the reconstructed surface, one with a rhombic unit cell and one with a rectangular unit cell, both of them having

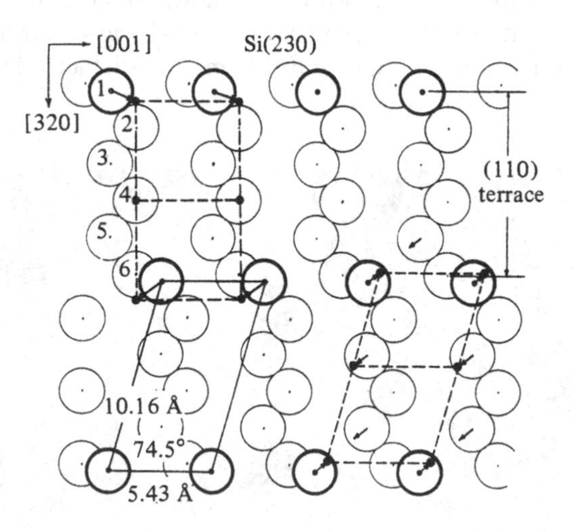

Fig. 4.18 Atomic structure of the (1 × 1) surface of the Si (230) surface which has a unit cell of size 5.43 by 10.16 Å, and possible atomic structures of the ($\frac{1}{2}$ × 1) reconstructed surfaces.

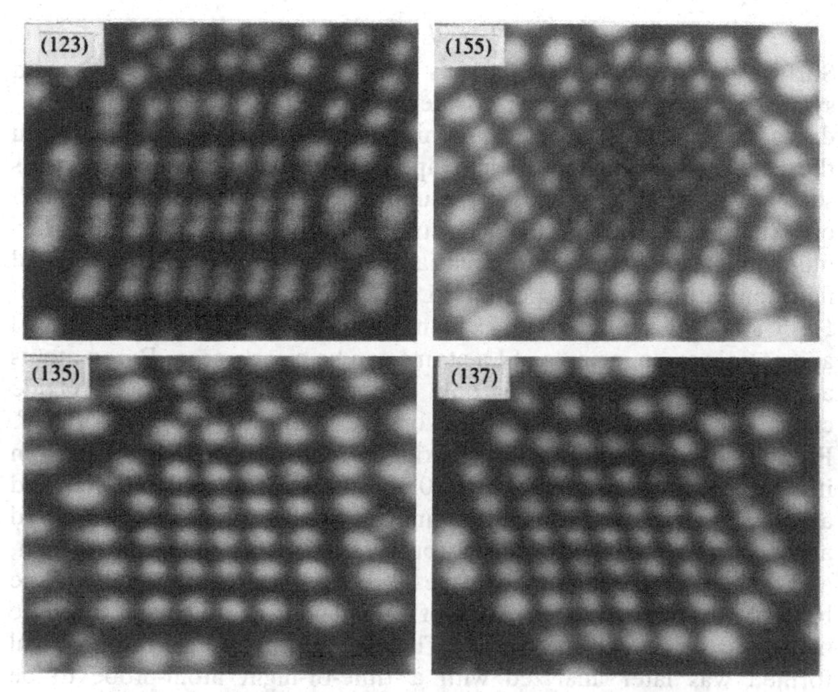

Fig. 4.19 Field ion images of some other surfaces of silicon. Identification of crystal planes and their atomic structures is most difficult since the entire Si emitter surface is reconstructed. An interesting plane is the (123), where atoms are found to be paired, presumably with the 2.35 Å Si–Si bonds, as found in the (1 × 1) structure.

the same (1 × 2) unit cell size but with two distinctively different shapes. The shapes of the unit cells agree with the FIM image structures. In Fig. 4.19, field ion images of {155}, {123}, {137} and {135} are shown. If one uses the average magnification of the FIM image then the {155} surface should have a c(3 × 1) structure and the {137} should have a (1 × 1) structure. The {123} surface is particularly interesting. The (1 × 1) structure should have atoms arranged in pairs with a unit cell of 8.34 Å by 5.92 Å and a rhombic angle of 61.6°. The field ion image indeed shows that atoms are paired. However, the paired bond direction differs completely from the (1 × 1) structure. At the present time, there are many uncertainties in determining the atomic structures of the reconstructed surfaces just based on the field ion images alone. Unfortunately, there are few other reported detailed studies of silicon high index surfaces in the literature, and many of the mysteries are unlikely to be solved in the very near future.

(*b*) *Surfaces of other materials*

Surfaces of some other materials such as metallic glasses, fivefold symmetry quasi-crystals, metal oxides, high temperature oxide superconductor compounds, LaB_6, and graphite, etc., have also been studied with the field ion microscope. Field evaporated surfaces of metallic glasses such as $Fe_{40}Ni_{40}B_{20}$ and $Pd_{80}Si_{20}$ usually show random image spots with only marginally recognizable ring structures of atomic layers along low index poles as a result of preferential field evaporation and preferential field ionization of alloy species.[67] LaB_6 can be imaged in a mixture of He-5% H_2.[68] Usually low index planes such as (001) are well developed and atomic arrangements of La atoms can be clearly seen. Boron atoms are hardly visible, appearing only as fuzzy cloud-like images. Oxide crystals seem to give much better ordered field ion images. Fortes & Ralph[69] studied the formation of oxide crystals of iridium by heating an iridium tip at 800°C in air for 20 minutes. Very well ordered and atomically well resolved field ion images, similar to images of ordered alloys, can be obtained. Obviously only the sublattice of Ir atoms is seen, and thus the resolution of the image appears exceptionally good. On the basis of a crystallographic analysis of the images, the stoichiometry of the oxide was concluded to be IrO_2. The composition of the oxide crystal formed was later analyzed with a time-of-flight atom-probe to be $IrO_{2.13\pm0.15}$.[69]

Particularly interesting and important oxide compounds are the high temperature oxide superconductor compounds known as '123' compounds,[70] etc. Some of these compounds have been successfully imaged by several groups of investigators, with Kellogg & Brenner[71] and Melmed *et al.*[72] giving first reports. Two methods have been used to prepare the samples. Melmed *et al.* simply crush mechanically a piece of high T_c compound, and then select pieces of better shaped sharp fragments, each of which has a smooth tapered section terminating in a sharp point with approximately circular symmetry. Each selected fragment is then attached with conducting epoxy to the end of a tapered tungsten wire, and the pointed fragment is imaged in the FIM in the conventional manner. Kellogg & Brenner, on the other hand, cut a piece of hot pressed high density high T_c compound with a low speed diamond blade saw, using alcohol as coolant, into strips about 0.4 mm wide and 1–2 cm long. It is then electrochemically etched in a solution of 10% perchloric acid in ethylene glycol mono-butyl ether. For FIM imaging, Ar, hydrogen and Ne can be used. Hydrogen seems to give better images. In Fig. 4.20 a hydrogen image of a $Ba_2YCu_3O_{7-\delta}$, obtained by Chen & Tsong[73] using the procedure described by Melmed *et al.*, is shown. The image shows parallel rows of presumably copper oxide layers. Melmed *et al.* believe

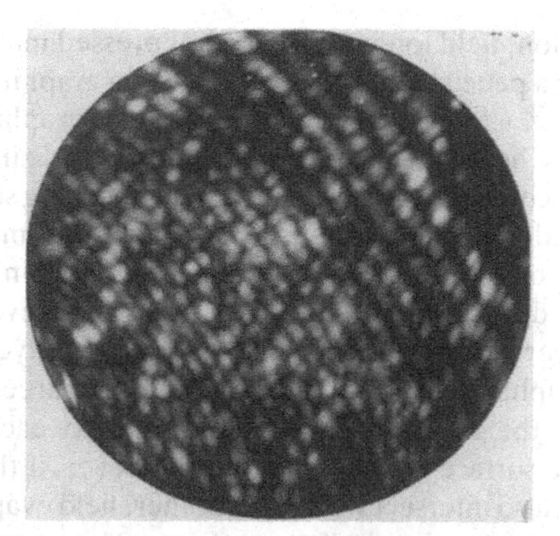

Fig. 4.20 8.5 kV hydrogen field ion image of a $Ba_2YCu_3O_{7-\delta}$ compound. Although the layered structure of the compound is evident, the quality of the image is not yet good enough to analyze the atomic arrangements on surfaces of this compound.

that for superconductor materials, images always show such layered structures. For non-superconductor materials, images show nearly random spot structures. The spacings of these layers are found to be approximately 11 to 14 Å, or about the same as the orthorhombic *c*-axis parameter of the Y-'123' oxide compound. They therefore argue that the distinctly different FI images of the material in superconductor and non-superconductor phases indicate the 2-dimensionality of the superconductivity. The field ion images of the superconductor phase are consistent with a material composed of conducting layers separated by insulating, and thus non-imaging, layers. While such a conclusion is not really based on sound scientific evidence and can be subjected to criticisms (for example Kellogg & Brenner found that both non-superconducting and superconducting materials give similar layered structures), the fact that good field images can be obtained for 'ceramic' compounds of such complexity should be considered an exciting and welcome new development in field ion microscopy. Atom-probe analyses of oxide superconducting materials have been reported by Nishikawa & Nagai,[74] and most other groups working on these materials. In general, the overall composition of a sample is found to agree with the expected stoichiometry of the sample, but it may vary noticeably over a range of 20 to 30 Å, which is a few times the perovskite lattice parameters. On the other hand, a careful consideration of statistical fluctuations inherent in the atom-probe data makes such a conclusion at best unreliable.

In this connection, field ion images of a hot pressed material such as a graphite, whether a pencil lead, a graphite rod or a graphite whisker, can also be obtained.[73] Chen & Tsong found that a graphite tip can be prepared by either heating a graphite fibre in an oxidizing flame or by electrochemical etching of a graphite fibre in a W-etching solution. These tips can be imaged with helium and layered field ion image structures similar to high T_c compounds can be obtained, as shown in Fig. 4.21. The layer spacings are difficult to estimate, but are at least several tens of Å. Each of these layers does not represent one atomic layer of the basal (0001) plane of graphite, but is more likely to be from three to four atomic layers. When all the layers are nearly parallel to each other, field evaporation of the surface is very sporadic. However, if these layers are very thin and they also intersect with one another, field evaporation is less sporadic, and becomes quite similar to silicon. Thus a study of carbon cluster ion formation[76] using pulsed-laser field evaporation has been successfully done in a manner similar to a study of silicon cluster ion formation.[77]

Materials of another type have also been successfully imaged in the FIM, namely the quasicrystals. Melmed & Klein[78] and Elswijk *et al.*[79] reported an FIM study of rapidly cooled Al–12 at.% Mn. The alloy was prepared in the form of long ribbons and ribbon fragments, a few mm in width, 1 to 2 cm in length and 3 to 4 × 10^{-3} mm thick. Slivers a few mm long, partially or completely broken from the edges of the very brittle large fragments, were clamped in a bent gold wire attached to the FIM specimen holder. Sharp tips were made by electropolishing in about 25% HCl in glycerin using a few volts a.c. in optical microscope and then

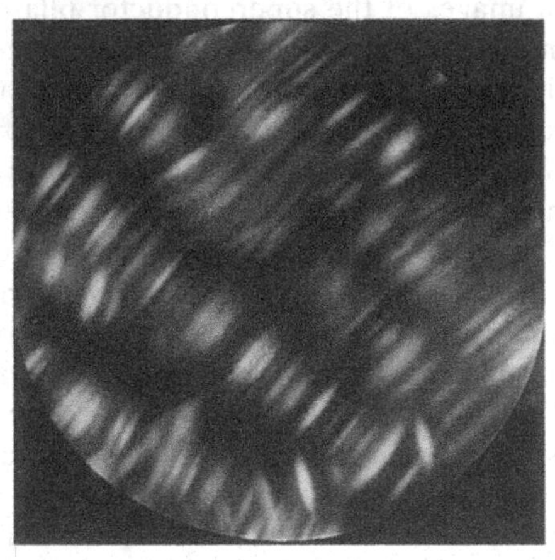

Fig. 4.21 4.5 kV helium field ion image of a graphite tip, etched from a piece of 0.5 mm pencil lead. Each stripe probably represents three to four layers of the basal plane. These stacks are about 35 to 40 Å apart.

rinsing in methanol. Imaging was done in hydrogen at 78 K or in hydrogen and neon at about 30 K. Low temperature neon imaging after initial field evaporation in hydrogen gave the best results. Micrographs were taken while the sample was slowly field evaporating. Collapsing rings were visible, but little short range order was found for the image. The overall symmetry of a specimen, over 1000 Å, is pentagonal, with 2-, 3- and 5-fold rotational symmetry poles in locations which are consistent with the long-range icosohedral symmetry determined from X-ray and electron diffraction studies. Additionally, a large amount of local disorder, replete with defects or antiphase boundaries, is found. There is no evidence for systematic twinning, which might account for the observed orientational symmetry.

4.1.5 Atomic structures of thin films

Thin film technology is becoming one of the important technologies today. While there are infinite varieties of thin film fabrication methods, most amorphous thin films seem to exhibit fractal-like atomic structures. Depending on the fabrication conditions, a thin film grows on the substrate into columnar structures with many voids interdispersed in the thin film.[80] These structures can be seen in the field ion microscope, and compositional variation can be analyzed with the atom-probe. In addition, formation of atomic clusters inside the thin film can be substantiated with the observation of a large fraction of cluster ions in field evaporation by the atom-probe.

FIM and atom-probe studies of thin films of Ni, Au, Pt, a-Ge:H, a-Si:H and WO_3, etc., on various substrates were reported by Krishnaswamy *et al.*[81] First, field ion tips each with a field evaporated surface were prepared. They are placed in an MRC model 8502 r.f. sputtering system. Tips were mounted on a recessed and shielded structure behind the sputtering surface which is bored with small holes about 1 to 2 mm in diameter. The very end of the tips came out of the holes to approximately the same level of the sputtering surface. Films were sputtered at about 20 mTorr Ar at an r.f. power of about 50 W. Thickness of a deposited thin film was controlled by both the r.f. power and the deposition time. Film thickness in the range of a few hundred to a few thousand Å were studied. These tips were then imaged with Ne in the field ion microscope, or analyzed in the flight-time-focused ToF atom-probe.

Depending on the deposition conditions, field ion images in general show patches of irregular shaped amorphous looking surfaces of 20 to 300 Å size. These patches are separated by interconnecting void structures. As a thin film is gradually field evaporated away, the area covered by the voids gradually reduces in size, and eventually reaches the substrate.

These FIM images are consistent with structures observed with transmission electron microscopes, and in addition FIM images also show clustering of atoms. When these surfaces are analyzed in the atom-probe, beside atomic ions of the thin film materials a large fraction of up to several per cent of cluster ions and compound ions are found. For example, cluster ion species found for Ni thin films deposited on a Mo substrate are $Ni_2O_3^{4+}$, $NiCO^{2+}$, $(Ni_2O_3)^{3+}$ $(NiH_2)^{3+}$, Ni_3^+ and Ni_6^+, etc. Similar cluster ions are found for Au and Pt thin films. For a-Ge:H films, a significant fraction of ions found are oxide and hydroxide ions. Some of these ion species are GeO^+, GeO_2^{3+}, GeO_3^+, $Ge_2H_2^+$, $Ge_3H_6^+$, $Ge_5H_9^+$, $Ge(OH)^+$, $Ge(OH)^{2+}$, $Ge(OH)^{3+}$ and Ge_2O^{2+}, etc. It is clear from these observations that atoms tends to form clusters in thin films. Thin films also tend to react with water in the atmosphere. In field evaporation, water molecules are desorbed in the form of hydride, oxide or hydroxide ions with the substrate atoms and atomic clusters. From the above discussions, it is clear that the FIM and atom-probe can now be applied to study a wide variety of problems for a wide variety of materials. Field ion microscopy is no longer confined to studying refractory metals, as many scientists still erroneously perceive.

4.2 Behavior of single atoms and clusters on solid surfaces

A fundamental understanding of many surface phenomena such as transport of atoms; surface reconstructions; nucleation and growth of surface layers and crystals; chemisorption, physisorption and desorption phenomena; and surface catalyzed chemical reactions, etc., requires knowledge of the behavior of single atoms and atomic clusters on the surface. Although there are many techniques capable of probing the behavior of single adatoms on solid surface, few give direct views like the field ion microscope can.

Atomic resolution of solid surfaces has of course already been achieved also by the electron microscope,[14,82] the scanning tunneling microscope[13] and the atomic force microscope.[83] However, field ion microscopy is at present the only microscopy capable of studying quantitatively the behavior of single adsorbed atoms on well defined and characterized solid surfaces. Direct observations of the atomic behavior of single adsorbed atoms on surfaces and quantitative study of the behavior are therefore still contributions unique to field ion microscopy even though the situation may change in the very near future. The capability of field ion microscopy to provide a surface of atomic perfection by low temperature field evaporation is unparalleled. As surfaces so prepared have (1×1) structures, so far all FIM studies of single atom behavior have been performed only on such surfaces, or non-reconstructed surfaces. In field ion microscopy, the number of adatoms deposited on a plane desired for an experimental study can be strictly controlled by repeated deposition and controlled field desorption,[84] something which cannot yet be done in other techniques.

The possibility of studying the behavior of single adsorbed atoms on solid surfaces was recognized by Müller[85] in 1957 in his early development of the field ion microscope. He demonstrated this capability of the FIM by depositing adatoms on a large, thermally annealed W (110) surface, and showed position changes of these adatoms when the surface was heated. A comprehensive study of surface diffusion of adatoms on tungsten surfaces with FIM was then reported by Ehrlich & Hudda[86] in 1966. They measured diffusion parameters of single adatoms on various surfaces of a tungsten emitter based on random walk analysis. Subsequent field desorption studies with adatoms were reported by Ehrlich & Kirk[87] and by Plummer & Rhodin.[88] From the strength of the desorption field required to desorb the adatoms from the surface, values of the binding energy of refractory metal adatoms on various planes of tungsten were derived. Unfortunately, a measurement by Tsong[89] of the effective polarizability of adatoms from a field dependence of the field desorption rate indicates that the polarization energy of adatoms is comparable to the binding energy, thus the binding energy data derived are not reliable and the method is no longer considered valid. It will be shown later that site specific binding energy of adatoms and surface atoms can be derived by ion kinetic energy analysis in field desorption using a high resolution pulsed-laser time-of-flight atom-probe. Bassett & Parsely[90] observed association of adatoms into clusters and dissociation of clusters into adatoms. Using the same method, nucleation of clusters into a surface layer, etc., can be studied. Tsong[84,91] observed long range adatom–adatom interactions and coupled motion of adatoms in the adjacent surface channels of the W (112) planes, etc.; from these observations it is now recognized that a strict control of the number of adatoms on a plane is important in deriving reliable data of various atomic processes of single adatoms on solid surfaces. A wide variety of single adatom experiments have since then been reported. These include random walk of an adatom under the influence of a chemical potential gradient,[92] adatom-substitutional impurity atom interactions,[93] correlation between adatom–adatom interactions and formation of adsorption layer superstructures,[94] and site dependent surface free energy of adatoms,[95] etc. There are of course still many atomic processes on surfaces of great interest which have not yet been studied with field ion microscopy. In Fig. 4.22 a few atomic processes of interest are illustrated.

A particularly stringent requirement of experiments of this type is the vacuum condition. At 10^{-6} Torr background pressure, approximately

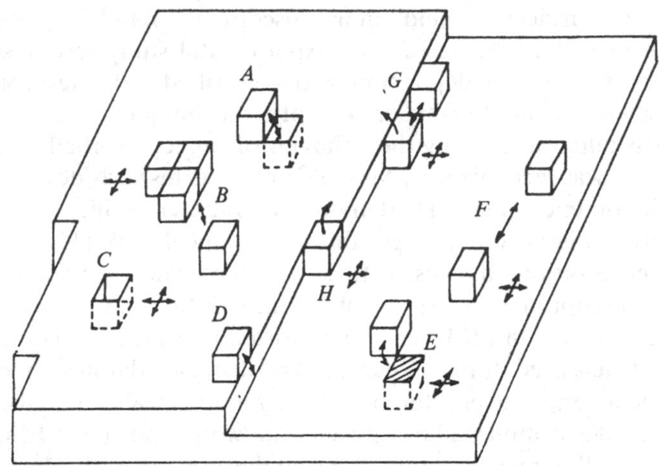

Fig. 4.22 On the surface of a solid, there are a wide variety of atomic processes. A: formation of a surface vacancy–adatom pair, or their recombination; B: association or dissociation of adatoms with an atomic cluster and cluster diffusion; C: diffusion of a surface vacancy, especially toward the lattice step; D: falling off a lattice step of an adatom; E: diffusion of a substitutional or interstitial impurity atom and its interaction with an adatom; F: diffusion of an adatom and its long range interactions with other adatoms; G: diffusion, dissociation and activation of a ledge atom; H: dissociation and activation of a kink atom into an adatom, a ledge atom, or an adatom on the layer above.

one monolayer of contaminating atoms will hit the surface every second. As these experiments always involve only a few adatoms on a surface of about 30 to 80 Å size, and the duration of the experiment is usually longer than a few hours, or of the order of 10^4 seconds, it is quite clear that vacuum at 10^{-11} Torr or lower is mandatory for the experiment to be free from the problems of contamination. Fortunately, when the image field is on, the surface is protected from contamination by residual gases of low ionization energies, as these gas molecules cannot penetrate the field ionization zone of the image gas, which is either helium or neon and therefore has a very high ionization energy, or a high image field strength.

Another stringent requirement is a careful degassing of the system. In performing the experiment, the tip normally has to be heated up to a temperature much higher than the imaging temperature with the imaging voltage turned off, and the emitter surface is no longer protected from contamination by the image field. Therefore before an experimental measurement the tip and its heating loop have to be carefully degassed. In addition, the adatom deposition source, usually a coil made of a wire of ~0.1 to 0.2 mm in diameter, has to be very carefully and thoroughly degassed before it can be used.

In a study of this type, data are composed of single atom events. Statistical fluctuations of the data have to be carefully considered and a careful statistical analysis of the data has to be done before the conclusions derived can be statistically meaningful and reliable. Although there are many precautions which have to be taken in experiments of this type, because of the very well defined nature of these experiments the technique has now been developed to a high degree of reliability and field ion microscope studies of the behavior of single atoms are among the best established of all field ion experiments.

4.2.1 Principle of surface diffusion by random walk of adatoms

Surface diffusion, a concept already used by Volmer & Easterman[96] for the atom transport needed in a crystal growth in 1921, is an important effect for many surface atomic processes. Let us consider only a case of crystal growth by condensation of atoms from the vapor phase. Irrespective of which of the well discussed three possible modes of crystal growth is occurring,[97] the atomic processes are quite similar. First the condensed atoms will move randomly on the surface. If they encounter a lattice step then they may be absorbed to it, causing the layer to grow in size. If there are no steps then adatoms may encounter each other during their random walks. They may combine into clusters. When a cluster reaches a critical size it may serve as a nucleation center for the growth of a two-dimensional layer or a three-dimensional microcrystal. Lattice defects and impurity atoms may also serve as trapping centers and nucleation centers. As surface diffusion is responsible for the transport of atoms on the surface, it is obviously important to all surface phenomena requiring transport and movement of atoms on the surface.

Surface diffusion can be studied with a wide variety of methods using both macroscopic and microscopic techniques of great diversity.[98] Basically three methods can be used. One measures the time dependence of the concentration profile of diffusing atoms, one the time correlation of the concentration fluctuations, or the fluctuations of the number of diffusion atoms within a specified area, and one the mean square displacement, or the second moment, of a diffusing atom. When macroscopic techniques are used to study surface diffusion, diffusion parameters are usually derived from the rate of change of the shape of a sharply structured microscopic object, or from the rate of advancement of a sharply defined boundary of an adsorption layer, produced either by using a shadowed deposition method or by fast pulsed-laser thermal desorption of an area covered with an adsorbed species. The derived diffusion parameters really describe the overall effect of many different atomic steps, such as the formation of adatoms from kink sites, ledge sites or even lattice sites, diffusion of surface atoms on terraces or along ledges, activation of atoms down or over lattice steps, etc. There are many different atomic processes involved in the mass transport, and the overall diffusion rate really reflects the rate of a rate limiting step in all these atomic processes. Surface diffusion parameters derived with macroscopic techniques are good for describing 'mass-transport diffusion' phenomena, but they are not really suitable for comparing with microscopic theories of surface diffusion where each of the steps must be treated as a separate phenomenon.

Surface diffusion parameters can of course be derived from a single crystal surface where diffusion of adatoms occurs by only one process, i.e. by random walk, or by random jumps of adatoms over surface potential barriers. But most atomic resolution microscope observations now indicate that atomically perfect surfaces rarely exist on a macroscopic scale even for the most carefully prepared surfaces. Lattice steps and defects are almost always present on such surfaces. Therefore surface diffusion parameters are best derived with a microscopic technique. The field emission microscope has been a powerful tool for studying surface diffusion, and has been employed by Drechsler,[99] Dyke,[100] Gomer,[101] Ehrlich,[102] Melmed,[103] Swanson,[104] Kleint,[105] Shrednik,[106] and their co-workers and many others in a variety of studies of mass-transport diffusion of metals and gases on mostly tungsten emitter surfaces. With the development of an autocorrelation analysis of field emission flicker noises,[107] surface diffusion parameters have been derived for adsorbed gases such as hydrogen, deuterium, oxygen, etc., diffusing on single crystal facets.[108] One of the interesting findings of this technique is that at very low temperatures diffusion of hydrogen and deuterium can occur by atomic tunneling.[109] In these field emission microscope studies of surface diffusion data are derived from time correlation of field emission flicker noises, which are collected under the influence of an applied field of ~0.3 V/Å and of many tunneling, field emitted electrons. These influences may render some of the conclusions drawn somewhat uncertain. For example, Naumovets[110] concludes from an experiment that the temperature independent diffusivity of hydrogen and deuterium observed at very low temperature in these field emission experiments is a result of excitations by the tunneling, field emitted electrons, not really by tunneling of the diffusing atoms. This uncertainty can of course be checked by measuring the low temperature diffusion coefficients as a function of field emission current. A great advantage of the field emission microscope is that diffusion of adsorbed gas species can be studied, whereas in field ion microscopy adsorbed gas species are in general not imaged and therefore cannot be studied with this technique.

In a field ion microscope study of surface diffusion, an adatom is deposited on an atomically perfect surface developed by low temperature field evaporation. The tip is cooled down to 20 to 80 K where the adatom is frozen for field ion imaging. To induce surface diffusion of the adatom, the image field is turned off, and the tip is heated to a desired temperature for a specified period of time. Once the tip is cooled down, the image voltage is raised again for field ion imaging. Atomic positions after these heating cycles are recorded and the displacements of the adatom as a function of the heating temperature are then measured from these micrographs. From these data, the mean square displacement as a function of inverse temperature is then calculated and plotted. During a heating cycle, there is no applied field or any other outside interaction. Thus atomic jumps occur by thermal effect alone without any other influence. At very low temperatures, the image field will not alter the atomic positions of the adatom. For studying other behaviors of single adatoms, similar procedures are followed. FIM studies of the behavior of single atoms are therefore performed under the best possible defined experimental conditions. Such great care makes these experiments unique in modern surface science.

(a) Unrestricted random walk

The size of a surface available for field ion microscope study of surface diffusion is very small, usually much less than 100 Å in diameter. The random walk diffusion is therefore restricted by the plane boundary. For a general discussion, however, we will start from the unrestricted random walk. First, we must be aware of the difference between the chemical diffusion coefficient and the tracer diffusion coefficient. The chemical diffusion coefficient, or more precisely the diffusion tensor, is defined by a generalized Fick's law as

$$\mathbf{J} = -\mathbf{D}\,\nabla c + c\langle \mathbf{v}\rangle_F, \tag{4.1}$$

where c is the number density of adatoms, $\langle \mathbf{v}\rangle_F$ is the average velocity of the adatoms under the influence of a driving force, or a chemical potential gradient, and $\mathbf{J}$ is the atom flux. In this definition of $\mathbf{D}$ interactions among diffusing atoms are included. $\mathbf{D}$ should therefore depend on the degree of coverage or the density of adatoms on the surface. In FIM single atom surface diffusion studies, only one adatom is present on the plane. The diffusion coefficient measured, D_s, is equivalent to D in zero coverage limit, and is called the tracer diffusion coefficient. It may also be called the single adatom diffusion coefficient, or the intrinsic diffusion coefficient. This coefficient may be defined as

$$2mD_s\tau = \langle |\mathbf{r}(\tau) - \mathbf{r}(0)|^2\rangle, \tag{4.2}$$

where m is the dimensionality of the diffusion which has a value of either 1 or 2 for surface diffusion, τ is the time period of each observation and $\mathbf{r}$ is the position vector of the adatom. The time average refers to an average of repeated observations with the time period τ.

The mean square displacement of an adatom at a temperature T is related to the potential barrier height the adatom has to be activated over to reach a neighbor site. The interaction between an adatom and an infinite substrate can be described by a periodic potential energy function which satisfies the following translational invariance:

$$U(\mathbf{r} + \boldsymbol{\rho}_n) = U(\mathbf{r}), \tag{4.3}$$

where $\mathbf{r}$ is the position vector of the adatom and $\boldsymbol{\rho}_n$ is a surface lattice vector. An equilibrium adsorption site is the site where the potential energy is minimum within a surface unit cell. Within each unit cell there is at least one site of minimum and maximum potential energy and multiple saddle points. If the potential barrier is narrow and low enough, light adatoms and admolecules such as H, D, H_2, D_2, He or Be may be able to tunnel from one site to another without the need of a thermal activation. The adatom cannot then be confined to a surface site for a long time by lowering the temperature of the surface. In FIM studies adatoms used so far have been confined to fairly heavy atoms. No atomic tunneling effect has yet been observed. To search for such an effect, one may try to study diffusion of Be on closely packed surfaces such as fcc (111) and bcc (110) surfaces. If the mean square displacement becomes independent of the temperature at a sufficiently low temperature, then tunneling effects dominate. At intermediate temperatures, both thermal activation and atomic tunneling effects should be important.[111]

For heavier atoms, jumps by thermal activation can occur much more readily. At a given temperature, an adatom has a certain probability of having an energy exceeding the potential energy at a saddle point, and may escape the confinement and translate along the surface. This translating adatom can lose a part of its energy by exciting phonons or by some other scattering mechanisms and be localized again to a new adsorption site. This constitutes a successful jump. Surface diffusion of adatoms consists of discrete random jumps of these atoms. The number of successful jumps in a time period τ is traditionally written as

$$\bar{N} = \tau \nu_0' \exp\left(-\frac{\Delta G}{kT}\right), \tag{4.4}$$

where

$$\Delta G = E_d - T\,\Delta S, \tag{4.5}$$

ν_0' is the frequency of the small oscillation, and ΔG and ΔS are, respectively, the difference in Gibbs free energy and entropy of the adatom at the saddle point and the equilibrium adsorption site. E_d is the activation energy of surface diffusion, or the barrier height of the atomic jumps.

Equation (4.4) can be more conveniently written as

$$\bar{N} = \tau \nu_0 \exp\left(-\frac{E_\mathrm{d}}{kT}\right). \tag{4.6}$$

ν_0 is simply the effective frequency factor, or the pre-exponential factor of the atomic jump rate equation. It accounts for the interaction between the adatom and the substrate, changes in the statistical weight, and the jump path.

Consider now a one-dimensional lattice of parameter l. The distance of each atomic jump depends on the rate of de-excitation once the adatom is excited and is translating along the lattice. This de-excitation process can be described by a characteristic life time τ' in the symmetric random walk, as in many other solid state excitation phenomena. The initial position of the adatom is taken to be the origin, denoted by an index 0. The adatom accomplishes a jump of distance il if it is de-excited within $(i - \frac{1}{2})l$ and $(i + \frac{1}{2})l$, where l is the lattice parameter, or the nearest neighbor distance of the one-dimensional lattice, and i is an integer. The probability of reaching a distance il in one jump is given by

$$\begin{aligned} p(il) &= (2\tau' v)^{-1} \exp\left(\frac{l}{2\tau' v}\right) \int_{(i-1/2)l}^{(i+1/2)l} \exp\left(-\frac{|x|}{\tau' v}\right) \mathrm{d}x \\ &= C \exp(-|i|a) \qquad \text{for } i = \pm 1, \pm 2, \ldots, \end{aligned} \tag{4.7}$$

where

$$a = \frac{l}{\tau' v} \qquad \text{and} \qquad C = \tfrac{1}{2}(\mathrm{e}^a - 1), \tag{4.8}$$

and v is the average speed of the translational motion. $(2\tau' v)^{-1} \exp(l/2\tau' v)$ is the normalization factor for $p(il)$ in the range from $-\infty$ to $-\frac{1}{2}l$ and $\frac{1}{2}l$ to ∞. $p(il)$ is the jump length distribution which, as can be seen from eq. (4.7), is an exceptionally decaying function. A random walk with an exponential jump length distribution has been considered by Lakatos-Lindenberg & Shuler.[112]

A one-dimensional random walk is not necessarily symmetric with respect to jumps toward the right and toward the left. If the chemical potential gradient is sufficiently weak we may still approximate the jump length distribution by an exponentially decaying function, but distinguish that toward the right from that toward the left.

$$p(il) = \begin{cases} C_\mathrm{R} \exp(-ia_\mathrm{R}) & \text{for } i = 1, 2, 3, \ldots, \\ C_\mathrm{L} \exp(ia_\mathrm{L}) & \text{for } i = -1, -2, -3, \ldots. \end{cases} \tag{4.9}$$

C_R, C_L, a_R and a_L are positive constants. The subscripts R and L refer, respectively, to atomic jumps in the right and left directions. a_R and a_L are the inverse of the characteristic jump lengths and C_R and C_L are normalization constants subject to

$$\sum_{|i|=1}^{\infty} p(il) = 1 \tag{4.10}$$

or

$$\frac{C_R}{e^{a_R} - 1} + \frac{C_L}{e^{a_L} - 1} = 1. \tag{4.11}$$

In a symmetrical random walk, $C_R = C_L = C$ and $a_R = a_L = a$. If the life time is much shorter than l/v, $a \gg 1$, and the jump length distribution becomes

$$p(il) = \tfrac{1}{2}\delta_{i\pm 1}. \tag{4.12}$$

This jump length distribution corresponds to the nearest neighbor, discrete random walk of jump length l.

In a one-dimensional random walk with excursion from the origin, the mth moment of the jump length distribution can be derived from

$$\langle(\Delta x)^m\rangle = (-j)^m l^m \left[\frac{\partial^m}{\partial\phi^m}\lambda^N(\phi)\right]_{\phi=0}, \tag{4.13}$$

where N is the number of jumps to have occurred in a time period τ, l is the nearest neighbor distance and $\lambda(\phi)$ is a generating function defined by

$$\lambda(\phi) = \frac{C_R}{\exp(a_R - j\phi) - 1} + \frac{C_L}{\exp(a_L + j\phi) + 1}, \tag{4.14}$$

and

$$\lambda(0) = 1.$$

We therefore have

$$\langle\Delta x\rangle = Nl\left[\frac{C_R\, e^{a_R}}{(e^{a_R} - 1)^2} - \frac{C_L\, e^{a_L}}{(e^{a_L} - 1)^2}\right], \tag{4.15}$$

$$\langle(\Delta x)^2\rangle = Nl^2\left[\frac{C_R\, e^{a_R}(e^{a_R} + 1)}{(e^{a_R} - 1)^3} + \frac{C_L\, e^{a_L}(e^{a_L} + 1)}{(e^{a_L} - 1)^3}\right] + \frac{(N-1)}{N}\langle\Delta x\rangle^2. \tag{4.16}$$

When $a_R \gg 1$ and $a_L \gg 1$, we have $p(il) \ll p(\pm l)$ for $|i| > 1$, and the walk becomes a nearest neighbor random walk. The mean displacement and mean square displacement are then given by

$$\langle\Delta x\rangle = Nl(p - q), \tag{4.17}$$

$$\langle(\Delta x)^2\rangle = Nl^2 + (N-1)\langle\Delta x\rangle^2/N, \tag{4.18}$$

where $p = C_R \exp(-a_R)$, $q = C_L \exp(-a_L)$ and $p + q = 1$. p and q are, respectively, the probabilities of jumping toward right and left.

For a symmetric, nearest neighbor discrete random walk, we obtain

$$\langle \Delta x \rangle = 0, \tag{4.19}$$

$$\langle (\Delta x)^2 \rangle = Nl^2. \tag{4.20}$$

By combining eqs (4.2), (4.6) and (4.20) we have

$$D_S = \frac{\langle (\Delta \mathbf{r})^2 \rangle}{2m\tau} = D_0 \exp\left(-\frac{E_d}{kT}\right), \tag{4.21}$$

$$D_0 = \frac{\nu_0' l^2}{2m} \exp\left(\frac{\Delta S}{k}\right) = \frac{\nu_0 l^2}{2m}, \tag{4.22}$$

where D_0 is the diffusivity, or is generally referred to as the pre-exponential factor of surface diffusion. Diffusion parameters in single adatom experiments can be derived by measuring the mean square displacement as a function of temperature, then plotting ln $(\langle (\Delta x)^2 \rangle / 2m\tau)$ against $1/T$, which should be a straight line. The slope of this plot, known as an Arrhenius plot, is $-E_d/kT$, and the intercept is ln D_0. In field ion microscope experiments, the size of the surface available for measurements is very small, only about 20 to 100 Å in diameter. The experimental data cannot be analyzed on the basis of the unrestricted random walk, and the effects of plane boundaries have to be taken into consideration.

(*b*) *Effects of plane boundaries*

FIM experiments show that a plane boundary acts mostly as a reflective barrier.[86] In a few cases plane boundaries are slightly absorptive.[95] Three effects will be considered here, namely on the mean square displacement, on the frequency of encountering a reflecting boundary, and on the mean adsorption time if the boundary is absorptive.

Consider here a one-dimensional lattice, or surface channel, with M equilibrium rest sites, numbered from 1 to M. The boundaries are assumed to be perfectly reflective. The probability that an atom initially at site i is found at site j after N jumps, $P_b(i, j, N)$, is given by summing the probabilities of displacements from site i to site j and all the images of site j.[113] The mirror images of site j are sites with indices $2kM + 1 - j$ and $2kM + j$ where $k = 0, \pm 1, \pm 2, \ldots$, etc. Thus for the symmetric random walk in a one-dimensional lattice of M sites and with reflective boundaries, $P_b(i, j, N)$ is given by

$$P_b(i, j, N) = \sum_k [P(i, 2kM + 1 - j, N) + P(i, 2kM + j, N)]. \tag{4.23}$$

$P(i, j, N)$ is the probability of finding the adatom, initially at site i, at site j after N jumps in an unrestricted random walk. The mean square displacement for a particular starting position, site i, is given by

$$\langle(\Delta x)^2\rangle_i = l^2 \sum_{j=1}^{M} (j-i)^2 P_b(i,j,N). \tag{4.24}$$

Averaging over all initial positions, one has

$$\langle\langle(\Delta x)^2\rangle\rangle_b = \frac{l^2}{M} \sum_{i=1}^{M} \sum_{j=1}^{M} (j-i)^2 P_b(i,j,N). \tag{4.25}$$

For data analysis, eq. (4.25) is impractically complicated. An approximate formula, which is derived by including only site j and its two nearest images and also approximating $P(i, j, N)$ by a continuous Gaussian form, has been given by Ehrlich[114] to be

$$\langle\langle(\Delta x)^2\rangle\rangle_b = \bar{N}l^2 \left[1 - \frac{4(2\bar{N}l^2)^{1/2}}{3\pi^{1/2}Ml}\right], \qquad \bar{N}l^2 = \langle(\Delta x)^2\rangle, \tag{4.26}$$

where $\langle(\Delta x)^2\rangle$ refers to the mean square displacement with the same number of atomic jumps, but performed on an infinite lattice. When an experiment is carried out on a finite plane, $\langle\langle(\Delta x)^2\rangle\rangle_b$ is obtained. For deriving diffusion parameters using Arrhenius analysis based on eq. (4.21), however, $\langle(\Delta x)^2\rangle$ has to be used. This quantity can be obtained from the experimental value of $\langle\langle(\Delta x)^2\rangle\rangle_b$ and eq. (4.26) using a numerical method.

The boundary effect of a two-dimensional plane is much more difficult to evaluate. Only approximate equations for circular planes are available. Two of them give a fairly good approximation if $\langle\langle(\Delta\rho)^2\rangle\rangle_b$ and $\langle(\Delta\rho)^2\rangle$ are not too large.

$$\begin{aligned}\langle\langle(\Delta\rho)^2\rangle\rangle_b &\approx \langle(\Delta\rho)^2\rangle \left\{1 - \frac{4}{3R}(2\langle(\Delta\rho)^2\rangle)^{1/2}\right.\\ &\quad \left.\times\left[1 + \frac{1}{2}\left(\frac{1}{2}\right)^2 + \frac{1}{3}\left(\frac{1\times 3}{2\times 4}\right)^2 + \frac{1}{4}\left(\frac{1\times 3\times 5}{2\times 4\times 6}\right) + \cdots\right]\right\}\\ &\approx \langle(\Delta\rho)^2\rangle[1 - 0.677(\langle\Delta\rho\rangle^2)^{1/2}/R],\end{aligned} \tag{4.27}$$

and

$$\begin{aligned}\langle\langle(\Delta\rho)^2\rangle\rangle_b &\approx \langle(\Delta\rho)^2\rangle\left[1 - \frac{8}{3\pi R}\left(\frac{\langle(\Delta\rho)^2\rangle}{\pi}\right)^{1/2}\right]\\ &= \langle(\Delta\rho)^2\rangle[1 - 0.479(\langle\langle(\Delta\rho)^2\rangle\rangle)^{1/2}/R].\end{aligned} \tag{4.28}$$

ρ is the distance variable and R is the radius of the circular plane. Equation (4.27) was derived by Tsong *et al.*[115] by taking an average of eq. (4.26) at all locations of the plane along all possible directions, and eq. (4.28) was given by Reed & Ehrlich.[116] If analytical solutions are difficult to find for boundary effects, they are rather easy to find with a computer

Monte Carlo simulation. Thus the finite sizes of field emitter surfaces pose no severe problems in data analyses.

Let us now consider the average number of times an atom encounters either one of the two boundaries in $\bar{N}$ jumps. Cowan,[117] from a clever argument, arrives at a conclusion that on a one-dimensional lattice of M sites this number should be

$$N_b = \bar{N}/M \tag{4.29}$$

For an elliptical plane of major and minor axes a and b, N_b can be shown to be given by[118]

$$N_b = \bar{N}l(1/a + 1/b)\pi. \tag{4.30}$$

For a circular plane, a and b are replaced with the radius of the plane R. These equations can be represented by

$$N_b = g\bar{N}, \tag{4.31}$$

where g is a geometrical factor. $g = 1/M$ for a linear lattice of M sites, $g \simeq l(1/a + 1/b)/\pi$ for an elliptical plane of major and minor axes a and b, and $g \simeq 2l/\pi R$ for a circular plane of radius R. The accuracy of eqs (4.29) and (4.30) has been ascertained by a Monte Carlo simulation.[118]

An adatom, when encountering a plane boundary, has a finite probability of jumping over the slightly higher potential barrier of the boundary. The probability, P_b, that within a heating period τ the adatom will step out of the boundary depends on the extra barrier height, ΔE_b, of the boundary, and can be calculated from the equation

$$\frac{P_b}{1 - P_b} \simeq N_b \exp\left(-\frac{\Delta E_b}{kT}\right) = g\tau\nu_0 \exp\left[-\frac{E_d + \Delta E_b}{kT}\right]. \tag{4.32}$$

Thus a plot of $\ln[P_b/(1 - P_b)]$ against $1/T$ will give an intercept of $\ln(g\tau\nu_0)$ and a slope of $-(E_d + \Delta E_b)/k$.

If the plane boundaries are absorptive, one would like to know what is the mean adsorption time of an adatom to the boundaries. The mean adsorption time and mean time to trapping are the same as the mean first passage time, and have been discussed by various investigators.[112] We should consider here a one-dimensional lattice of M sites. The boundary sites 1 and M are absorptive. We also assume that the random walk has an exponential jump length distribution as discussed earlier. The first passage time, expressed in the average number of jumps, $n_{ab}(i)$, required to pass a boundary for the first time from the starting site, i, has been given by

$$n_{ab}(i) = \frac{(1 - e^{-a})^2}{(1 + e^{-a})}(M - i)(i - 1) + e^{-a}\left[(M - 1)\left(\frac{1 - e^{-a}}{1 + e^{-a}}\right) + 1\right]. \tag{4.33}$$

Averaged over all starting positions, i yields

$$\langle n_{ab}\rangle = \frac{1}{M-2}\sum_{i=2}^{M-1} n_{ab}(i)$$
$$= \frac{(1-e^{-a})^2}{(1+e^{-a})}\frac{M(M-1)}{6} + \frac{e^{-a}(1-e^{-a})}{(1+e^{-a})} \times (M-1) + e^{-a}. \quad (4.34)$$

For a large plane where $a \gg 1/M$, one has

$$\langle n_{ab}\rangle = M^2(1-e^{-a})^2/6(1+e^{-a}). \quad (4.35)$$

The corresponding equations for the one-dimensional, discrete, nearest neighbor random walk are obtained by setting $a \rightarrow \infty$. They are

$$n_{ab}(i) = (M-i)(i-1), \quad (4.36)$$
$$\langle n_{ab}\rangle = M(M-1)/6. \quad (4.37)$$

When the size of the plane is large, $M \gg 1$, and we have

$$\langle n_{ab}\rangle \simeq M^2/6. \quad (4.38)$$

These equations can also be used to derive diffusion parameters. For example, within the nearest neighbor jump approximation, we have from the last equation,

$$\ln \tau_{ab} = \ln\left(\frac{M^2}{6\nu_0}\right) + \frac{E_d}{kT}, \quad (4.39)$$

where τ_{ab} is the mean time to adsorption by either one of the two boundaries. Thus a plot of the logarithm of the mean adsorption time against $1/T$ should be a straight line having a slope of E_d/kT and an intercept of $\ln(M^2/6\nu_0)$. This alternative method has not yet been used in field ion microscope studies of surface diffusion, although the method appears to be very attractive for those systems where the boundaries are either absorptive or have a lower barrier so that every time an adatom encounters a boundary it simply falls off the step.

(c) *Random walk diffusion with a driving force*

An adatom, during its random walk diffusion, may be subjected to the influence of a driving force.[119] When a random walk is completely symmetric, or when there is no driving force present, there can be no net mass flow in one direction, as can be easily seen from eqs (4.1) and (4.19). Rigorously speaking, truly symmetric random walks are rare. An adatom is almost always subjected to the influence of a force, or a chemical potential gradient, even though when the potential energy change within a distance l, the shortest jumping length of the random walk, is much smaller than E_d, the average barrier height in and against the direction of the chemical potential gradient, the random walk may well still be approximated by a symmetric random walk. A driving force may arise from adatom–adatom interactions, adatom–lattice step interactions, adatom–lattice defect interactions, a temperature gradient and a field gradient, etc. Even a concentration gradient of adatoms can produce a driving force as a result of the adatom–adatom interactions. The kinetics of a random walk under the influence of a driving force, however, can be treated without considering the specific nature of the driving force. The symmetric random walk may be considered a special case of the more general class of random walks with a driving force, but the driving force is zero. We will consider only the case where the driving force is very weak, and the temperature is sufficiently low so that a random walk of an adatom can still be described by a discrete, nearest neighbor random walk.

Under the influence of a driving force, the random walk of an adatom is no longer symmetric. The mean displacement given by eqs (4.17) and (4.19) no longer vanishes. Consider here a nearest neighbor discrete random walk on a one-dimensional lattice of infinite length. The adatom is under the influence of a very weak driving force, or the chemical potential gradient is very small. The force is pointing toward the right, which is taken to be the positive x-direction. Let us assume that within a time period τ, the adatom moves from a site i_0 to a site i_f, resulting in a net displacement of $(i_f - i_0)l$. At a certain instant of time the atom is located at site i. The net frequency with which the atom jumps toward right in site i is given by

$$\nu_i \approx \left[\nu_0 \exp\left(-\frac{E_d - f_i l/2}{kT}\right) - \nu_0 \exp\left(-\frac{E_d + f_i l/2}{kT}\right)\right], \tag{4.40}$$

where f_i is the average driving force in the vicinity of site i. The average time needed for the atom to displace one step to the right is

$$\tau_i = \frac{1}{\nu_i} \approx \nu_0^{-1} \exp\left(\frac{E_d}{kT}\right)\left(2\sinh\frac{f_i l}{2kT}\right)^{-1}. \tag{4.41}$$

The total time it takes for the atom to move from site i_0 to site i_f is

$$\tau \approx \nu_0^{-1} \exp\left(\frac{E_d}{kT}\right) \sum_i \left(2\sinh\frac{f_i l}{2kT}\right)^{-1}. \tag{4.42}$$

And the average velocity of the atom under the action of the driving force is

$$\langle \dot{x} \rangle = \frac{(i_f - i_0)l}{\tau} \approx (i_f - i_0)\nu_0 l \exp\left(-\frac{E_d}{kT}\right) \times \left\{\sum_i \left[2\sinh\frac{f_i l}{2kT}\right]^{-1}\right\}^{-1}. \tag{4.43}$$

Under a weak and constant driving force, this equation reduces to[120]

$$\langle \dot{x} \rangle \approx 2\nu_0 l \exp\left(-\frac{E_\mathrm{d}}{kT}\right) \sinh\left(\frac{fl}{2kT}\right), \tag{4.44}$$

or

$$\langle x \rangle \approx 2\bar{N} l \sinh\left(\frac{fl}{2kT}\right). \tag{4.45}$$

Experimental observation of single atom diffusion under the influence of a driving force, produced by an electric field gradient, was made by Tsong & Walko,[92] and later studied in greater details in connection with a measurement of adatom polarizability, which will be discussed in Section 4.3.3.

4.2.2 FIM studies of single adsorbed atoms

(*a*) *Measurement of diffusion parameters*

In field ion microscope studies of single atom diffusion, a field ion emitter is first degassed in ultra-high vacuum by heating the tip to about 700°C for a long time, usually more than several minutes, to ensure the cleanliness of the tip and the supporting loop, and this is repeated if necessary. An image gas is then admitted to the system, and the surface is processed by low temperature field evaporation. Once the surface plane of interest is prepared, the image field is turned off and a few adatoms are deposited on the surface by heating a source coil. The source coil has to be thoroughly degassed before it can be used. The image field is turned on again for field ion imaging. Nowadays, surface diffusion parameters are always derived with only one adatom on a plane. Any excess adatoms can be easily removed by controlled field evaporation. Once the image is recorded, the field is turned off, and the surface is heated to a desired temperature for a specified period of time. After the heating, the surface is cooled again and the image field is turned on once more to record the new position of the adatom. This process is repeated many times, usually over one hundred times for each temperature. From these field ion micrographs, mean displacements at different temperatures can be derived. In Fig. 4.23(*a*) and (*b*) we show a few field ion micrographs obtained for a one-dimensional random walk of one Re adatom on a W (123) surface and of one W adatom on the ledge of a W (112) surface, and a two-dimensional random walk of one Si adatom on a W (110) surface.

In a surface diffusion experiment one has to determine the adsorption site and the displacement of an adatom within a heating period from two field ion images, one taken before the heating period and one after. An accurate determination of the image magnification and of distances from

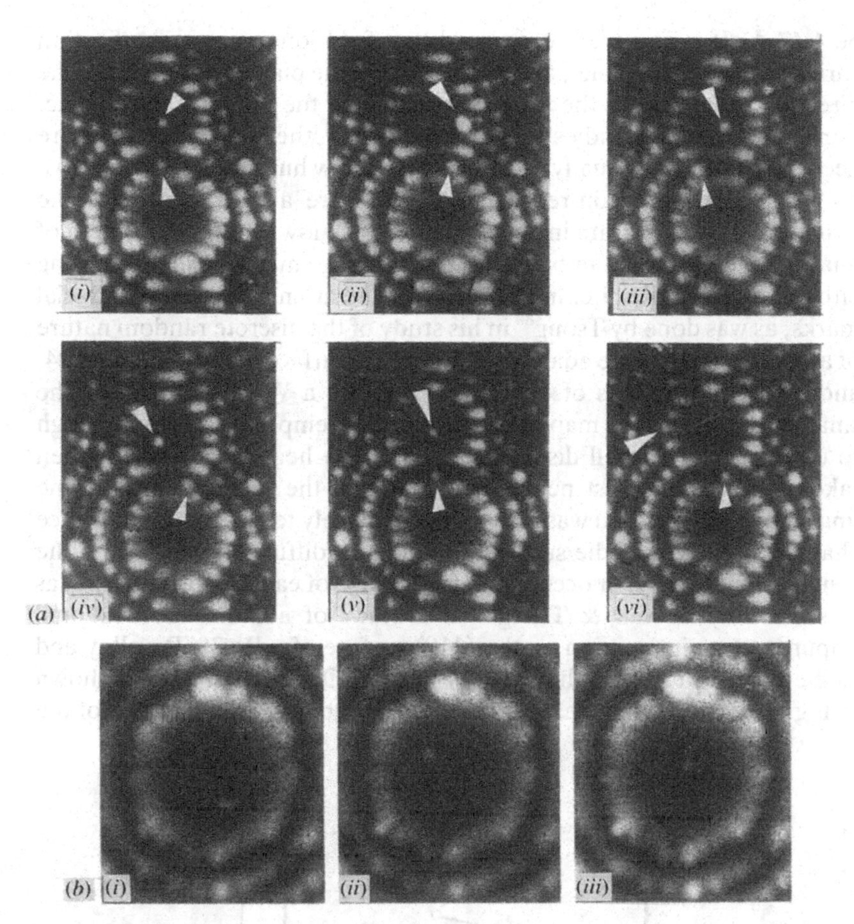

Fig. 4.23 (*a*) Some field ion images showing the one-dimensional random walk of a Re adatom on a W (123) surface, and that of a ledge atom of a W (112) surface. At (*vi*), the Re adatom falls off the lattice step and is absorbed into a kink site as can be seen by following the arrow. The discreteness of the random walk was established through a careful mapping of adatom locations on these and other images of 1972. (*b*) Some helium field ion images showing two-dimensional random walk of a Si adatom on a W (110) surface. The image brightness of the Si adatom is only about 3% of those of other adatoms. To bring out the image of Si atoms, these micrographs are taken with ~30 times normal exposure times.

field ion images is one of the most difficult problems in field ion microscopy, as has already been discussed in Section 4.1.2 in connection with a study of surface reconstruction of silicon high index surfaces where all the surfaces seem to have reconstructed. In early days, the distance calibration was usually based on the separation between two neighboring atomic rows of a W (112) surface in a field ion image which was taken to be 4.47 Å. It is now well recognized that field ion image magnification varies from plane to plane, and even on the same plane it varies from one direction to another if the atomic structure of the plane is anisotropic. Fortunately, in FIM studies of adatom behavior, the heating temperature needed is quite low, usually no higher than a few hundred K, where most of the surfaces are non-reconstructed and are also stable. Thus the structure of the substrate in an experiment is known and the problem of image magnifications can be solved by an image mapping method using either internal marks, i.e. images of substrate atoms, or external fiducial marks, as was done by Tsong[84] in his study of the discrete random nature of atomic jumps of a Re adatom on a W (113) surface, shown in Fig. 4.24, and of the movements of one Re adatom on a W (110) surface. The smallest distance in the map, when the heating temperature is low enough to observe only a small displacement in a few heating periods, is then taken to be the nearest neighbor distance of the substrate. The same image mapping method was also used extensively to establish the surface channel structures of the substrate in adatom diffusion, and to find the visitation frequency, or occupation frequency, of each adsorption site, as was done by Cowan & Tsong[93] in a study of adatom–substitutional impurity atom interaction on the (110) surface of a W–3%Re alloy and single atom diffusion on the W (112) surface. One of their maps is shown in Fig. 4.25. In the map, each dot represents an observed location of the

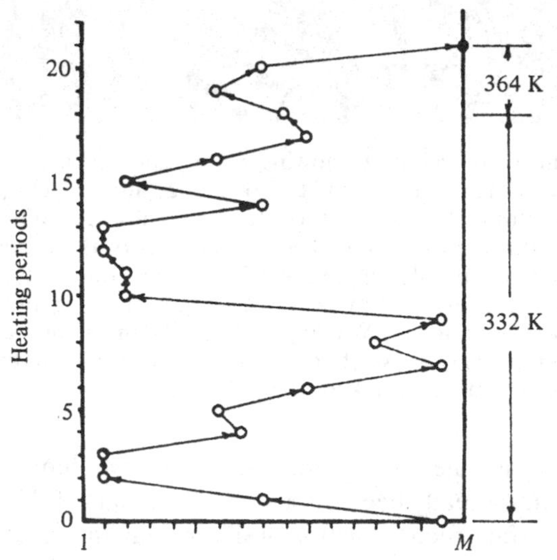

Fig. 4.24 A careful mapping of the adatom position of a Re adatom on a W (123) surface (some of these images are shown in Fig. 4.23*a*) demonstrates the discrete random nature of the atomic jumps in adatom diffusion.

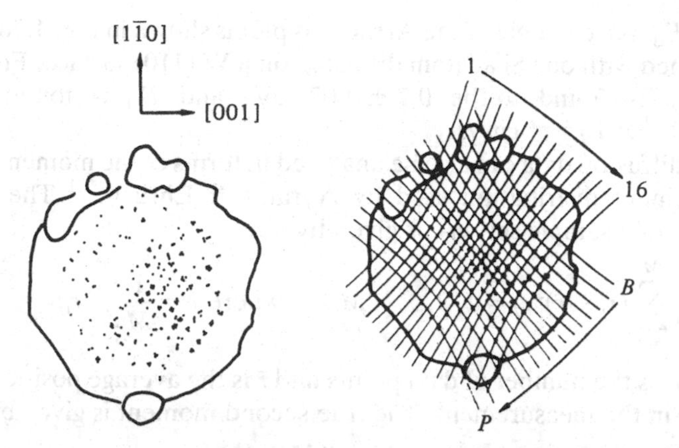

Fig. 4.25 Map of the locations of an adatom on a W (110) surface after about 300 heating periods. Each dot represents an observed location of the adatom. Only a fraction of the about 300 locations are shown. These dots are clustered, and the clusters are found to register with slightly curved grid lines parallel to the [1$\bar{1}\bar{1}$] and [1$\bar{1}$1] surface channel directions. Using this map, the length calibration can then be done accurately from the known size of the surface channels.

adatom (only a fraction of the dots are shown in the figure). Clusters of dots are found to register with the intersections of the slightly curved grid lines parallel to the [1$\bar{1}\bar{1}$] and [1$\bar{1}$1] surface channel directions. The distribution of dots within a cluster is found to have a circular symmetry. From the map and the atomic structure of the substrate, it is concluded that the adsorption site of an adatom is the lattice site, and the grid points are the lattice points of the (110) plane of the W surface. With the availability of an image digitizing capability of the microcomputer, the accuracy of the image mapping has been greatly improved, as has been demonstrated by Fink & Ehrlich.[95] From a published work of Fink & Ehrlich, it is estimated that positions of an adatom on a mid-size plane can be determined to an accuracy of better than 0.3 to 0.4 Å, better than the ~0.5 Å achievable with a photographic mapping method.[93]

Once the displacements of an adatom as a function of tip temperature are determined, the mean square displacement $\langle\langle(\Delta x)^2\rangle\rangle_b$ can be calculated. Using a numerical method, $\langle(\Delta x)^2\rangle$ is then derived using eq. (4.26), (4.27) or (4.28). Usually the temperature range of an experiment is chosen such that the displacements within a heating period are very small, and therefore the boundary correction is not large. A plot of the mean square displacement per unit time against inverse temperature, known as an Arrhenius plot, is made, which can be analyzed with a linear regression method. According to eq. (4.21), the intercept of the straight line should give the diffusivity D_0, and the slope should give the activation energy E_d. An example of the Arrhenius plot is shown in Fig. 4.26, which is obtained with one Si adatom diffusing on a W (110) surface. From this plot, E_d is found to be 0.7 ± 0.07 eV, and D_0 is found to be $3.1 \times 10^{-4} \times 10^{\pm 1.28}$ cm²/s.

The diffusion data can also be analyzed in terms of the moments of the displacement distribution used by Ayrault & Ehrlich.[121] The second moment of a sample set, m_2, is given by

$$m_2 = \frac{1}{M}\sum_{j=1}^{M}(r_j - \bar{r})^2, j = 1, 2, \ldots, M \qquad \text{where } \bar{r} = \frac{1}{M}\sum_{j=1}^{M} r_j, \tag{4.46}$$

where M is the number of data points and $\bar{r}$ is the average position of the adatom in the measurement. The true second moment is given by

$$\mu_2 = Mm_2/(M-1). \tag{4.47}$$

The variance of μ_2 is given by

$$\sigma_2 = \frac{1}{M(M-1)}[(M-1)\mu_4 - (M-3)\mu_2^3], \tag{4.48}$$

where

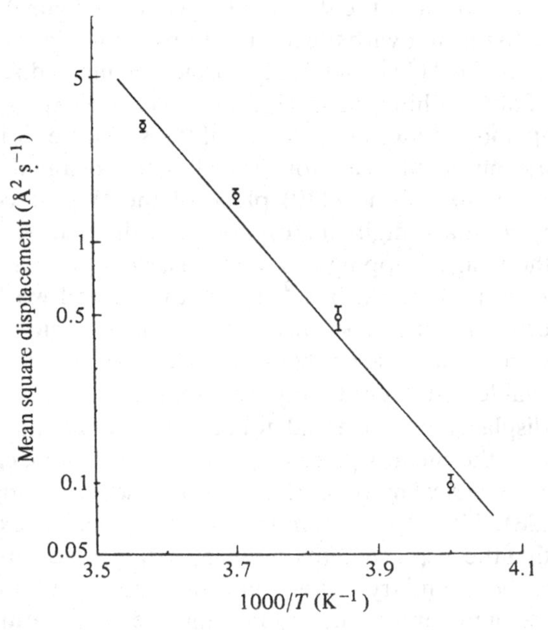

Fig. 4.26 An Arrhenius plot for diffusion of a Si adatom on the W (110) surface, derived from ~360 heating periods of observation with one Si adatom. From the slope of the plot one obtains $E_d = 0.70 \pm 0.07$ eV. From the intercept one obtains $D_0 = 3.1 \times 10^{-1} \times 10^{\pm 1.28}$ cm²/s.

$$\mu_4 = \frac{M}{M-1}\frac{1}{1-\frac{3}{M}+\frac{3}{M^2}}\left[m_4 - \frac{6}{M-1}\left(1-\frac{3}{2M}\right)m_2^2\right], \tag{4.49}$$

and

$$m_4 = \frac{1}{M}\sum_{j=1}^{M}(r_j - \bar{r})^4. \tag{4.50}$$

Over the years, surface diffusions of many systems have been studied and a considerable amount of data accumulated. In Table 4.1 are listed some of the data available in the literature. The activation energies of adatoms on flat surfaces range from ~0.16 eV for Rh adatoms on the Rh (111) to ~1.0 eV for Re adatoms on the W (110), and may exceed 2 eV for W adatoms in the W (111). These values are very low compared with those measured with macroscopic techniques and early field emission microscopy experiments. In these measurements, the activation energies obtained are all close to 3 eV, which are not really the barrier heights seen by adatoms on flat surfaces. These activation energies are valid only for mass-transport diffusions. A mass-transport diffusion is really a combination of many atomic processes such as dissociation of atoms from plane edges, diffusion on terraces, stepping up or down the lattice steps, and atomic cluster association, dissociation and diffusion, etc., as have already been illustrated in Fig. 4.22. An advantage of FIM studies is that, in principle, each of all these elementary surface atomic processes can be studied separately, and an atomistic understanding of macroscopic phenomena is possible.

The activation energy of surface self-diffusion varies from one crystal plane to the other. In general, on fcc metal surfaces, the smaller the corrugation of the surface the lower is the activation energy. Thus self-diffusion of Rh on the Rh (111) surface have one of the lowest activation energies, ~0.16 eV, measured. This is, however, found to be not the case for a bcc metal, tungsten. The activation energy is found to be considerably higher on the smooth (110) surface than on the (112) and (113), as can be seen from the data listed in Table 4.1. It is not at all clear whether this is a general property of bcc metals or only for tungsten. This peculiar behavior cannot be understood from Lennard–Jones or Morse type potential calculations using a static, rigid lattice model. The low activation energy of diffusion on the W (112) and (123) surfaces can however, be explained by assuming that surface channel atoms open up for a diffusing atom.[122] Unless it can be shown that the 'open-up' atomic configuration, i.e. including the adatom in the surface channel, has a lower surface free energy, an amount of energy has to be supplied to open up the channel and it is difficult to see how this can lower the activation energy of surface diffusion.[123] It will be clear from further discussions that

Table 4.1 *Diffusion parameters*

Diffusing species	E_d eV	D_0 cm^2/s	E_d eV	D_0 cm^2/s	E_d eV	D_0 cm^2/s
Plane	W {110}		W {211}		W {321}	
Ta	0.78	4.4×10^{-2} [b]	0.49	9.0×10^{-8} [b]	0.67	1.9×10^{-5} [b]
	0.70	— [f]				
W	0.92	2.6×10^{-3} [a]	0.53	3.0×10^{-8} [a]	0.87	3.7×10^{-4} [b]
	0.86	2.1×10^{-3} [b]	0.56	3.8×10^{-7} [b]	0.84	1.2×10^{-3} [b]
	0.87	— [f]	0.76	3.0×10^{-4} [h]	0.82	1.0×10^{-4} [h]
	0.90	6.2×10^{-3} [k]	0.73	1.2×10^{-4} [n]		
W_2	0.81	7.0×10^{-4} [i]	0.37	2.0×10^{-11} [g]		
	0.92	1.4×10^{-3} [k]	0.82	7.0×10^{-4} [n]		
Re	1.04	1.5×10^{-2} [b]	0.88	1.1×10^{-2} [b]	0.89	4.8×10^{-4} [b]
	1.01	— [d]	0.86	2.2×10^{-3} [m]	0.88	— [c]
	0.91	— [f]				
Re_2			0.78	4.5×10^{-4} [m]		
Ir	0.78	8.9×10^{-5} [b]	0.58	2.7×10^{-5} [b]		
	0.70	1.0×10^{-5} [f]	0.53	5.0×10^{-7} [l]		
Ir_2			0.68	9.0×10^{-6} [i]		
Pt	0.61	$\sim 10^{-4}$ [b]				
	0.63	— [f]				
	0.67	— [o]				
Pt_2	0.67	— [o]				
Pt_3	0.79	— [o]				
Pt_4	0.87	— [o]				
Mo			0.56	2.4×10^{-6} [b]		
			0.57	9.3×10^{-7} [i]		
Mo_2			0.26	2.3×10^{-12} [j]		
Plane	Rh {111}		Rh {311}		Rh {110}	
Rh	0.16	2.0×10^{-4} [e]	0.54	2.0×10^{-3} [e]	0.60	3.0×10^{-1} [e]
Plane	Rh {331}		Rh {110}			
Rh	0.64	1.0×10^{-2} [e]	0.88	1.0×10^{-3} [e]		
Plane	Pt {113}		Pt $\{011\}_{\parallel}$		Pt $\{011\}_{\perp}$	
Pt	0.69	10^{-6} [p]	0.84	8×10^{-3} [p]	0.78	1×10^{-3} [p]
Ir	0.74	— [p]	0.80	10^{-5} [p]	0.80	— [p]
Au	0.56	3×10^{-1} [p]	0.63	10^{-7} [p]		

Table 4.1 (*cont.*)

Diffusing species	E_d eV	D_0 cm²/s	E_d eV	D_0 cm²/s	E_d eV	D_0 cm²/s
	Pt {113}					
Pt	0.84	4×10^{-4} [p]				
Plane	Ni {113}		Ni $\{110\}_{\parallel}$		Ni {331}	
Ni	0.30	2×10^{-6} [q]	0.23	10^{-9} [q]	0.45	2×10^{-3} [q]
Plane	Ni {001}		NI $\{110\}_{\perp}$			
Ni	0.33	— [q]	0.63	— [q]		

Notes:
[a] Ehrlich & Hudda (1966)
[b] Bassett & Parsely (1969)
[c] Tsong (1972)
[d] Tsong (1973)
[e] Ayrault & Ehrlich (1974)
[f] Tsong & Kellogg (1975)
[g] Graham & Ehrlich (1973)
[h] Graham & Ehrlich (1975)
[i] Tsong *et al.* (1975)
[j] Sakata & Nakamura (1975)
[k] Cowan & Tsong (1975)
[l] Reed & Ehrlich (1975)
[m] Stolt *et al.* (1976)
[n] Cowan & Tsong (1977)
[o] Bassett (1976)
[p] Bassett & Weber (1978)
[q] Tung & Graham (1980)

atomic interactions on metal surfaces do not show a monotonic distance dependence. The monotonic Lennard–Jones and Morse potentials are therefore not realistic for explaining surface atomic processes, as was also cautioned by Flahive & Graham.[123] Obviously the activation energy of surface diffusion should depend on the chemical specificity of the diffusing adatoms, as can be seen from data listed in Table 4.1. An interesting behavior is found for surface diffusion of 5-d transition metal adatoms on the tungsten (110) surface where E_d shows a systematic change with the numbers of *d*-electrons in the adatoms, as can be seen in Fig. 4.27.[124,125]

If the activation energies of surface diffusion vary over a wide range from one element to another and from one surface to another, the diffusivity, or the pre-exponential factor, D_0, does not. Within the very limited accuracy of the field ion microscope measurements, usually no better than one order of magnitude because of the narrow temperature range within which a measurement can be conveniently done, all measured values of D_0 are consistent with eq. (4.22) by taking $\Delta S = 0$, and $\nu_0 = kT/h$ where h is the Planck constant, or D_0 are about a few 10^{-3} cm²/s. This has been pointed out repeatedly by the author[126] since there is

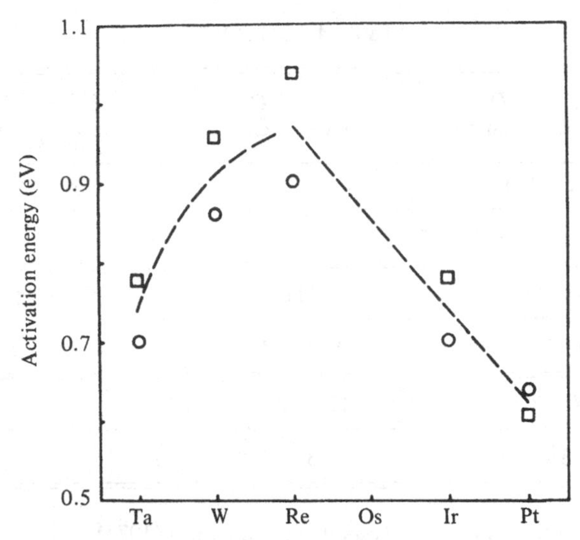

Fig. 4.27 Activation energies of surface diffusion of 5-d transition metal adatoms on the W (110) surface. Squares are data obtained with a few adatoms on a plane by Bassett and circles are data with one adatom on a plane by Tsong & Kellogg.

no reason why the diffusivity should vary over a wide range for different diffusion systems; the frequency spectra of small oscillations of metal adatoms on the surface should not vary more than an order of magnitude, and will be difficult to detect with the limited accuracy of the FIM measurements of D_0. A very careful measurement has recently been done for metallic systems by Wang & Ehrlich.[127] They do not find a significant difference in D_0 for different diffusion systems either. This author believes that to find a significant difference in the diffusivity one should study diffusion systems consisting of very light adatoms and surfaces of metals of very large atomic numbers, such as Be adatoms on W surfaces etc. Such systems are of particular interest because diffusion by atomic tunneling may occur. Recently some efforts have also been directed toward calculating the diffusivity using renormalization theory.[128] No comparison with FIM data has been reported yet.

(*b*) *Atomic jumps and displacement distributions*

At a very low temperature where an adatom jumps only occasionally, about one atomic jump in every few seconds, field ion microscope studies conclude that the surface diffusion of adatoms is consistent with a discrete nearest neighbor random walk. However, in molecular dynamic simulations of diffusion phenomena, which are carried out only for high temperature diffusions where atomic jumps are very rapid, i.e. an atomic jump occurs in the subnanosecond range, atomic jumps are less clearly defined and long distance jumps dominate the diffusion process. In fact, even in field ion microscopy, the validity of a discrete nearest neighbor random walk has been scrutinized, as will be discussed later in this section. In any case, it should be of interest to study the effect of jump length distribution of a random walk as a function of temperature. Experimentally, however, it is impossible to observe directly a jump length of an adatom with an atomic resolution microscopy. This will require a time resolution of the order of 10^{-13} s. In field ion microscopy, only a displacement of an adatom in each heating period is measured, and thus only the displacement distribution can be obtained.

Our analysis described in Section 4.2.1(*a*) indicates that the jump length distribution, in general, should be an exponentially decaying function. The average jump length, $\bar{l}$, should depend on the surface temperature. At low temperatures, adatoms translate with an energy just sufficient to overcome the potential barrier, and the jumps should be dominantly nearest neighbor jumps. At high temperatures, adatoms can translate with an energy considerably larger than the barrier height. Their translational motion may reach far beyond the nearest neighbor sites. We will show here first that under such a circumstance the Arrhenius plot will no longer be a straight line over a large temperature range; thus those analyses based on a linear Arrhenius behavior will not be valid over a large temperature range. For simplicity, we will consider only a one-dimensional lattice of infinite length, and the random walk is also symmetric. From eqs (4.15) and (4.16), valid for a random walk with an exponential jump length distribution, we should have

$$\langle(\Delta x)^2\rangle = \bar{N}l^2\frac{(1+\mathrm{e}^{-a})}{(1-\mathrm{e}^{-a})^2} = \bar{N}l^2\frac{(2-p)}{p^2}$$
$$= \bar{N}\bar{l}^2\left(2-\frac{l}{\bar{l}}\right), \tag{4.51}$$

where

$$p = 1-\mathrm{e}^{-a} \tag{4.52}$$

is the probability that a translating adatom will be de-excited within the first potential well it encounters, and $\bar{l}$ is the average jump length which is related to p and a by

$$\bar{l} = l/p = l/(1-\mathrm{e}^{-a}). \tag{4.53}$$

If $\bar{l}$, or p, is a constant, then an Arrhenius plot will still be a straight line but with a minor modification, i.e. ν_0 should be replaced by $\nu_0(2-p)/p^2$. In general neither $\bar{l}$ nor p will be constant over a wide temperature range; $\bar{l}$ should increase with temperature. Thus the diffusion coefficient, D, defined by

$$D = \frac{\langle(\Delta x)^2\rangle}{2\tau} = D_0 \left(\frac{\bar{l}}{l}\right)^2 \left(2 - \frac{l}{\bar{l}}\right) \exp\left(-\frac{E_d}{kT}\right) \tag{4.54}$$

will increase more rapidly with temperature than a straight line in the Arrhenius plot. The non-linear effect should become important when the temperature is close to the melting point of the substrate crystal, where molecular dynamic simulations all show long jump lengths of diffusing atoms.

Mass-transport self-diffusion studies by Bonzel & Latta[129] of Ni, and by Binh & Melinon[130] of W, have indeed observed such a deviation from the linear Arrhenius behavior at above ~75% of the melting point of the bulk crystal. Bonzel & Latta interpret their observation as an indication of the formation of atomic clusters of various sizes and the diffusion of these 'non-localized' atomic clusters. The non-localization means that the random walk is no longer discrete, or the jump length is rather long. Binh & Melinon, on the other hand, interpret their observation as arising from the formation of a viscous fluid layer and a co-operative motion of atoms at a temperature above 75% of the melting point of the W crystal. Both of these authors adhere to a linear Arrhenius analysis, and as a result the high temperature data are interpreted to have very high activation energies and prefactors. For tungsten, the activation energy changes from 2.85 eV to as high as 5.57 eV, whereas D_0 changes from 0.24 cm^2/s to 1.08×10^4 cm^2/s. The physical meaning of such a high activation energy is obscure. It is simply an artifact of insisting on the validity of a simple linear Arrhenius behavior of the diffusion coefficient. It is much more natural to interpret these observations in terms of a temperature dependent jump length distribution, so that there is no need to change the activation energy. For both the Ni and W data, in reality, the diffusion coefficients, $D = D_0 \exp(-E_d/kT)$, at $T_m/T \simeq 1.1$ are only about 10 times larger than calculated values using the low temperature values of E_d and D_0. This factor can easily be accounted for by eq. (4.54) using the same activation energy but with an average jump length of $3l$ to $4l$. This idea is admittedly not transparent in explaining a mass-transport diffusion phenomenon where many atomic processes are involved in the transport of atoms. The important point here is that one should not insist on the validity of a simple linear Arrhenius analysis based on a temperature independent jump length of atoms at temperatures close to the melting point of the crystal.

An Arrhenius analysis, when properly applied, can provide diffusion parameters and clarifies the energetics of surface diffusion, but does not give any information about the geometrical aspect of these atomic jumps and how these jumps are related to the atomic structure of the substrate. Such information can be derived from displacement distributions. A method for deriving the displacement distribution function from a given jump length distribution has been summarized by Barber & Ninham.[131] The essential steps are as follows:

Define

$$\lambda(\boldsymbol{\phi}) = \sum_{\mathbf{r}} p(\mathbf{r}) \exp(i\boldsymbol{\phi}\cdot\mathbf{r}), \tag{4.55}$$

then

$$P_n(\mathbf{r}) = \left(\frac{1}{2\pi}\right)^d \int_{-\pi}^{\pi} \lambda^n(\boldsymbol{\phi}) \exp(-i\boldsymbol{\phi}\cdot\mathbf{r})\, d\boldsymbol{\phi}, \tag{4.56}$$

and finally

$$P(\mathbf{r}, \bar{N}) = \sum_{n=0}^{\infty} \frac{\bar{N}^n}{n!} e^{-\bar{N}} P_n(\mathbf{r}). \tag{4.57}$$

Here d is the dimensionality, $\bar{N}$ is the average number of jumps during the time period τ, $P_n(\mathbf{r})$ is the probability of a displacement $\mathbf{r}$ following n jumps and $p(\mathbf{r})$ is the probability of a displacement $\mathbf{r}$ in one jump, or the jump length distribution. The solution for the exponential jump length distribution has been solved by Cowan[117] and can be found also in reference 126. The solution reduces to a displacement distribution for a nearest neighbor random walk, orlginally given by Montroll,[132] and by Ehrlich,[133] if one lets $a \rightarrow \infty$. which is

$$P(j, \bar{N}) = \exp(-\bar{N}) \sum_{m=0}^{\infty} (\tfrac{1}{2}\bar{N})^{2m+j}[m!(m+j)!]^{-1}$$

$$= \exp(-\bar{N}) I_j(\bar{N}), \tag{4.58}$$

where $I_j(\bar{N})$ is a hyperbolic Bessel function.

An experimental measurement of the one-dimensional displacement distributions has been reported for self-diffusion of W on the W (112) plane by Ehrlich & Fudda,[86] and Re on W (112).[134] Their result agrees with eq. (5.57) to within statistical uncertainties. However, a later result by Ehrlich[135] agrees better with a model having 10% of the atomic jumps extended to the second nearest neighbor distance.

In two-dimensional random walks with nearest neighbor atomic jumps, the displacement distribution is given by

$$P(i, j, \bar{N}) = \exp(-\bar{N}) \cdot I_i(\bar{N}_\zeta) \cdot I_j(\bar{N}_\xi), \tag{4.59}$$

$$\bar{N} = \bar{N}_\zeta + \bar{N}_\xi, \tag{4.60}$$

where the two-dimensional lattice is represented by two intersecting axes ζ and ξ, and i and j are displacements in units of the shortest jump lengths along the ζ and ξ axes. Experimental measurements of two-dimensional displacement distributions have been reported by Tsong & Casanova[136] for self-diffusion of W on the W (110) surface, which has an atomic structure shown in Fig. 4.28. Two questions are addressed in this study. They are the jump directions and the jump length distributions. At 299 and 309 K, the experimental data fit remarkably well with the nearest neighbor random walk of eqs (4.59) and (4.60) if the jump directions are taken to be along the ⟨111⟩ surface channel directions, or the ζ and ξ directions of Fig. 4.28. These data are shown in Fig. 4.29, and a detailed comparison with eqs (4.59) and (4.60) is given in Table 4.2. A two-dimensional displacement distribution of a W adatom on a W (110) surface taken at 330 K by Cowan,[117] on the other hand, fits best with a theoretical result based on an exponential jump length distribution

$$p(il) = \tfrac{1}{2}[\exp(a) - 1]\exp(-aj), \qquad j = 1, 2, 3, \ldots, \tag{4.61}$$

with $a = 2.04 \pm 0.16$. This result indicates that at 330 K, 13 ± 2% of the atomic jumps go beyond the first nearest neighbor distance. As the amount of data is still small, further study is needed to establish the temperature dependence of the average jump distance $\bar{l}$. It appears, though, that at low temperatures, diffusion of adatoms proceeds by a nearest neighbor random walk, whereas at higher temperatures, jumping beyond the nearest neighbor distance becomes important; thus the jump length distribution is temperature dependent. Experimental data available for higher temperatures are, however, preliminary. The effect of a plane boundary will make analysis of these measurements quite difficult at high temperatures where boundary effects become very important. Monte Carlo simulations may be a good way for analyzing high temperature data on displacement distributions.

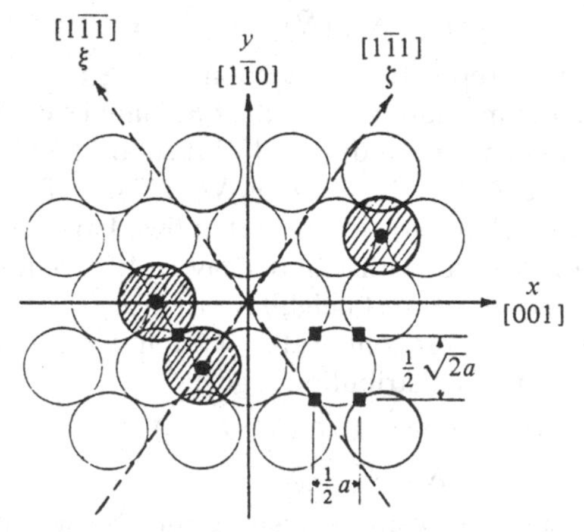

Fig. 4.28 The atomic structure of the W (110) surface. Single adatoms sit at the lattice sites. Closely packed diatomic clusters have their centers of mass sitting in the bridge sites.

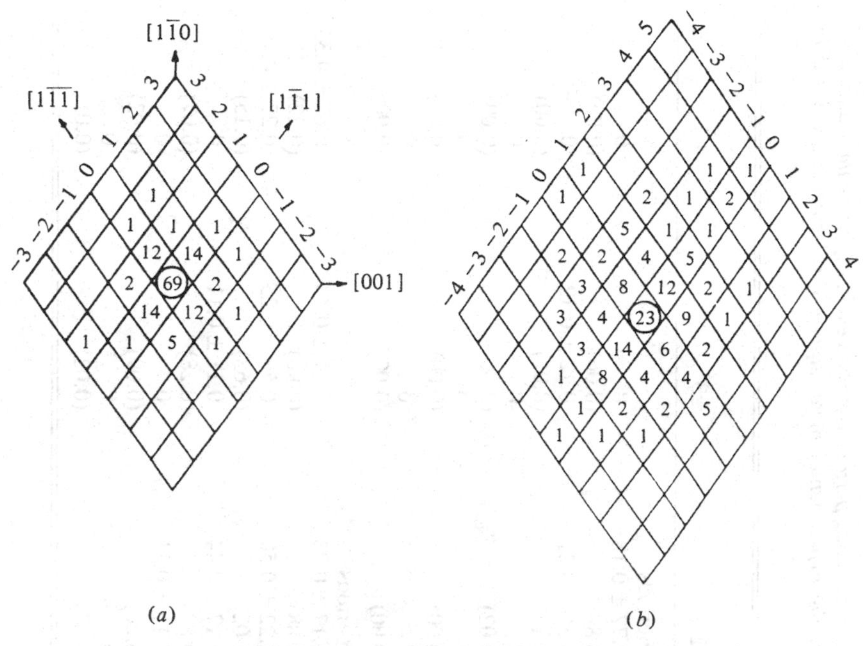

Fig. 4.29 Two-dimensional displacement distributions obtained with one W adatom on a W (110) plane. Each heating period is 60 s. (*a*) is taken at 299 K and (*b*) is taken at 309 K.

(c) *Surface anisotropy*

Adatom diffusion, at least under the low temperature of field ion microscope measurements, almost always follows the direction of the surface channels. Thus adatoms on the W (112) and Rh (110) surfaces diffuse in one direction along the closely packed atomic rows of the surface channels. Such one-dimensional surface channel structures and random walks can be directly seen in the field ion images, and thus the diffusion anisotropy is observed directly through FIM images. Unfortunately, for smoother surfaces such as the W (110) and the fcc (111), no atomic or surface channel structures can be seen in field ion images. But even in such cases, diffusion anisotropy can be established through a measurement of the two-dimensional displacement distributions, as discussed in the last section. Because of the anisotropy of a surface channel structure, the mean square displacements along any two directions will be different. In fact this is how diffusion anisotropy on the W (110) surface was initially found in an FIM observation.[120]

Diffusion anisotropy on the W (110) surface was realized when the mean square displacements of a Re adatom on a W (110) plane along the [100] and [110] directions, or x- and y-directions shown in Fig. 4.28, were

Table 4.2 *Two-dimensional displacement distribution function: W on W {110}; heating periods were 60 s each; experimental values were averaged over equivalent directions. Theoretical values calculated from the experimental mean square displacments are listed in parentheses.*

$\Delta\zeta$ \ $\Delta\xi$	0	±1	±2	±3	±4
			299 K, 138 heating periods		
0	50.00 ± 6.02 (50.11)	9.42 ± 1.31 (9.39)	0.72 ± 0.18 (0.89)	0 (0.06)	0 (0.00)
±1	9.42 ± 1.31 (9.39)	1.81 ± 0.57 (1.76)	0.36 ± 0.21 (0.17)	0.12 ± 0.12 (0.01)	0 (0.00)
±2	0.72 ± 0.18 (0.89)	0.36 ± 0.21 (0.17)	0 (0.02)	0 (0.00)	0 (0.00)
±3	0 (0.06)	0.12 ± 0.12 (0.01)	0 (0.00)	0 (0.00)	0 (0.00)
±4	0 (0.00)	0 (0.00)	0 (0.00)	0 (0.00)	0 (0.00)
			309 K, 150 heating periods		
0	15.33 ± 3.20 (9.34)	6.67 ± 1.05 (6.56)	3.17 ± 0.73 (2.88)	1.17 ± 0.44 (0.90)	0.67 ± 0.33 (0.22)
±1	6.67 ± 1.05 (6.56)	3.50 ± 0.76 (4.61)	2.22 ± 0.50 (2.02)	0.56 ± 0.25 (0.63)	0.22 ± 0.16 (0.15)
±2	3.17 ± 0.73 (2.88)	2.22 ± 0.50 (2.02)	1.17 ± 0.44 (0.89)	0.11 ± 0.11 (0.28)	0 (0.07)
±3	1.17 ± 0.44 (0.90)	0.56 ± 0.25 (0.63)	0.11 ± 0.11 (0.28)	0 (0.09)	0 (0.02)
±4	0.67 ± 0.33 (0.22)	0.22 ± 0.16 (0.15)	0 (0.07)	0 (0.02)	0 (0.01)

Data from Ref. 136

found to differ by a factor of 2.09.[120] A simple analysis based on atomic jumps along the ⟨111⟩ surface channel directions gave a ratio exactly equal to 2, which differs from the experimental value by less than 5%, well within the expected statistical uncertainty of the measurement. This can be easily understood as follows. As the jumping directions have been established to be along the ⟨111⟩ directions, one nearest atomic jump will result in either of the four displacements with their components along the [001] and [110] directions, or along the x- and y-directions listed below:

$$\begin{aligned} x_1 &= -a/2 & y_1 &= \sqrt{2}/2a; \\ x_2 &= -a/2 & y_2 &= -\sqrt{2}/2a; \\ x_3 &= +a/2 & y_3 &= \sqrt{2}/2a; \\ x_4 &= +a/2 & y_4 &= -\sqrt{2}/2a; \end{aligned} \tag{4.62}$$

where a is the lattice constant. The jumping frequencies to these four sites should be the same, and are given by

$$\nu_1 = \nu_2 = \nu_3 = \nu_4 = \tfrac{1}{4}\nu_0 \exp(-E_d/kT). \tag{4.63}$$

Hence

$$\langle(\Delta x)^2\rangle = \tau\Sigma\nu_j x_j^2 = \tfrac{1}{4}\tau a^2\nu_0 \exp(-E_d/kT), \tag{4.64}$$

$$\langle(\Delta y)^2\rangle = \tau\Sigma\nu_j y_j^2 = \tfrac{1}{2}\tau a^2\nu_0 \exp(-E_d/kT), \tag{4.65}$$

and

$$R = \langle(\Delta y)^2\rangle/\langle(\Delta x)^2\rangle = 2. \tag{4.66}$$

The factor R may be called a diffusion anisotropy factor for the surface. For diffusion of a W on the W (110), Tsong & Casanova find a diffusion anisotropy factor of 1.88 from a set of data taken at 299 K, and of 2.14 from a set of data taken at 309 K. The average is 2.01, which agrees with the theoretical value, 2, to within 0.5%. A more detailed general analysis has since then been reported,[137] and diffusion anisotropy on the W (110) surface has also been observed in field emission experiments.[138] It should be noted, however, that the same ratio can be expected if an adatom jumps instead in the [001] and [110] directions with an equal frequency. Thus a measurement of surface diffusion anisotropy factor alone does not establish uniquely the atomic jump directions. The atomic jump directions can, of course, be established from a measurement of the displacement distribution function in two directions as discussed in the last section. Such measurements can only be done with atomic resolution field ion microscopy.

(d) Exchange diffusion

Low temperature diffusions of adatoms are found to be almost always along the surface channel directions. Thus for the bcc (112) and (123), and the fcc (110), (113) and (133) surfaces, surface diffusions are, in general, one-dimensional, and for the bcc (110) and the fcc (111) and (001) diffusions are two-dimensional. A few exceptions have however been observed. For example, on the Pt (110), while Au adatoms still diffuse in one dimension along the surface channel direction, surprisingly Pt and Ir adatoms diffuse in two dimensions.[139] A mechanism involving an interchannel diffusion of adatoms on fcc (110) has been proposed by Bassett & Weber which postulates the formation of a surface vacancy, and a subsequent displacement of the adatom into the vacant site, as shown in Fig. 4.30. In other words, the cross-channel diffusion of an adatom is accomplished by an exchange of adsorption site of a surface atom by the adatom. This mechanism was examined by Wrigley & Ehrlich[140] using an atom-probe FIM. In their experiment, one W atom was deposited on an Ir (110) surface. The chemical identity of the adatom remaining on the surface after a cross-channel diffusion was then determined by the atom-probe. The adatom was found to be Ir, establishing that the cross-channel diffusion occurs by exchange of the original W adatom with an Ir lattice atom.

This author can envision a few interesting consequences of the exchange diffusion. First, as diffusion parameters are derived from many heating periods of observations with one adatom on a plane, these derived parameters for the cross-channel diffusion really represent those for self-diffusion irrespective of the chemical species of the originally deposited adatom. Second, it may be possible to incorporate exactly one foreign atom into a surface which can be used for studying adatom–substitutional impurity atom interaction, provided of course that the incorporated atom does not diffuse into the bulk at the temperature of the experiment. Some unsolved and interesting questions are why cross-channel diffusion occurs for some adatoms but not others, and why it occurs on some fcc metals but not others, and how atomic interactions and atomic sizes play a role in the process. On the basis of the study of surface reconstruction by Kellogg and by Gao & Tsong, this author would also like to point out that FIM experiments on surface diffusion reported so far, including those on the Pt (110) and Ir (110), are performed on non-

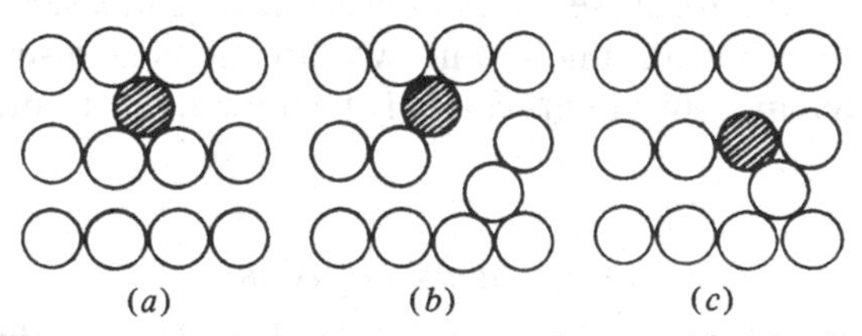

Fig. 4.30 Map showing the exchange of an adatom with a lattice atom in cross channel adatom diffusion on the fcc {110} plane.

reconstructed surfaces despite one conclusion to the contrary which is by no means an accepted view. There is, however, no reason why such a measurement could not be performed on reconstructed surfaces such as the (1 × 2) surfaces of Ir and Pt (110).

Molecular dynamic simulations[141] of single atom diffusion on the fcc (110) surface can reproduce the observed exchange diffusion at a much higher temperature than can the FIM experiments; i.e., at the temperatures where surface reconstruction should occur. However, no surface reconstruction is observed in the simulations. The simulations do not explain why no exchange diffusion occurs for Au adatoms on the Pt (110) either. One simple explanation is that Au atoms are too big to be accommodated to the Pt lattice without producing a large distortion of the lattice, and thus costing too much in strain energy. Also, if molecular dynamic simulations are going to be useful in understanding the behavior of atoms on surfaces, it will have to be shown why no exchange diffusion occurs on the bcc (112) and (123) surfaces which have surface channel structures resembling that of the fcc (110). FIM experiments have not observed such diffusions on these surfaces.

(*e*) *Boundary height*

When an adsorbed atom diffuses on a surface, it may encounter the boundary of the plane. Field ion microscope experiments show that a plane boundary is usually reflective to a diffusing adsorbed metal atom at low temperatures. The potential energy barrier height a diffusing adatom must overcome at the boundary is therefore higher than the diffusion barrier of the surface. It is also observed that at higher temperatures a diffusing atom has a finite probability of falling down the plane, and be absorbed into a kink site or a ledge site of the lattice. Thus adatom is incorporated into the substrate layer, as shown in Fig. 4.23*a*(iv). This adsorption indicates that a kink site has a lower energy than a terrace site. A study of these different energies is important in our understanding of mass transport diffusion of solid materials, and possibly also of surface reactivity.

The extra barrier height of a reflective boundary is very low. An estimate of ΔE_b for a Re adatom on the W (123) plane puts the extra barrier height at only about 0.05 eV.[84] Bassett and co-workers[142] tried to measure this energy for W, Re, Pt and Ir on the W (110) plane using different methods, and found it too small to be measured reliably using an Arrhenius plot. Because of the limited temperature range within which FIM measurements can be conveniently done, the accuracy in activation energy measurements is usually limited to about ±10%, or about ±0.06 to ±0.1 eV, comparable to the extra barrier height of the reflective plane boundary. To overcome such a difficulty, Wang & Tsong[143] approxi-

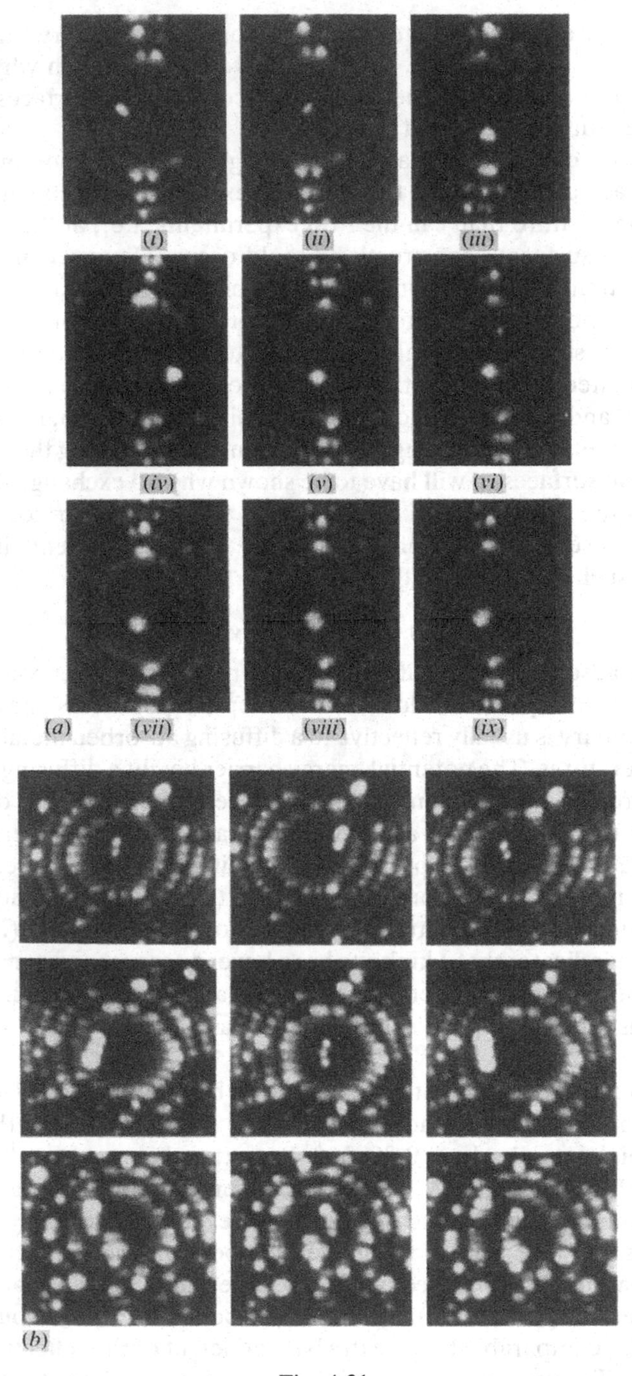

Fig. 4.31

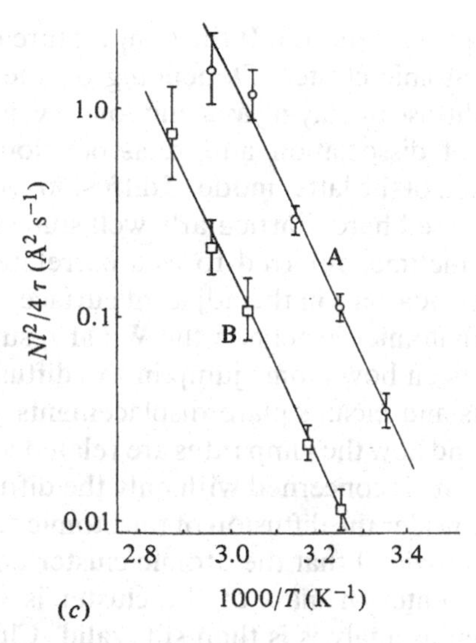

Fig. 4.31 (*a*) Formation of multiple atom clusters of W on the W (110). (*i*), (*ii*) W_2 diffusion and in two orientations; (*iii*), (*iv*) W_3 diffusion and with two different atomic structures; (*v*) W_4; (*vi*) a W_5 and a W adatom; (*vii*)–(*ix*) when the adatom combines with the W_5, a W_6 is formed. Below ~390 K, W_6 has a highly symmetric David-star-like metastable structure. By heating the surface above 390 K, a stable structure is formed which lines up with either of the two ⟨111⟩ channel directions of the substrate. (*b*) Co-operative walks of two to four W adatoms in the adjacent surface channels of the W (112) surface. Note that two nearest-neighbor adatoms can only assume two bond lengths, one of ~4.47 Å and one of ~5.32 Å, as these 1971 micrographs show.[84] (*c*) Arrhenius plots for diffusion of single W adatom (A) and single W_2 cluster (B) on the W (110) surface. From these plots E_d and D_0 are derived to be 0.90 ± 0.07 and 0.92 ± 0.15 eV, and $6 \times 10^{-1} \times (13)^{\pm 1}$ cm^2/s and $1 \times 10^{-3} \times (100)^{\pm 1}$ cm^2/s for W and W_2, respectively.

mated the pre-exponential factor of the diffusion rate equation by kT/h, and derived ΔE_b from experimentally measured P_b and E_d using eq. (4.32). They found the extra barrier height for W, Re and Ir adatoms on the W (110) surface to be, respectively, 199 ± 9 meV, 145 ± 6 meV and 195 ± 12 meV. These energies are about 20 to 25% of the activation energies of single atom diffusion.

4.2.3 Random walk of small atomic clusters

When diffusing adatoms encounter each other, in most cases they will associate into an atomic cluster if the surface temperature is low enough. Figure 4.31(*a*) and (*b*) show formation and diffusion of clusters of various sizes on the W (110) and W (112) surfaces and (*c*) gives an example of

Arrhenius plots in cluster diffusion. If the temperature is high, atoms can dissociate from an atomic cluster. Depending on the temperature, an atomic cluster may diffuse by staying as a unit, or may, in addition, diffuse by a combination of dissociation and re-association processes.[144] A quantitative treatment of the latter mode of diffusion is more complicated and will not be discussed here. Particularly well studied systems are the coupled motion, sometimes referred to as a correlated or cooperative walk, of two to three adatoms in the adjacent surface channels of the W (112) surface, and diatomic clusters on the W (110) surface. Subjects of great interest have been how atoms jump in the diffusion process, how mean displacements and mean square displacements are related to the atomic jump rates, and how the jump rates are related to adatom–adatom interactions, etc. If one is concerned with only the diffusion parameters, then one may still consider the diffusion of the atomic cluster as a random walk of a particle, provided that the atomic cluster does not change its structure, thus the center of mass of the cluster is well defined. The conventional Arrhenius analysis is then still valid. Clusters of this type include diatomic clusters of refractory metals and Pt clusters of different sizes on the W (110) surface.[145] In these clusters, atoms always line up along the ⟨111⟩ directions of the substrate. Diffusion parameters of W_2, Ir_2, Mo_2, and Pt_2 to Pt_4 are also listed in Table 4.1. As is clear from this table and from Fig. 4.31(*c*), the activation energies and pre-exponential factors of surface diffusion of these clusters differ only very slightly from those for single adsorbed atoms. If displacement of a cluster is achieved by jumps of individual atoms in the cluster, then the small difference in the activation energies simply means that the energy needed to stretch bonds to achieve the atomic jump is very small. In this connection, early experiments[146] reporting a much smaller D_0 for dimmer diffusion on the W (112) surface is an error of neglecting the center of mass changes during the diffusion. A proper account of this effect will be discussed in Section 4.2.5(c).

In a molecular dynamic simulation[147] of bulk atomic diffusion by a vacancy mechanism, two atoms may occasionally jump together as a pair. The temperature of the simulation is close to the melting point of the crystal. In FIM studies of single atom and atomic cluster diffusion, the temperature is only about one tenth the melting point of the substrate. All cluster diffusion, except that in the (1×1) to (1×2) surface reconstruction of Pt and Ir (110) surfaces already discussed in Section 4.1.2(b), are consistent with mechanisms based on jumps of individual atoms.[148,149] In fact, jumps of individual atoms in the coupled motion of adatoms in the adjacent channel of the W (112) surface can be directly seen in the FIM if the temperature of the tip is raised to near 270 K.[150]

Although the effect of the image field cannot be isolated in this observation, the mode of atomic jumps observed is thought to be real since, near the center of the plane, the field gradient is zero and there is no driving force. In the absence of an applied field, time lapsed images taken at a very slow atomic jump rate, where one atomic jump is observed for every several micrographs taken can also establish with some certainty that a displacement of the cluster is achieved by jumps of individual atoms. Atomic jumps for the coupled motion of chain-like atomic clusters on the W (112) are relatively simple to figure out because the smallest displacement of a cluster involves only one jump of one of the atoms. Also, atoms in a cluster are distinguishable as they are confined to separate atomic channels of the surface.

Atomic jumps in random walk diffusion of closely bound atomic clusters on the W (110) surface cannot be seen. A diatomic cluster always lines up in either one of the two ⟨111⟩ surface channel directions. But even in such cases, theoretical models of the atomic jumps can be proposed and can be compared with experimental results. For diffusion of diatomic clusters on the W (110) surface, a two-jump mechanism has been proposed by Bassett[151] and by Cowan.[152] Experimental studies are reported by Bassett and by Tsong & Casanova.[153] Bassett measured the probability of cluster orientation changes as a function of the mean square displacement, and compared the data with those derived with a Monte Carlo simulation based on the two-jump mechanism. The two results agree well only for very small displacements. Tsong & Casanova, on the other hand, measured two-dimensional displacement distributions. They also introduced a correlation factor for these two atomic jumps, which resulted in an excellent agreement between their experimental and simulated results. We now discuss briefly this latter study.

The smallest possible displacements, produced by the least possible number of atomic jumps of the two atoms, are called elementary displacements. All other displacements are then different combinations of these elementary displacements. Three elementary displacements have been identified which are called α, β and γ displacement steps. These steps produce, respectively, a center of displacement of $\frac{1}{2}a\mathbf{i}$, $\frac{1}{2}\sqrt{2}a\mathbf{j}$ and $\frac{1}{2}a\mathbf{i} + \frac{1}{2}\sqrt{2}a\mathbf{j}$, where $\mathbf{i}$ and $\mathbf{j}$ are unit vectors along the [001] and the [110] directions as shown in Fig. 4.32. Each of these elementary displacements can be produced by two nearest neighbor jumps of either of the two atoms as can be seen in the figure. The first atomic jump places the cluster in a metastable state with its bond direction aligned either along the [001] or the [110] direction. The second jump returns the cluster to its normal bond configuration. Since out of a few thousand observations, not a single metastable bond state has been observed, the jumping rate of the second

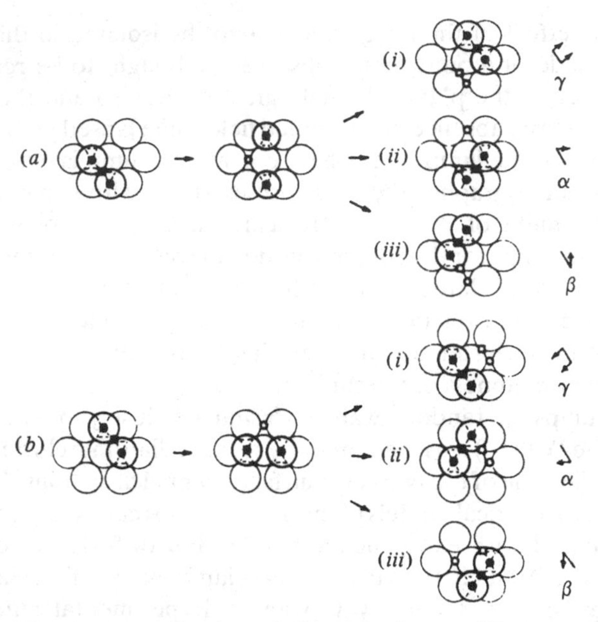

Fig. 4.32 Two-jump mechanism of diatomic cluster diffusion on the W (110) surface with the [110] and [001] intermediate bond configurations.

atomic jump is at least 10^4 times faster than the first atomic jump. The diffusion rate is therefore limited by the rate of occurrence of the first atomic jump.

There are two possible directions for both the atoms in both the first and second atomic jumps. If the jumping direction is completely random and the two atoms have the same probability of performing a jump, then these atomic jumps are said to be uncorrelated. A correlation factor, f, has been introduced for the two atomic jumps, which is defined as the extra probability that the atom making the first jump will also make the second jump in the forward direction. The rest of the probability, $(1 - f)$, is then shared equally for either of the two atoms jumping in either of the two directions. Two experimental displacement distributions measured at 299 K and 309 K fit best with a Monte Carlo simulation with $f = 0.1$ and $f = 0.36$, respectively. The correlation factor increases with diffusion temperature as can be expected. It is interesting to note that when $f = 1$, only α and β steps can occur.

4.2.4 Atom–substrate interactions

Several interactions are fundamentally important for understanding the dynamics of atomic processes on surfaces. These are:

1. interaction between an adsorbed atom and the substrate of different atomic structures, or site specific binding energy of an adatom;
2. interactions between two adatoms;
3. interaction between an adatom and an impurity atom in the substrate;
4. interaction between an adatom and plane edges, lattice defects of various kinds etc.

These interactions are vital to understanding the dynamics of many surface phenomena, such as cluster formation and diffusion, crystal growth, surface reactivity, adsorption and desorption, and many others.

Experimental studies of adatom interactions focus on two quantities, namely the binding energy and the interatomic force, or the distance dependence of the potential energy. These are two different quantities, although in the past they have been occasionally mixed up in some studies. In many FIM studies where the term force is used, concern is in reality only with binding energy at a certain bond distance or a certain site. We will describe briefly here some FIM studies of adatom interactions with metallic substrates. In Section 4.2.5 adatom–adatom and adatom–substitutional impurity atom interactions will be discussed.

(a) Site specific binding energy of surface atoms by evaporation field

One fundamental interaction of a surface atom is that with the substrate. The energy needed to take a kink site atom, i.e. a surface atom with exactly one half the coordination numbers of a bulk atom, away from the surface to the vacuum level at zero temperature may be called the sublimation energy. It can easily be shown analytically that if atomic interactions are pair-wise additive then this energy should be equal to the cohesive energy of the solid, or the energy needed to take a bulk atom away to the vacuum level at zero temperature.[154] In fact, even if atomic interaction is not pairwise additive, the same conclusion can still be drawn from a simple argument. The energy needed to take a kink site atom (all kink site atoms are equivalent since all of them have the same atomic configurations) away to the vacuum level should be the total cohesive energy of the solid divided by the number of atoms in the solid, or exactly the same as the cohesive energy of an atom in the solid. The energy needed to take a surface atom other than a kink site atom to the vacuum level is called simply the binding energy of that atom, and should depend on the atomic configuration of the atom–substrate system.

In solid states, cohesive energy is usually derived from a thermodynamic measurement, such as the temperature dependence of the vapor pressure of a solid. In surface science, the binding energy is usually derived by a temperature programmed thermal desorption measurement. Most of these measurements are done at high temperatures where neither the atomic strutures of the substrates nor the adsorption sites of the desorbed atoms are well defined. Thus it is meaningless to talk about how the binding energy of an atom depends on its atomic configuration in the solid, or on the surface, in these measurements. In field ion microscopy, atoms can be field desorbed at low temperatures from well defined sites on a well defined atomic structure of the surface. Thus in principle it is possible to derive the site specific binding energy of surface atoms.

It was recognized quite early in the development of the FIM that the binding energies of adatoms on a surface could be determined from their desorption fields.[85] Experimental measurements were first reported by Ehrlich & Kirk[87] for W on W surfaces and by Plummer & Rhodin[88] for 5-d transition metal adatoms on W surfaces. A few other measurements, such as Ga and Sn,[155] and Si on W surfaces,[156] have also been reported. It is now well recognized, however, that the method is experimentally poorly defined and theoretically uncertain, and the results obtained are at best unreliable. For historical reasons, however, we will briefly discuss this method and then explain why it is poorly defined. In the next section we will discuss a completely new method where the binding energy of surface atoms is derived from a measurement of the ion kinetic energy distribution in field desorption.

As discussed in Section 2.2.2, in the image hump model of field evaporation, the activation energy for field evaporating an atom as an n-fold charged ion is given by

$$Q_n(F) = Q_0 - (ne)^{3/2}F^{1/2}, \tag{4.67}$$

where

$$Q_0 = \Lambda' + \tfrac{1}{2}\alpha' F^2 + \sum_{i=1}^{M} I_i - n\phi_0. \tag{4.68}$$

Λ' is the binding energy of the surface atom in the applied field F, ϕ_0 is the emitter work function of the surface, and α' is the difference between the polarizability of the surface atom and that of the ensuing ion. In the desorption field measurement one simply assumes that field evaporation occurs when $Q_n(F) = 0$. Thus

$$\Lambda' \simeq (ne)^{3/2}F^{1/2} + n\phi_0 - \sum_{i=1}^{n} I_i. \tag{4.69}$$

Binding energies of a wide variety of adatoms on various planes of tungsten surfaces have been derived by using eq. (4.69), but assuming that $\Lambda' = \Lambda$ and $\alpha' = 0$, or $\Lambda' + \tfrac{1}{2}\alpha' F^2 = \Lambda$. The results obtained have been summarized in a review article.[37] As these values are unreliable, they will not be reproduced here. Now we discuss why the method is ill defined in principle as well as in experimental measurements.

Several serious uncertainties of the method are as follows:

1. The 'desorption field' is an ill defined quantity as this field depends on the rate of field evaporation the measurement is carried out. A d.c. evaporation field differs from the field needed in ns pulsed-field evaporation by as much as ~15% as shown in Fig. 2.10*b*. The field evaporation rate depends on the local properties of the atom and the substrate under the influence of the applied high electric field and these influences are largely unknown.

2. Although the detected charge states in field evaporation of metals are now known from atom-probe studies, they are unknown for adatoms. In addition, many investigators now believe that the multiply charged ions observed in field evaporation are produced by post field ionization of field evaporated ions. Thus the value of n in eq. (4.69) is uncertain for adatoms also.

3. The polarization energy is as large as the binding energy. It is not known how the polarization energy can be correctly accounted for in these experiments. The activation energy of field evaporation is surely significantly changed by this energy. Thus while the method is interesting in a historical sense, it is considered neither a useful nor a valid one.

(*b*) *Site specific binding energy of surface atoms by ion kinetic energy analysis*

The binding energy of surface atoms can be directly derived from a measurement of the kinetic energy of low temperature field desorbed ions. This method is still in an early state of development but will be discussed because of its directness. As atoms cannot move prior to their desorption, the binding energy derived is site specific. In Sec. 3.2.6, it had already been discussed that it is now possible to measure the ion energy distribution and the critical ion energy deficit in low temperature field desorption with an accuracy of 0.1 to 0.2 eV out of a total ion energy of 10 to 20 keV using a pulsed-laser ToF atom-probe. The advantage of this method is that the total energy of a system is constant regardless of how the system is changed from one state to another and how the potential and kinetic energies interchange during this state change. To find the energy needed to transform a system from the initial state to the final state, all the energy changes of the intermediate states can be omitted from consideration. A surface atom just before its desorption has a potential energy of $-(\Lambda' + \frac{1}{2}\alpha_s F_0^2)$ and a kinetic energy of Q. It is removed from the surface to a field free region in the form of an n^+ ion. In a retarding potential energy analyzer, just before the ion is collected, it has a potential energy of $\Sigma I_i - n\phi_{col}$ and zero kinetic energy. The energy needed for the field desorption manifests in the critical ion energy deficit ΔE_c, which is the difference in the total energy of the final and the initial states, or

$$\Delta E_c = \Lambda' + \tfrac{1}{2}\alpha_s F_0^2 + \Sigma I_i - n\phi_{col} - Q. \qquad (4.70)$$

The question is whether $(\Lambda' + \frac{1}{2}\alpha_s F_0^2)$ is equal to Λ or not. When a positive field is applied, the electronic charges around a surface atom will be displaced and the potential energy of the system is lowered by $\frac{1}{2}\alpha_s F_0^2$. These charges are also responsible for the binding of the atom with the surface. Thus the binding energy of the surface atom will be lowered by presumably the same amount. Within the experimental accuracy of a preliminary atom-probe measurement, the data show that they are most likely equal. After replacing ϕ_{col} with ϕ_{av}, we have

$$\Lambda \approx \Delta E_c + n\phi_{av} - \sum_{i=1}^{n} I_i + Q. \qquad (4.71)$$

The activation energy Q can in general range from about 0.1 eV to 1 eV, depending on the temperature and field at which the evaporation is carried out, as already discussed in Section 2.2.4(a). In low temperature field evaporation, it is only about 0.1 to 0.3 eV, a small term in the equation. A similar equation was used to derive the binding energy of Ag atoms on a silver emitter surface using a retarding potential ion energy analyzer.[157] In that experiment, the critical energy deficit of Ag^+ ions was measured when the tip temperature was changed all the way from 80 to 800 K. The critical energy deficit decreased as the tip temperature was raised and correspondingly the field was lowered. This can easily be seen from eqs (4.70) where Q increases with the tip temperature. When the value of the critical energy deficit measured at 80 K, thus Q is only about 0.2 eV, is used, the binding energy of Ag atoms is found to be 2.94 eV, in good agreement with the sublimation energy, or equivalently the cohesive energy at 0 K, of Ag, which is 2.96 eV. The important message here is that the binding energy of surface atoms can be derived from eq. (6.71) provided that the critical ion energy deficit is measured at very low temperature at a very high field evaporation rate, so that Q in the equation is small, and can either be neglected from the equation or corrected with a small factor. Measurements under such conditions have been done using a pulsed-laser ToF atom-probe.

In a preliminary pulsed-laser atom-probe measurement,[158] the critical ion energy deficits of Rh and W atoms, field evaporated from kink sites, are measured at a rate of $\sim 10^{10}$ layers per second at a temperature around 100 K. Using the system constants, i.e. the flight path constant and the time delay constant, calibrated from the onset flight times of pulsed-laser field desorbed inert gas ions discussed in Section 3.2.6, the critical ion energy deficits for W^{3+} and Rh^{2+} desorbed from W (110) and Rh (001) lattice steps can be calculated from their onset flight times. These energy deficits are listed in Table 4.3. The binding energy of these atoms in the kink sites of their lattices, calculated from eq. (4.71) by assuming $Q = 0.2$ eV, are also listed in the table. The binding energies of kink site atoms, or sublimation energy, of W and Rh are found to be 9.08 eV and 6.12 eV, respectively, from this measurement. It is most surprising to see that these values agree with the cohesive energies of W and Rh to within 0.4 eV, or within ~5%. The binding energy of kink site atoms should be equal to the cohesive energy at 0 K even if metallic bonds are not pair-wise additive. It appears possible now to measure both the cohesive energy of a solid and the binding energy of surface atoms on specified adsorption sites with a low temperature field desorption method.

This method can of course be used to measure the binding energy of adsorbed atoms on various crystal planes, although the measurement will be very time consuming. Since the detection efficiency of a Chevron channel plate is only about 60% and a reliable ion energy distribution needs at least a few hundred ions, and an element usually has a few isotopes, one would have to study about two thousand deposited adatoms from each plane just to derive the binding energy of one adatom–surface system. The experiment has not yet been done. The method can also be used to study binding energies of different constituent atoms in an alloy or a compound, and also the binding energies of atoms of the same species but in different lattice sites of a crystal, for example oxygen atoms in two different sites of 123 high T_c superconducting compounds, etc. For semiconductors, photo-excitation seems to occur in photon stimulated field evaporation even when the power of the laser pulses is very low, as already discussed in Section 2.2.6. Therefore there are still many uncertainties in using this method to derive the binding energies of surface atoms in semiconducting materials.

Table 4.3 *Critical energy deficits for various metal ions*

Ionic species	Crystal plane	Measured $\Delta E_c/n$ eV	E_s eV	Measured Λ eV
W^{3+}	W (110)	15.12 ± 0.17	8.90	9.08 ± 0.30
Rh^{2+}	Rh (100)	11.23 ± 0.04	5.75	6.12 ± 0.30

4.2.5 Interactions between surface atoms

(a) Adatom–adatom interactions

In this section, we will consider adatom–adatom interactions and interaction between an adatom and a lattice defect such as a plane edge or a substitutional impurity atom in the surface layer.[159] Consider first adatom–adatom interactions. Many surface atomic processes are dictated by adatom–adatom interactions, or the force laws governing the interaction between adatoms. Some of the obvious ones are the formation of adsorption layer superstructures, surface atomic reconstructions, two-dimensional phase transitions of surface layers, surface reactivities, adsorption and desorption, and chemical surface diffusions, etc. Although there are a considerable number of works on chemisorptions on single crystal faces, few studies provide detailed information on adatom–adatom interactions. Macroscopic techniques have been very successful in studying two-dimensional island formation, adsorption layer superstructure formation, and structural and other phase transformations.[160,161] From these studies, people are able to estimate the interaction energies at different neighbor sites, and also whether these interactions are attractive or repulsive and so on. Such information can also be derived from studies of how the heat of adsorption varies with the degree of coverage of an adsorbed layer, and desorption and migration kinetics, etc.[162,163] All these studies are indirect, and really do not provide information on how the pair potential energy between two adatoms depends on their separation over a large distance range. It is also difficult to distinguish pair effects from many body effects. The field ion microscope, with its capability to image single adatoms and to provide atomically perfect surfaces, can be used to derive the pair potential by a direct method. Before we discuss FIM experiments, a few words about theoretical developments are appropriate.

In general, if two adatoms are sufficiently close to one another so that a significant overlap of their wave functions occurs, a strong interaction can be expected between them. For an attractive interaction, it is essentially a chemical bond and the strength should be comparable to the chemical bond of the adatom with the substrate. For larger inter-adatom separations, the interaction should fall off exponentially and should become negligibly small when the distance is more than a few Å. This interaction, expected from our knowledge of molecular and solid state physics, has however not been observed in FIM experiments with metal surfaces. FIM studies find that adatom–adatom interaction at the strongest bond separation is of the order of 0.1 eV or less for all adatom pairs studied to date, including a Si–Si pair where formation of covalent bond might be expected. In addition, the interaction may extend to beyond several Å. It is also well known that most gas molecules dissociate upon chemisorption on 5-d transition metal surfaces. Chemical bonds are therefore drastically modified on metal surfaces, possibly by the breaking of symmetry of the crystal at the surface, and the strong adatom–substrate interactions.

If chemisorbed species are polar, or dipole moments are induced by adatom–substrate interactions, there will be a dipole–dipole interaction between two adatoms. The interaction is given by

$$U_{d\text{-}d}(R) = 2\mu_a\mu_b/R^3 \qquad (4.72)$$

where μ_a and μ_b are the surface dipole moments of the two adatoms. The factor of 2 was introduced by Kohn & Lau[164] to account for the fact that although outside a metal the image charge acts as though it were at the image point, in reality it is distributed on the surface. The interaction energy is 625 meV times the two dipoles in Debyes over R^3 in Å.[165] For like adatoms, the interaction is repulsive.

Two adatoms will always interact with each other weakly via the van der Waals interaction. This interaction, while important in physisorption, is a relatively minor factor in chemisorption. The leading term is

$$U(R) \simeq -C/R^6, \qquad (4.73)$$

where C is a constant proportional to the square of the polarizability, and is roughly 30 eV·Å^6 for Ar, N_2 and O_2, and ~150 eV·Å^6 for Xe. The interactions considered so far are direct interactions since these interactions propagate directly through space and the substrate does not play a direct role. Adatoms can also interact with each other by a mutual perturbation of the substrate. This 'indirect' interaction can have different origins. The electronic indirect interaction arises from the fact that the atomic wavefunctions of the adatoms can tunnel through the narrow potential barrier to the metal and couple with metal wavefunctions.[166,167,168] If the coupling places the two atomic wavefunctions in phase, the interaction is attractive. If it is out of phase, the interaction is repulsive. From the oscillatory nature of the intermediate wavefunctions, the electronic indirect interaction should be an oscillatory function of inter-adatom distance. The asymptotic form of the interaction is given by

$$U_{in}(R) \propto \frac{\cos(2k_F R)}{(2k_F R)^n}, \qquad (4.74)$$

where k_F is the Fermi wavevector and n is a positive constant ranging from 1 to 5, depending on the shape of the Fermi surface in the direction of the interaction. Other indirect interactions originated in the lattice, such as the phonon field and the elastic distortion, have also been discussed.[169] These interactions are usually very weak, of the order of only 10^{-6} eV. FIM measurements give the total effect of all these possible interactions. These interactions are already very weak, and therefore it is difficult to separate one interaction from another in an experimental measurement.

Adatom–adatom interactions will affect the behavior of single atoms on the surface. Thus one may be able to derive the interaction between two adatoms by studying their behavior in the presence of one another. The pair potential energy between two adatoms may be defined as the difference between the potential energies of an adatom on the surface with and without the presence of another adatom on the same surface. Since the substrate contains tens of thousands of atoms, the term 'pair potential' here really means the potential of mean force, which may be called 'an effective pair potential', or simply the pair potential. In doing so, it is understood that we are including all the possible changes in the substrate lattice due to the adatoms, which cannot be directly observed, into the adatom–adatom interaction, which can be directly observed. There are basically two kinds of experiments from which the pair potential can be derived. They are equilibrium and kinetic experiments. In the former kind, pair potential energies are derived from the relative frequencies of having different bond distances of the two adatoms. In the latter kind, the transition rate from one bond length to another bond length is measured and then related to the barrier height at the saddle point. So far, FIM studies of adatom–adatom interactions focus on the W (110) and (112) surfaces. Below we briefly describe some of these studies.

(*b*) *Adatom–adatom interaction on W* (110) *surfaces*

The tungsten (110) surface is one of the best studied of all surfaces, especially in field emission and field ion microscopy for many reasons. It is a very stable surface without surface reconstruction or phase transformation. It is also inert to contaminations. For the study of adatom–adatom interactions, it is a very smooth plane with the largest density of adsorption sites available of any W surface. Lesser restrictions are imposed on the adatom–adatom separation. As the surface is structurally very smooth, wave mechanical interference effects are least affected by the surface atomic structure.

The simplest of adatom–adatom interaction studies is a measurement of dissociation time as a function of temperature, which gives the binding energy, or adatom cohesive energy, at the bond separation of strongest binding, E_c. The average dissociation time is given by[170]

$$t_{dis} = k^{-1} \simeq h/kT \exp\left[(E_d + E_c)/kT\right]. \tag{4.75}$$

E_d is the activation energy of surface diffusion. The adatom cohesive energies of various adatom pairs on various planes of tungsten range from 0.1 to 0.6 eV. They show strong chemical specificity. On the W (110),

Bassett[170] found the adatom cohesive energies of diatomic clusters to be $\sim 0.62 \pm 0.15$ eV for Ta_2, $\sim 0.31 \pm 0.10$ eV for W_2, ~ 0 for Re_2, $\sim 0.15 \pm 0.10$ eV for Ir_2, and $\sim 0.15 \pm 0.05$ eV for Pt_2, $\sim 0.15 \pm 0.10$ eV for Pd_2, and $\sim 0.31 \pm 0.10$ eV for Ni_2. Similar results are obtained for Mo clusters on W surfaces.[171] The dimer binding energy of 5-d transition metal adatoms on the W (110) surface thus shows a trend opposite to that of the activation energy of surface diffusion of single 5-d adatoms on the same surface, as shown in Fig. 4.27. In general the adatom cohesive energy increases monotonically with the cluster size. Thus the dissociation temperature of linear Ir clusters increases from ~320 K for Ir_2 to ~420 K for Ir_4, and to 430 to 470 K for larger clusters.[172] In some cases, this does not occur. Cowan & Tsong find that W_2 is much more stable than W_3. W_3 often dissociates around 350 K whereas W_2 does not dissociate until near 400 K. Thus the W–W interaction has to be non-monotonic in distance dependence, and at least one of the bond energies in a W_3 cluster is negative, or the interaction is repulsive.

Kinetic experiments, while useful in estimating the adatom cohesive energy of a cluster, become very complicated if pair energies at more than one bond state, or information on the inter-adatom potential, are desirable. The distance dependence of pair interaction can be more easily derived by an equilibrium experiment.[173] The principle is very simple. At equilibrium, the relative frequencies of observing the two adatoms at various bond states or bond separations at a given temperature are related to their pair energies according to the Boltzmann factors. Thus

$$\frac{p_e(\mathbf{r}_i)}{p_e(\mathbf{r}_j)} = \frac{p_0(\mathbf{r}_i)\exp\left[-\dfrac{U(\mathbf{r}_i)}{kT}\right]}{p_0(\mathbf{r}_j)\exp\left[-\dfrac{U(\mathbf{r}_j)}{kT}\right]}, \tag{4.76}$$

where $p_e(\mathbf{r})$ is the probability of observing a bond represented by $\mathbf{r}$, $U(\mathbf{r})$ is the pair energy of the bond, and $p_0(\mathbf{r})$ is the probability of observing such a bond of the surface if the interaction between the adatoms is turned off, $p_0(\mathbf{r})$ is a geometrical factor determined solely by the adsorption site structure and the shape of the finite size substrate. We therefore have

$$p_e(\mathbf{r}_i)/p_0(\mathbf{r}_i) = K\exp[-U(\mathbf{r}_i)/kT], \tag{4.77}$$

where K is an undetermined proportionality constant. The pair energy is given by

$$-U(\mathbf{r}_i) = E_c(\mathbf{r}_i) = kT\ln\left[\frac{p_e(\mathbf{r}_i)}{p_0(\mathbf{r}_i)}\right] - kT\ln K, \tag{4.78}$$

$E_c(\mathbf{r})$ is the pair binding energy at bond vector $\mathbf{r}$. If the distance range of

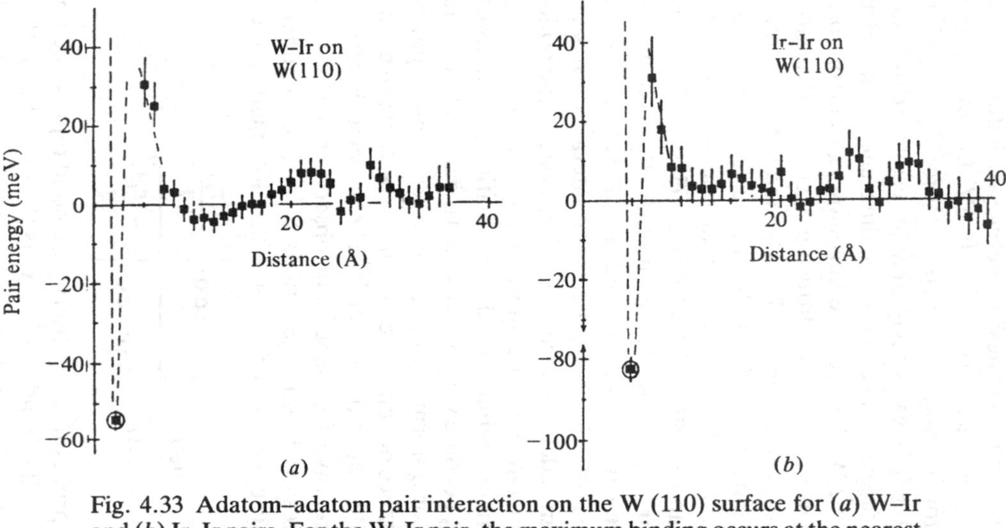

Fig. 4.33 Adatom–adatom pair interaction on the W (110) surface for (*a*) W–Ir and (*b*) Ir–Ir pairs. For the W–Ir pair, the maximum binding occurs at the nearest neighbor bond of the substrate, while for the Ir–Ir pair it occurs at twice the nearest neighbor bond distance. All the pair cohesive energies are less than 100 meV.

the measurement is sufficiently large such that the interaction becomes negligibly small, then a reference level of the pair potential can be found. When one takes the pair energy to be 0 at a very large adatom–adatom separation, one has $K = 1$. This method thus involves measuring the relative frequencies of observing adatoms at different bond distances (vectors), theoretically calculating the relative probabilities of adatoms to assume these bond separations (vectors) if the adatom–adatom interaction is turned off, and then calculating $E_c(\mathbf{r})$ as a function of $\mathbf{r}$ according to eq. (4.77).

In their studies,[174] Tsong & Casanova make two approximations. First the W (110) surface is assumed to be isotropic and thus no attempt is made to find a directional dependence of the adatom–adatom interaction. In other words, the distance dependence of pair energy is measured without considering the direction of the adatom–adatom bonds. The bond length as well as the bond direction of the strongest binding energy state is identified, however. Second experimental data are analyzed in histograms of 1 Å intervals and an average procedure is used to reduce statistical fluctuations of the data. The average procedure is needed in spite of great efforts to make about 1000 observations for each set of data. Except for the closest bond, no identification of the bonds is attempted. The statistical weights, $p_0(\mathbf{r})$, are calculated numerically using an expression based on the assumption that adatoms sit only on lattice sites, and it is also plotted in a histogram of 1 Å width like that used in the data analysis. We will refer the details of this approximation method to original articles. In Fig. 4.33 (*a*) and (*b*), pair interactions of W–Ir and Ir–Ir pairs on the W (110) surface are shown. There are a few interesting features of the adatom–adatom interactions. First, all the adatom pairs studied in their experiment, except Re–Re, exhibit an attractive interaction at a close bond, which is the nearest neighbor bond, ~2.74Å parallel to the ⟨111⟩ direction, of the W (110) surface for W–Re and W–Ir interactions, and twice the nearest neighbor bond, ~5.48 Å parallel to the ⟨111⟩ direction, for Ir–Ir interaction. Second, the adatom cohesive energy is less than 0.1 eV in all cases, or less than one tenth of what can be expected from a chemical bond (see Table 4.4). Third, there is a repulsive region extending to a separation of nearly 10 Å, and beyond that a small oscillatory tail of about 10 meV or less amplitude can be detected. The large repulsive potential observed cannot be explained in terms of a dipole–dipole interaction between the adatoms as this energy is less than 1 meV. The non-monotonic behavior of adatom–adatom interaction is consistent with what can be expected from an electronic indirect adatom–adatom interaction, but no detailed calculations of specific systems are available to compare with the experimental data. The smallness of the adatom–adatom interaction, only about 10% of what should be expected

Table 4.4 *Adatom–adatom interaction derived from pair distribution functions*

Adatom pair	Distance of strongest bond	Bond energy meV	Expected dipole repulsive energy meV
Re–Re	2.5–3.5 Å	?	0.35
W–Re	2.5–3.5 Å	99.0 ± 0.7	0.61
Ir–Ir	5.0–6.0 Å	82.0 ± 2.5	~0
W–Ir	2.5–3.5 Å	53.2 ± 3.6	~0

from a chemical bond, has been explained in an elaborate theoretical calculation by Feibelman,[175] who finds the pair binding energy is only about 60 meV for an Al–Al interaction on an aluminum surface.

The pair potential energy of Re–Re interaction is more difficult to derive. While the oscillatory tail is more clearly established than the other pairs studied, the repulsive potential in the region between 4 and 6 Å is too high to be overcome by the adatoms at a 330 K heating of the surface. Hence not even a single bond of >6 Å is observed in a total of 1045 heating periods of observations. The strong repulsive region is recognized in the first study of adatom–adatom interactions,[91] and is further confirmed by this study if one realizes that without the existence of a repulsive potential barrier in the region between 4 to 6 Å one should observe at least 35 bonds of a separation less than 6 Å. Closely packed Re_2 clusters, however, can occasionally be observed if the heating temperature is above 400 K. In fact, Re clusters of up to 4 atoms have been observed at 400 K heating. Unfortunately, at such a high temperature, a loss of adatoms frequently occurs, and a quantitative study of pair distribution function is difficult. Thus the adatom cohesive energy at closest bond has not been established. Some investigators simply believe that the adatom cohesive energy of a Re–Re pair is zero.[170]

In the experiment discussed above, no directional dependence of the pair interaction is attempted. Pair interactions are simply assumed to be isotropic on the W (110) surface. The pair interaction, in general, should depend both on the direction of the adatom–adatom pair bond and on the bond length. Thus pair energies should therefore be measured for each possible pair bond. A preliminary study in this direction has been reported by the same authors for Si–Si interaction on the W (110) surface.[94] Si–Si interaction is of particular interest since (1) Si atoms interact with one another in solid state by forming covalent bonds rather than metallic bonds; it would be interesting to see how the interaction of Si adatom pairs on a metal surface is different from that of metal adatom pairs; (2) semiconductor–metal interfaces are technologically important and electronically interesting. Understanding diffusion and interaction behavior will help us understand many interface phenomena of great interest.

There are a few technical difficulties in this study. To avoid any problem of contamination, helium is used for the image gas. Si adatoms are found to be able to resist the image field of helium, but the image intensity is only 3 to 5% that of metal adatoms. It is particularly difficult to see a Si adatom if it is near the center of the plane. Accidental field desorption of adatoms cannot be avoided. A compromise is made by depositing 3 to 6 adatoms on a W (110) surface, and then observing the relative frequencies of having different pair bond configurations. Fortunately another Si adatom is rarely present near a pair bond if the bond length does not exceed ~10 Å. Thus the relative pair energies for bond lengths less than the sixth nearest neighbor bond, 6.32 Å in the [001] direction, can be reliably derived. The pair binding energies $E_c(r)$ derived are listed in Table 4.5.

The Si–Si interaction again shows a non-monotonic distance dependence. The largest binding energy is found when two Si adatoms are separated by the third nearest neighbor bond of the substrate, or 4.47 Å in the [110] direction. The binding energy is only about 60 meV, thus no formation of a covalent bond is definitely established. The repulsive energy at the second nearest neighbor bond, 3.16 Å in the [001] direction, is only about 35 meV, or much weaker than those in metal adatom–adatom interactions. As pair energies are obtained only for adatom–adatom distances of less than 7 Å, the reference level, or zero level, of the adatom–adatom potential energy cannot be obtained. This level is estimated from comparing these experimental observations with a Monte Carlo simulation, and is shown in Fig. 4.34. As will be described in Section 4.2.6, formation of a two-dimensional superlattice of a silicon adsorption layer on the W (110) surface was anticipated from this pair potential and was subsequently found. Thus the energetics of the formation of an adsorption layer superstructure is clarified from this study.

As time progresses and the techniques are better perfected, adatom–adatom interactions are also studied in greater detail. With the availability of microcomputers the image mapping can now be done with a digital method and greater accuracy thus achieved. It is now possible to study the directional dependence of pair interactions without making any approximation in the data analysis. Thus the frequencies of having the two adatoms of assuming different bonds can be directly derived from field ion images by a digital mapping method. The theoretical pair distribution for two non-interacting atoms on a plane of the same shape, $p_0(\mathbf{r}_i)$, can also be obtained from a Monte Carlo simulation and then used for deriving $E_c(\mathbf{r}_i)$ at different bond lengths and directions. A recent

Table 4.5 *Frequencies of observing various bonds in Si–Si interaction on the W* {110} *at* 300 *K and relative pair energies*

Bond directions	Bond lengths	Frequency of observations	Statistical weight	Relative binding energies
$[1\bar{1}1]$	2.74 Å or $(\sqrt{3}/2)\,a$	44	4	0 (ref. level)
[001]	3.16 Å or a	5	2	−38.00 ± 12 meV
$[1\bar{1}0]$	4.47 Å or $\sqrt{2}\,a$	150	2	50.00 ± 3 meV
$[1\bar{1}1]$	5.48 Å or $\sqrt{3}\,a$	30	4	±10.00 ± 5 meV
[001]	6.32 Å or 2 a	27	2	5.00 ± 5 meV
$\alpha \sim 65°$	5.24 Å or $(\sqrt{11}/2)\,a$	7	4	−47.52 ± 10 meV
$[1\bar{1}0]$	Two 4.47 Å bonds chain	10	—	—
$[1\bar{1}0]$	Three 4.47 Å bonds chain	1	—	—

Angle α in column 1 is the angle between the $[1\bar{1}0]$ direction and the bond direction of the cluster.
a is the lattice constant of W.
Data from ref. 156.

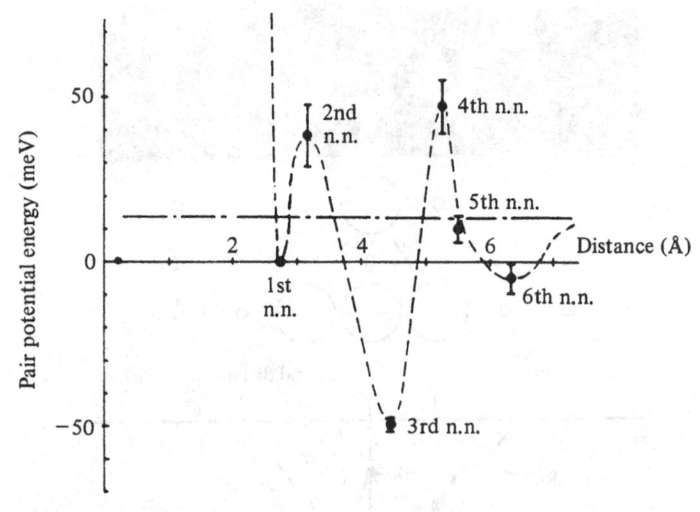

Fig. 4.34 Relative pair energies between two Si adatoms at six pair bond distances of the W (110) substrate. The broken curve is a freehand drawing merely for the purpose of showing the non-monotonic distance dependence of the Si–Si pair interaction. The centered line is the asymptotic value determined by comparing these experimental data with a Monte Carlo simulation.

measurement of W–Re and Ir–Ir interactions on the W (110) plane distinguishes pair potential along [001], [110] and [111] directions.[176] Again, non-monotonic dependences of these pair potentials are found. The adatom cohesive energies at strongest bonds are again of the order of 100 meV, or about one tenth the strength of a chemical bond. However, the potential energy of adatom–adatom interaction does depend on the crystallographic direction.

(c) *Adatom–adatom interactions on the W (*112*) surface*

It is found that adatoms in the adjacent surface channels of a W (112) plane will combine into a linear chain when these adatoms encounter one another.[177] This chain can diffuse by coupled motion, or cooperative walk, of these adatoms. Adatoms in the chain are found to assume only two bond lengths, 4.47 Å parallel to the [110] and 5.32 Å parallel to the [311] or the [131], usually referred to as the straight and staggered bonds, or configurations 0, +1 and −1, respectively. Figure 4.35 shows these three configurations of a W_2 cluster. The chain-like clusters move by coupled atomic jumps, or correlated changes of bond lengths, of the adatoms. For a diatomic cluster, the adatom–adatom potential energies at three bond separations can be derived by a combined measurement of a pair distribution function and transition rates from one configuration to

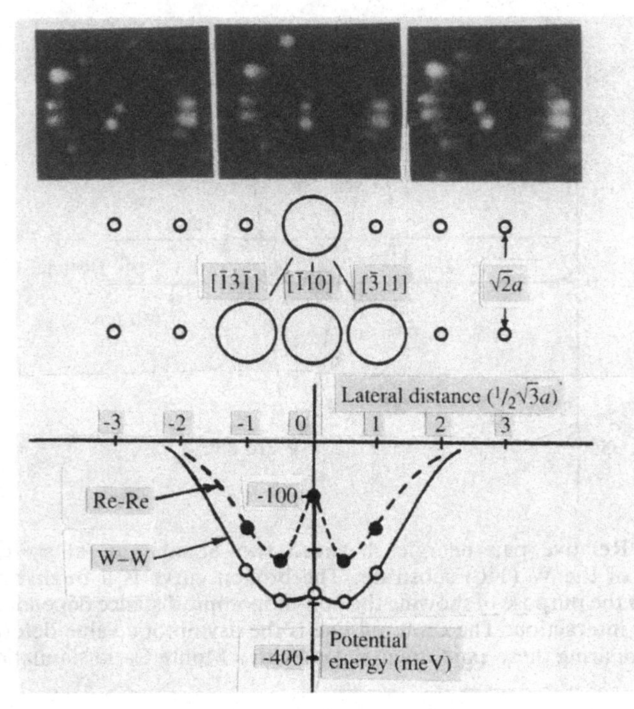

Fig. 4.35 The field ion images show two W adatoms forming a straight bond, or configuration 0, and staggered bonds, or configurations ±1 on a W (112) surface. Also shown are pair energies at four bond distances for Re–Re pairs, from Ehrlich and co-workers, and W–W pairs, from Cowan and Tsong, interacting across the atomic channels of the W (112) plane.

another.[178] The forward transition rate, i.e., transition from configuration 0 to configuration 1, is given by

$$k^+ = 4\nu_0\left(\frac{M-1}{M}\right)\exp\left(-\frac{\Delta E^+}{kT}\right), \tag{4.79}$$

where $\Delta E^+ \equiv E_d + E_c(\mathbf{r}_0) - E_c(\mathbf{r}_{1/2})$; $\mathbf{r}_0$ and $\mathbf{r}_{1/2}$ are, respectively, bond vectors at configuration 0 and at the saddle point between configuration 0 and configuration 1. The factor 4 comes from the fact that a change in the configuration can occur by jumps of either of the two adatoms of the cluster in either of the two possible directions. M and $(M-1)$ represent, respectively, the configuration weights of the two bonds on a linear lattice of M adsorption sites. The backward transition rate, i.e. the transition rate from configuration 1 to 0, is given by

$$k^- = 2\nu_0 \exp\left(-\frac{\Delta E^-}{kT}\right), \tag{4.80}$$

where $\Delta E^- \equiv E_d - E_c(\mathbf{r}_{1/2}) + E_c(\mathbf{r}_1)$, and $\mathbf{r}_1$ is the bond vector of configuration 1. The thermodynamic equilibrium probabilities $p(\mathbf{r}_0)$ and $p(\mathbf{r}_1)$ of finding the two atoms in configuration 0 and 1 are given, respectively, by eqs (4.78) and (4.79) to be

$$p(\mathbf{r}_0) = \frac{k^-}{k^+ + k^-}, \tag{4.81}$$

$$p(\mathbf{r}_1) = \frac{k^+}{k^+ + k^-}. \tag{4.82}$$

The time dependence of the probabilities of finding a diatomic cluster in configurations 0 and 1 after time t, starting from the same configuration, are respectively

$$p(\mathbf{r}_0, t) = \frac{k^+k^-}{k^+ + k^-}\left\{\frac{1}{k^+} + \frac{1}{k^-}\exp\left[-(k^+ + k^-)t\right]\right\}, \tag{4.83}$$

$$p(\mathbf{r}_1, t) = \frac{k^+k^-}{k^+ + k^-}\left\{\frac{1}{k^-} + \frac{1}{k^+}\exp\left[-(k^+ + k^-)t\right]\right\}. \tag{4.84}$$

The mean square displacements for the center of mass of the diatomic cluster in a series of walks starting from configurations 0 and 1 are, respectively, given by,

$$\langle(\Delta x)^2\rangle^{(0)} = \frac{k^+l^2}{2(k^+ + k^-)}\left\{k^-t + \frac{(k^+ - k^-)}{2(k^+ + k^-)}\left[1 - \exp\left[-(k^+ + k^-)t\right]\right]\right\}, \tag{4.85}$$

$$\langle(\Delta x)^2\rangle^{(1)} = \frac{k^-l^2}{2(k^+ + k^-)}\left\{k^+t + \frac{(k^+ - k^-)}{2(k^+ + k^-)}\left[1 - \exp\left[-(k^+ + k^-)t\right]\right]\right\}, \tag{4.86}$$

For long periods, such that $t \gg 1/(k^+ + k^-)$, the mean square displacement for the center of mass becomes

$$\langle(\Delta x)^2\rangle = \frac{k^+k^-}{2(k^+ + k^-)}l^2t = 2Dt. \tag{4.87}$$

Thus $\langle(\Delta x)^2\rangle$ becomes independent of the starting configuration of the cluster. The same equation is obtained if one averages over all configurations in a large number of events. The activation energy of diffusion of the center of mass of the cluster, ΔE_{cm}, as derived from an Arrhenius plot, is related to ΔE^+ and ΔE^- by

$$\Delta E_{cm} = \frac{k^-\Delta E^+ + k^+\Delta E^-}{k^+ + k^-} = p(\mathbf{r}_0)\Delta E^+ + p(\mathbf{r}_1)\,\Delta E^-. \tag{4.88}$$

E_c at $\mathbf{r}_0$, $\mathbf{r}_1$, $\mathbf{r}_{1/2}$ can be derived by measuring ΔE_{cm}, $p(\mathbf{r}_0)$, and $p(\mathbf{r}_1)$, or the activation energy of surface diffusion of the center of mass of the cluster, and the relative abundances of the straight and the staggered bonds. Stolt *et al.*[179] find that for two Re atoms on the W (112), $E_c(\mathbf{r}_0) - E_c(\mathbf{r}_1) = -35$ meV, $E_c(\mathbf{r}_1) \simeq 90$ meV, and $E_c(\mathbf{r}_0) - E_c(\mathbf{r}_{1/2}) = 100$ meV. For two Ir adatoms on the W (112) plane, $E_c(\mathbf{r}_0) = 147$ meV. For Ir_2, only configuration 0 is observed. Thus no E_c at other bond distances can be obtained.

A slightly different method, which recognized the need to consider the mean square displacement of the center of mass of the cluster, and used an approximation in the derivation, was originally used by Cowan & Tsong[180] for analyzing coupled motion of W_2 clusters on the W (112) surface. Although the method is slightly simpler and also easier to visualize the physical significance of properly accounting for the center of mass diffusion, we will omit discussing it here since an exact treatment is available as already mentioned. Interested readers can find the method in the references cited. Data obtained in that experiment are: $E_c(\mathbf{r}_1) - E_c(\mathbf{r}_{1/2}) = -53$ eV, $E_c(\mathbf{r}_0) - E_c(\mathbf{r}_1) = 37$ meV, and $E_c(\mathbf{r}_0) = 280$ meV.

In Fig. 4.35, images of a W diatomic cluster on a W (112) surface having three configurations, 0, -1 and $+1$, are shown. The pair potential energies at four bond vectors, $\mathbf{r}_0$, $\mathbf{r}_{1/2}$, $\mathbf{r}_1$, and ∞ for W–W and Re–Re interactions on the W (112) are also shown. Aside from the fact that Re_2 is more stable in configuration 1 than in configuration 0 and W_2 is just reversed, the two potential curves are quite similar in shape and magnitude. No oscillatory tail, or a non-monotonic distance dependence, of the adatom–adatom interaction is apparent. The difficulty of studying adatom–adatom interaction on this plane is obvious. No pair energies below 4.47 Å can be obtained on this plane because of the surface channel structure which limits the shortest possible bond length the two atoms can assume to be greater than 4.47 Å. Also, the rough atomic channel structure of the surface may dampen the indirect electronic interactions. This system, while not particularly suited for studying adatom–adatom interactions, is very fruitful for studying one-dimensional cluster diffusions since analytical stochastic equations are available for data analysis of the cluster diffusion.[181]

(*d*) *Adatom–substitutional impurity atom interactions*

As an example of adatom–lattice defect interactions on a surface, we will describe here a study of the interaction between an adatom and an impurity atom in a surface layer. An impurity atom, like any other lattice defect, in the surface layer can perturb the periodicity of the surface both electronically and elastically. Such a perturbation will change the potential energy of a diffusing adatom on the surface. A study of adatom–

impurity atom interaction, and other adatom-lattice defect interactions, is of considerable experimental and theoretical interest for the following reasons:

1. If adatom–impurity atom interaction is attractive, then the impurity atom can act as a trapping center. A diffusing adatom may be trapped. In heterogeneous catalysis, the reaction rate may be changed by the trapping effect of impurities as also by lattice defects and lattice steps and so on.
2. In crystal growth by vapor condensation, an attractive impurity atom may act as a nucleation center by trapping diffusing adatoms when the number of trapped adatoms reaches a critical number.
3. Many theories of adatom–adatom interactions and other interactions are based on a model Hamiltonian developed by Anderson for dilute alloys. Such a Hamiltonian may also be used to explain adatom–impurity atom interactions.

Despite a possible wide significance of such a topic, there is only one reported study of adatom-substitutional impurity atom interaction, where the interaction of a W adatom with substitutional Re atoms in a W lattice is studied by using a W–3% Re alloy as the substrate.[182] The planes used in FIM studies of adatom behavior are usually quite small containing only a few hundred atoms. Thus a plane of a W–3% Re alloy is likely to contain a few Re substitutional atoms. The perturbation to the overall electronic and elastic properties of the substrate lattice should still be relatively small. Therefore the interaction of a single substitutional impurity atom with a diffusing adatom can be investigated.

As in a study of adatom–adatom interactions, two types of measurements may provide information about the interaction. They are an equilibrium experiment and a kinetic experiment. The former measures the frequencies of visit at a given temperature of an adatom, deposited on the (110) surface of the alloy, to all the adsorption sites of the surface. The latter measures the jumping rates from one site to neighbor sites at a given temperature, or as a function of temperature. Using an image mapping method, it is found that when a W adatom is allowed to diffuse on a W (110) surface the site occupation numbers vary considerably from one site to another. The variation, however, is consistent with what can be expected from random events and a Monte Carlo simulation of the random walk diffusion process. On a (110) surface of the W–3% Re alloy, however, the site occupation numbers are found to be clustered, indicating that the diffusing W adatom is very often trapped by substitutional Re atoms in the surface layer. From analysis of these site occupation numbers and transition rates from one site to others, a potential energy diagram of the adatom in the vicinity of a substitutional Re atom is derived. It is found that at the nearest-neighbor sites, when the adatom is at a lattice site and a bridge site, the interaction is attractive. At the second nearest neighbor site, or when the adatom is on another lattice site, the interaction is highly repulsive (for locations of these sites, see Fig. 4.36). The repulsive region extends to several Å. If one ignores a possible directional dependence of the adatom–substitutional interaction, an interaction potential as shown in Fig. 4.36 is obtained. r_T is when the adatom is in a lattice site and r_B is when the adatom is in a bridge site. The interaction is again very weak, less than ~90 meV. It is most interesting to point out here that this interaction potential is almost identical to the one found for adatom–adatom interaction of a W–Re pair on the W (110) surface (see Fig. 9 of reference 174). The adatom cohesive energy is found to be 99 ± 1 meV at the nearest neighbor distance, and the potential is highly repulsive at the second nearest neighbor distance. This repulsive region extends to several Å also. It appears that one may be able to do this type of study on an fcc (110) surface, such as the Pt (110) surface, in a much better controlled manner. First, an impurity atom, for example an Ir adatom, is incorporated into the surface by an exchange diffusion. An Au adatom, which does not cross the surface channel in its diffusion, is then deposited on the same surface channel of the Pt (110) surface. Site occupation numbers and transition rates within the one-dimensional surface channel can then be derived by a diffusion experiment. This experiment, while requiring great skill, can at least be done in principle.

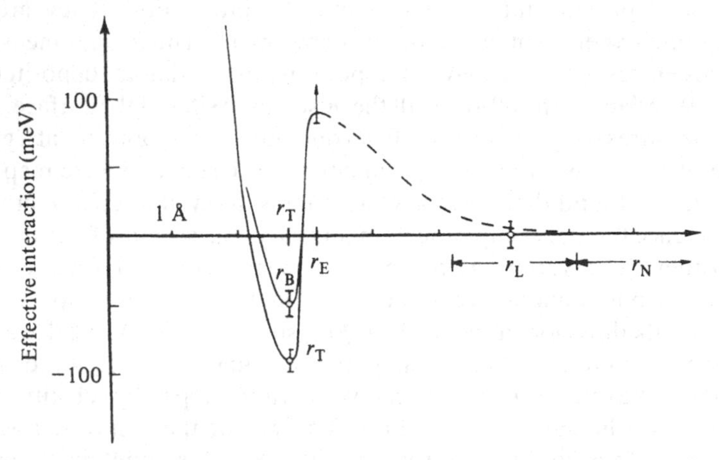

Fig. 4.36 Intraction potential between a W adatom and a substitutional Re atom inside the (110) surface layer. This interaction is almost identical to W–Re adatom–adatom interaction on the W (110) surface.

4.2.6 Formation of an adsorption layer

Fundamental to the growth of a crystal is how atoms interact and nucleate into a surface layer and eventually into a three-dimensional crystal. FIM experiments find that when adatoms on a surface encounter each other in their random walk diffusion, in most cases they will combine into a cluster. On the W (110) plane,[84] a W diatomic cluster is linear with its axis always lining up with either one of the ⟨111⟩ surface channel directions of the substrate. However, two-dimensional growth starts already when the number of adatoms in a cluster exceeds three. Thus W_3 clusters are triangular in shape even though more than one structure exists. For W_6 clusters two structures have been observed; one has an image the shape of a David star and is a metastable structure which can be formed at a temperature below ~390 K, while the other has a less symmetric shape, but is a stable structure which can be formed by annealing the metastable W_6 cluster above 390 K, as already shown in Fig. 4.31(*a*). For Re adatoms on the W (110) surface, a Re_2 cluster, having its axis parallel to the ⟨111⟩, can occasionally be formed if the heating temperature is above 400 K. Unfortunately at this temperature the Re_2 cluster is unstable and cannot be studied. Re_3 and Re_4 clusters can also be formed above 400 K; they are also stable and have two-dimensional structures, similar to W clusters. In contrast, Bassett[183] finds that Pt and Ir adatoms on the W (110) always form linear chains lining up along the ⟨111⟩ surface channel directions. These chains are highly stable, and they may intersect with one another without transforming into two-dimensional islands. Ni and Pd adatoms on the W (110) also form one-dimensional chain clusters when the cluster size is small. When the cluster size exceeds a critical number, 4 for Ni and 8 for Pd, two-dimensional structures become more stable. Thus growth of two-dimensional islands becomes possible above these critical numbers. On the Ir (100) surface, Schwoebel & Kellogg[184] find that the critical number is 6. For 5 or fewer atoms the stable configuration is that of a linear chain oriented in a [110] direction. For 6 or more atoms, the stable configuration is a two-dimensional island. Two-dimensional metastable structures of clusters of less than 5 atoms and one-dimensional metastable structures of clusters of more than 6 atoms can be formed at low temperatures. But upon annealing they always return to the stable structures. Figure 4.37 gives two examples. In (*a*), an island of 6 atoms equilibrated at 460 K is shown. An atom is then field evaporated from the island at 77 K. Upon annealing to 450 K, the atoms reconfigure to a linear chain as shown in (*c*). In (*d*), a one-dimensional, 6-atom chain is formed by aggregating adatoms at 315 K. However, upon annealing to 450 K, a stable two-dimensional island is formed. This is quite similar to the structure transformation found for W_6 on the W (110), although in such a case both structures are two-dimensional. The fact that the shape of a fairly large island can be changed by adding or subtracting an atom from the island indicates that the atomic interaction is long ranged. It is possible to derive information on adatom–adatom interaction by studying the stable structures of islands of different sizes, or to try to correlate formation of an island, or formation of a surface layer with the adatom–adatom interaction.

The energetics in the formation of a superstructure of a surface layer can be understood from FIM studies. A correlation between adatom–adatom interaction and the atomic structure of an adsorption layer has

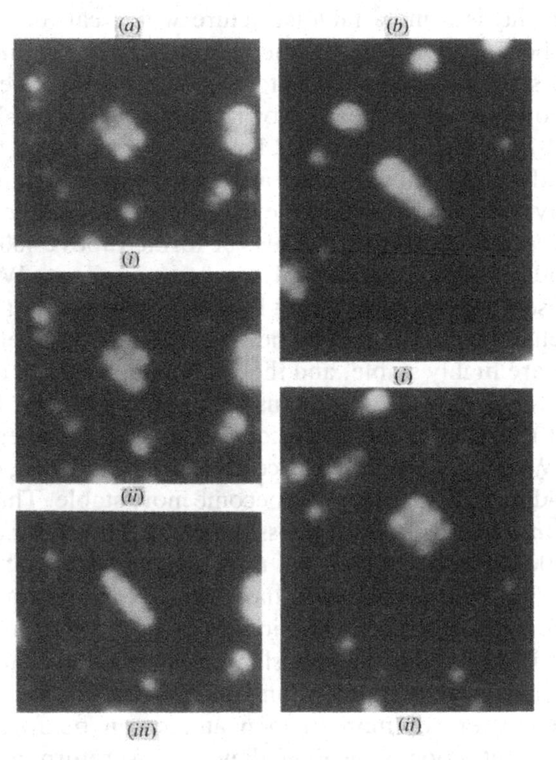

Fig. 4.37 (*a*) On the Ir (001) surface, for an island of fewer than 6 Ir atoms, a two-dimensional structure is metastable, whereas the linear chain structure is stable. (*i*) An island of 6 Ir atoms equilibrated at 460 K; (*ii*) one atom is field evaporated from the 6-atom island; (*iii*) upon heating to 450 K, the two-dimensional five-atom island reconfigures to a linear chain. (*b*) For an island of 6 or more Ir atoms, the linear chair is metastable whereas the two-dimensional island is stable. (*i*) 6 Ir atoms in a linear chain, equilibrated at 315 K; (*ii*) upon heating to 450 K, the island transforms into a two-dimensional structure. From Schwoebel & Kellogg[184].

been found for Si on the W (110),[174] as already mentioned in Section 4.2.5*b*. From the pair energy data measured for Si–Si pair interactions on the W (110) surface, shown in Fig. 4.34, one can easily see that this interaction has two relative minima, one at the third nearest neighbor bond, 4.47 Å parallel to [110], and one at the sixth nearest neighbor bond of the substrate, 6.32 Å parallel to [001]. One can therefore anticipate that when a large number of Si atoms are deposited on a W (110) surface, the equilibrium structure of the adsorption layer or the island should be a conventional p(2 × 1) structure of the W (110). The unit cell of this structure has the same two mentioned bonds of the two relative minima in the Si–Si interaction potential. The two-dimensional binding energies per atomic pair for different adlayer structures, calculated from the Si–Si pair energy data, are listed in Table 4.6. Indeed the p(2 × 1) structure has the lowest free energy, and thus should be the stable adlayer structure. An experimental result is shown in Fig. 4.38. When about one quarter of a monolayer of Si atoms is deposited on a W (110) surface, upon annealing to 300 K for 50 s an ordered structure of the adlayer is formed which indeed has the anticipated p(2 × 1) structure.

A Monte Carlo simulation has been carried out which uses the pair energies shown in Fig. 4.34.[185] This simulation also finds the equilibrium structure, at less than a quarter monolayer of Si adatoms, to be the p(2 × 1) structure. At a larger adatom coverage other structures can be expected,[186] but have not yet been studied experimentally. For Ir adatoms on the W (110) surface, a minor adjustment of pair energy data shown in Fig. 4.33*b* produces[156] both stable and metastable island structures of Ir on the W (110) surface found by Fink & Ehrlich.[187] Since the accuracy of FIM pair potential energy measurements is still limited, comparing observed adlayer structures with a Monte Carlo simulation is a complementary method of deriving adatom–adatom interaction.

Table 4.6 *Relative binding energies per adatom in a two-dimensional Si adlayer of various structures on the tungsten* {110}

Adlayer structures	Binding energies
1 × 1	−98 meV
2 × 1	−62 meV
c(2 × 2)	−3 meV
$\left(\frac{2\sqrt{2}}{\sqrt{3}} \times \frac{4}{\sqrt{3}}\right)$R35.26°, or p(2 × 1) in LEED literatures	55 meV

Data from ref. 156.

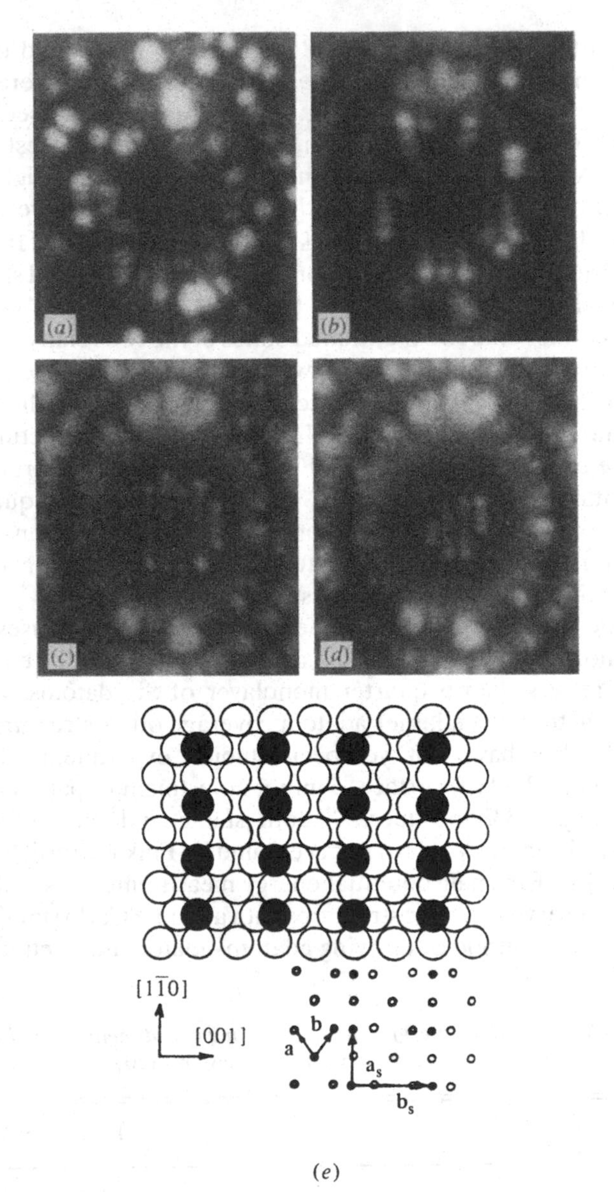

(*e*)

Fig. 4.38 (*a*) 'As deposited' Si adsorption layer on W (110) surface. (*b*) Ordered superstructure, $p(2 \times 1)$, is formed after annealing the layer at 300 K for 50 s. (*c*) and (*d*) Gradual field evaporation of the Si layer reveals the atomic structure of the adsorption layer more clearly. (*e*) Schematic of the $p(2 \times 1)$ structure of the Si adsorption layer.

The Si adlayer superstructure identified in the FIM study is in agreement with a LEED study by Boiko *et al.*[188] They find that at coverages less than 0.25, silicon adatoms form a $p(2 \times 1)$ superstructure on the W (110). At higher coverages, a series of superstructures are formed. Of course, when the degree of coverage exceeds 0.25, the $p(2 \times 1)$ structure can no longer accommodate all the adatoms, and a new structure of lowest free energy at the given degree of coverage will be formed. FIM pair energy data are not yet accurate enough to account for these structures. Also, when the degree of coverage is high, adatom separations become very small, and many body effects may become important. Einstein[189] finds that as much as 25% correction has to be made for a triatomic cluster with a nearest neighbor bond configuration. It is possible that many body effects are important only if adatom distances become very small. We have to recognize that the majority of physical phenomena have been understood with the explicit assumption that pair potential energies are additive. FIM experiments, of course, can easily be adapted to study quantitatively many body effects by comparing the radial distribution functions obtained with two and three adatoms on a plane and vice versa. Data analyses will be very involved and experiments will be very tedious. But even at this stage of the studies, FIM works have been able to contribute to a better understanding of the energetics of superstructure formation of surface layers.

4.2.7 Studies of the behavior of single atoms by other microscopies

Observations of surface diffusion of single atoms and molecules and mass transport diffusion have also been reported from electron microscope and scanning tunneling microscope studies. In transmission electron microscopy, resolution of better than 2 Å is now quite common. However, this resolution can be achieved only in the lattice imaging mode, or imaging arrangements of atomic rows in a very thin film. Thus jumping of atomic rows (or single atoms?) in the (1×1) to (1×2) reconstruction of the Au (110) surfaces, at edges of very thin Au films, can be directly seen.[13] It is impossible to distinguish whether single atoms or atomic rows (of what sizes?) are jumping, and neither can one be sure that these jumps are not produced by the bombardment of electrons. Takayanagi *et al.*[190] showed migration of Au atoms aggregating into small atomic clusters on carbon films. For direct imaging of single atoms and molecules, a scanning transmission electron microscope (STEM) with a field emission gun, first developed by Crewe and co-workers, has been very successful.[191] The resolution, ~2.5 Å, is comparable to that of the FIM. To achieve an adequate image contrast, however, only high-Z atoms adsorbed on low-Z thin substrate film can be studied.

An observation of motion of single atoms and single atomic clusters with STEM was reported by Isaacson *et al.*[192] They observed atomic jumps of single uranium atoms on a very thin carbon film of ~15 Å thickness or less. Coupled motion of two to three atoms could also be seen. As the temperature of the thin film could not be controlled, no Arrhenius plot could be obtained. Instead, the 'Debye frequency', kT/h, was used to calculate the activation energy of surface diffusion, as is also sometimes done in field ion microscopy. That the atomic jumps were not induced by electron bombardment was checked by observing the atomic hopping frequencies as a function of the electron beam intensity.

Quantitatvely more reliable amounts of data were later collected by Utlaut.[193] Activation energies are again derived by using the 'Debye frequency' as the prefactor. The activation energies for Ag, $CdCl_2$, In, Au, $AuCl_3$, UO_2Cl_2 and UO_2Ac_2 and their dimers are measured; these energies are, surprisingly, all in a narrow range from 0.83 to 0.90 eV. Utlaut also measured the radial distributions of single atoms (or molecules) for Ag, Cd, In and UO_2Cl_2, which were similar to those reported in an early FIM work for Re adatoms on the W (110) surface.[173] His result gives the potentials of mean force at the closest bond for In and UO_2Cl_2 to be, respectively, 150 and 90 meV, comparable to those obtained from FIM works. The potentials are long ranged and weak, and seem to show non-monotonic distance dependence as is also found in the early FIM work. Since the particle density is very low, the potentials of mean force should approximate well the pair energies, especially at small interatomic separations.

In scanning tunneling microscopy, surface diffusion of Au on a nanometer scale is found to occur at room temperature by Jaklevic & Elie[194] when a Au surface is intentionally spot damaged by the STM scanning tip. New atomic layers are formed around the damaged spots. Tsong *et al.*[195] also found rapid three-dimensional changes of clusters and nanostructures of Pt and Au surfaces and thin layers of W, Pt and Au, deposited on carbon films and mica, at room temperature in atmosphere without artificially inducing lattice defects on the surfaces and thin films. Ganz *et al.*[196] claimed to have seen rearrangement of Au atoms on a graphite surface at room temperature in UHV. At the present moment, none of these diffusion studies is quantitative, and it is very difficult to tell when the situation will change. In electron microscope studies, atomic behavior is always studied for heavy atoms deposited on a very thin carbon film of only a few atomic layers. It is very difficult to see how the phonon spectrum of such a thin film can be similar to the bulk. Also, what do the temperature and thermodynamic mean with a film of a few atomic layers in thickness? In FIM studies of surface diffusion, the activation energies are very specific both to the chemical species of the adatoms and substrates, and also to the atomic structures of the substrates. In the work of Utlaut, the activation energies measured for all adatom species are, to within the experimental uncertainties, all identical at ~0.88 eV. It is not clear whether this is due to the effect of thin carbon films or to that of electron bombardment, even though the latter effect was ruled out by these authors. In FIM and STM studies of surface diffusions, samples used should represent well those of a bulk. As STM is very sensitive to thermal drift, it is difficult to see how the sample temperature can be controlled for quantitative studies of single atom behavior on surfaces. On the other hand, STM is capable of operating in air or liquid with atomic resolution, and behavior of single atoms and atomic clusters in different environments can be studied.

4.3 Some properties of adsorbed atoms

4.3.1 Charge distribution and electronic density of states

FIM can also be used to study properties, such as the surface induced dipole moment and the effective polarizability of some surface atoms, kink site atoms and adsorbed atoms etc. The charge distribution of a surface atom is obviously completely different from that of a free atom because of its interaction with the surface and in addition surface atoms are partially shielded by itinerant charges of the surface. The charge distribution of a surface atom can be described by the magnitudes of the electric multipoles of the atom.

When an atom of a spherical charge distribution is brought near a solid surface, the atom–surface interaction will deform the electronic charge distribution or a charge transfer may occur between the atom and the surface. In general both processes will occur. A dipole moment will then be induced on the atom, or the atom–substrate system.[197] The magnitude of the induced dipole moment depends on the nature of the interaction, and also on how polarizable the atom is. For physisorbed atoms, deformation of electron orbitals should be the main effect, whereas for chemisorbed atoms the charge transfer effect should be more important. FIM studies of adatoms are concerned with mostly chemisorbed atoms. We will consider how the charge distribution of an adatom is changed by the atom–surface interaction and the effect of the applied field.[198,199] In Fig. 4.39(*a*), a simplified view of a generally accepted theory of chemisorption is shown. The presence of the adatom–surface interaction broadens an originally sharp energy level ε_a of the free atom into a band of width $\sim 2\Gamma$, and also shifts the center of the band to ε'_a. The band can be described by a density of states function $\rho_a(\varepsilon)$. The net charge on the adatom is given by

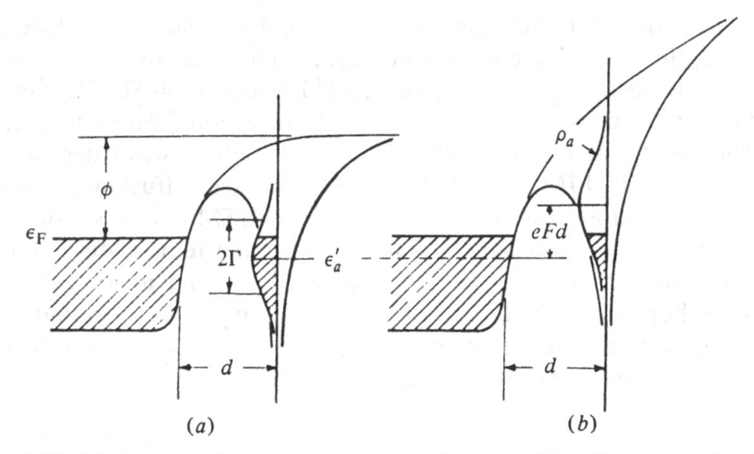

Fig. 4.39 (*a*) A generally accepted view of chemisorption. The originally sharp atomic energy level is broadened and shifted by adatom–surface interactions. (*b*) In an applied positive electric field F, the localized adatom band is shifted upward by an amount eFd.

$$q = e\left(n_0 - \int_{-\infty}^{\varepsilon_F} \rho_a(\varepsilon)\, \mathrm{d}\varepsilon\right), \tag{4.89}$$

where ε_F is the Fermi level and n_0 is the occupation number of the ε_A level in the free atom. The surface induced dipole moment of the atom is then given by

$$\mu_0 = qd, \tag{4.90}$$

with d representing the distance between the centers of the positive and negative charges.

The amount of the charge transfer can be changed by an applied field as can be seen in Fig. 4.39(*b*). When an electric field is applied to the metal surface, the atomic band will be shifted upward by eFd. The dipole moment of the atom is now field dependent, and is given by

$$\mu(F) = ed\left(n_0 - \int_{-\infty}^{\varepsilon_F - eFd} \rho_a(\varepsilon)\, \mathrm{d}\varepsilon\right), \tag{4.91}$$

where the electronic density of states has been modified by the applied field. A new definition of the polarizability of metallically chemisorbed atom may be defined as

$$\alpha(F) = \frac{\mathrm{d}\mu(F)}{\mathrm{d}F} = -\frac{\mathrm{d}}{\mathrm{d}F}\left(ed \int_{-\infty}^{\varepsilon_F - eFd} \rho_a(\varepsilon)\, \mathrm{d}\varepsilon\right). \tag{4.92}$$

The charge transfer induced polarizability of a surface atom is closely related to the electronic density of states of the atom as can be seen by considering

$$\alpha(F) = \lim_{\Delta F \to 0} \left(-ed \int_{-\infty}^{\varepsilon_F - e(F+\Delta F)d} \rho_a(\varepsilon)\, d\varepsilon + ed \int_{-\infty}^{\varepsilon_F - eFd} \rho_a(\varepsilon)\, d\varepsilon \right) \Big/ \Delta F$$

$$\approx e^2 d^2 \rho_a(\varepsilon_F - eFd). \tag{4.93}$$

Thus

$$\rho_a(\varepsilon_F - eFd) \approx \alpha(F)/e^2 d^2. \tag{4.94}$$

This is a surprisingly simple result where the electronic charge density of states of a surface atom at eFd below the Fermi level is related to the polarizability of the atom measured at that field. Therefore it is possible to investigate the adatom electronic band by measuring the dipole moment and the polarizability of the adatom as a function of the electric field.

4.3.2 Surface induced dipole moment of adsorbed atoms

The dipole moment of adsorbed atoms can be derived by measuring the change in work function as a result of the presence of the adatoms. In macroscopic measurements, the states of the deposited adatoms are unknown. They may combine into clusters of different sizes and some of them may be absorbed into lattice steps. One can also use the field emission microscope for this purpose. However, similar uncertainties exist. To achieve a well characterized surface and a known number of individual adsorbed atoms on a surface, a combined experiment with field ion microscope observations and field emission Fowler–Nordheim (F–N) plots has been most successful.[198,200]

The experimental procedure is illustrated in Fig. 4.40. An atomically flat small surface is developed by low temperature field evaporation. With an external gimbal system, the tip orientation is adjusted until the (110) surface falls into a Faraday cap. The helium image gas is pumped out, and a F–N plot is derived by reversing the polarity of the tip. With the tip voltage turned off again, a desired number of adatoms are deposited on the (110) surface. The number of adatoms on the surface is examined in FIM mode by admitting the image gas into the system and applying the image voltage to the tip. This number is controlled by repeated deposition and controlled field desorption, as discussed earlier. The system is reversed again to the field emission mode, and a F–N-plot is measured again, now with the presence of a known number of adatoms. From the change in the slope of the two F–N-plots, shown in Fig. 4.40(*e*), the dipole moment of a deposited adatom can be calculated.

The field emission current from an area A of a field electron emitter surface of work function ϕ_0 at a field F is given by

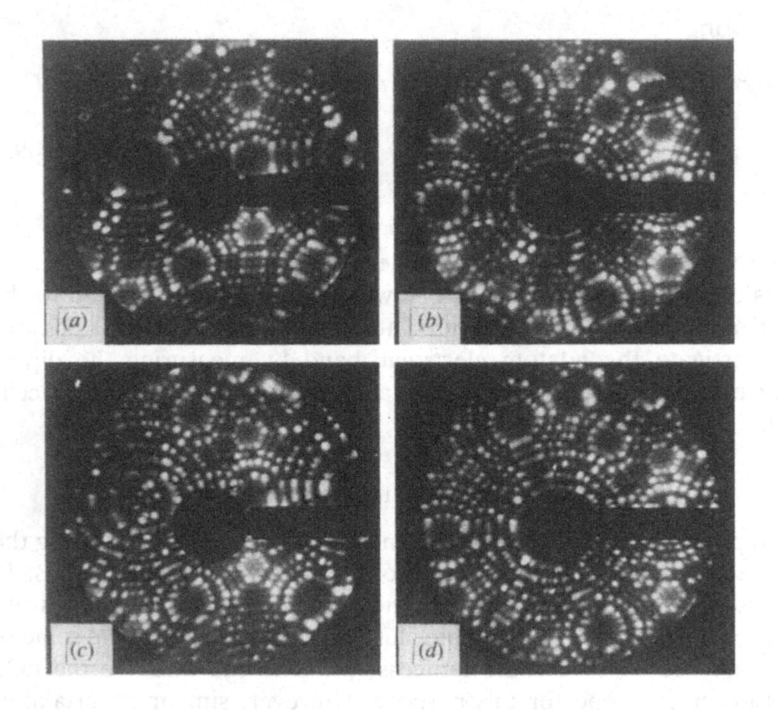

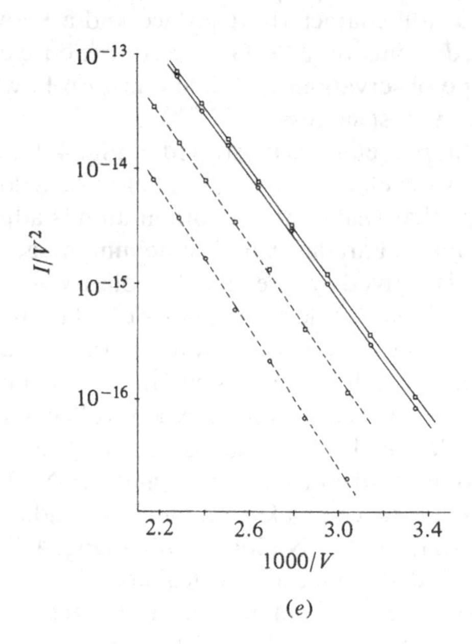

(e)

$$I = \frac{BA(\beta V)^2}{\phi_0} \exp\left(-\frac{b\phi_0^{3/2}}{\beta V}\right), \tag{4.95}$$

where β is a constant defined by $F = \beta V$, V is the applied voltage, and B and b are constants independent of the applied voltage. An F–N-plot, ln (I/V^2) against $1/V$ gives a straight line with a slope of

$$m_0 = -b\phi_0^{3/2}/\beta. \tag{4.96}$$

When the same surface has deposited on it n adatoms per unit area, each with a dipole moment μ and polarizability α, the work function in the presence of an applied field βV changes to

$$\phi_n = \phi_0 - 4\pi n\mu + 4\pi n\alpha\beta V. \tag{4.97}$$

If the number density of adatoms is not too large and the applied field is in the field electron emission range, one can show that, in general, $4\pi n\mu \ll \phi_0$, $4\pi n\alpha\beta V \ll \phi_0$ and $\alpha\beta V \ll \mu$. The slope of the F–N plot is therefore given by

$$m_n \approx -\frac{b}{\beta}(\phi_0 - 4\pi n\mu)^{3/2}. \tag{4.98}$$

Thus

$$\mu = \frac{\phi_0}{4\pi n}\left[1 - \left(\frac{m_n}{m_0}\right)^{2/3}\right]. \tag{4.99}$$

By comparing the slopes of F–N-plots, as shown in Fig. 4.40(*e*), with and without adatoms on the surface, the dipole moments of adatoms are then derived. Table 4.7 lists some of the data obtained by different investigators. These values are in fair agreement with those obtained by other macroscopic techniques.[201] One must recognize, however, that values obtained with macroscopic techniques are measured without the knowledge of whether these are single adatoms, or whether these adatoms have already been combined into clusters of different sizes, while some of them may in fact have been absorbed into lattice steps. FIM experiments have

Fig. 4.40 Procedures used in measuring the dipole moment of adatoms by an FIM–FEM combination experiment. (*a*) A clean W surface is prepared by low temperature field evaporation and characterized by the FIM image. (*b*) The (110) plane is lined up with a Faraday cap using an external gimbal, the image gas is pumped out and the polarity of the tip is changed. A FN plot is then taken. (*c*) Adatoms are deposited on the (110) surface and their number found by the field ion image, in this case 9 adatoms. (*d*) A FN plot is taken again from the same surface area but now with the presence of 9 adatoms on the surface. (*e*) Two sets of data collected from two different (110) planes are shown (circles are for clean W (110), and squares are with 10 adatoms). The solid lines are for W adatoms and the broken lines are for Ta adatoms.

Table 4.7 *Experimental results for the adatom dipole moments and polarizabilities*

Adatoms	Dipole moments[a] Debye	Polarizability[b] $Å^3$
Ta	0.5	11.9
W	0.2; 0.7;[c] 1.0[d]	6.8, 14.4[d]
Re	0.1	4.0
Ir	0	3.3
Pt	0	2.7

[a] From ref. 199. [c] From ref. 200.
[b] From ref. 198, 199. [d] From ref. 203.

shown clearly that adatoms will combine into clusters at the room temperature where these experimental measurements are done. Such ambiguity does not exist in the FIM–FEM combination experiments just described.

4.3.3 Effective polarizability of surface atoms

The effective polarizability of surface atoms can be determined with different methods. In Section 2.2.4(*a*) a method was described on a measurement of the field evaporation rate as a function of field of kink site atoms and adsorbed atoms. The polarizability is derived from the coefficient of F^2 term in the rate vs. field curve. From the rate measurements, polarizabilities of kink site W atoms and W adatoms on the W (110) surface are determined to be 4.6 ± 0.6 and 6.8 ± 1.0 $Å^3$, respectively. The dipole moment and polarizability of an adatom can also be measured from a field dependence of random walk diffusion under the influence of a chemical potential gradient, usually referred as a directional walk, produced by the applied electric field gradient, as reported by Tsong *et al.*[150,198,203] This study is a good example of random walk under the influence of a chemical potential gradient and will therefore be discussed in some detail.

It is well known that in an applied field the field strength near the center of a plane is smaller than near the edge of the plane. The field gradient across a plane can be measured from the desorption field of adatoms deposited at different locations of the plane. For example, the field gradient on the largest obtainable W (110) plane, developed by slow field evaporation at 10^{-1} to 10^{-2} layer/s, has been measured.[156,202] The field strength differs by as much as ~14% from the center to the edge of the plane. The field gradient is nearly constant except near the center and edge of the plane. The potential energy of an adatom on a plane is shown schematically in Fig. 4.41. Although we do not know a detail form of the periodic surface potential, one may, to a first approximation, include only the first harmonic term $U(\rho, 0) = \frac{1}{2}E_d(1 - \cos 2\pi\rho/l)$, where ρ is the radial distance from the center of the plane. Because of the symmetry of this potential, the jumps of an adatom are randomly directed. In an applied field the polarization energy must be included:

$$U(\rho, F) \approx U(\rho, 0) - \mu_0 F(\rho) - \tfrac{1}{2}\alpha F(\rho)^2, \tag{4.100}$$

where the field strength is given approximately by $F(\rho) \simeq F_c + \beta' F_c$, F_c is the field at the center of the plane and β' is the field gradient. The driving force due to the non-uniform field can be found from the gradient of $U(\rho, F)$ to be

$$f \simeq \mu_0 \beta' + \alpha \beta' F_c \tag{4.101}$$

if the higher power terms of β' are neglected. Substituting f into eq. (4.45) one obtains

$$\langle \rho \rangle_F \approx 2\bar{N} l \sinh\left(\frac{\mu_0 \beta' l + \alpha \beta' F_c l}{2kT}\right), \tag{4.102}$$

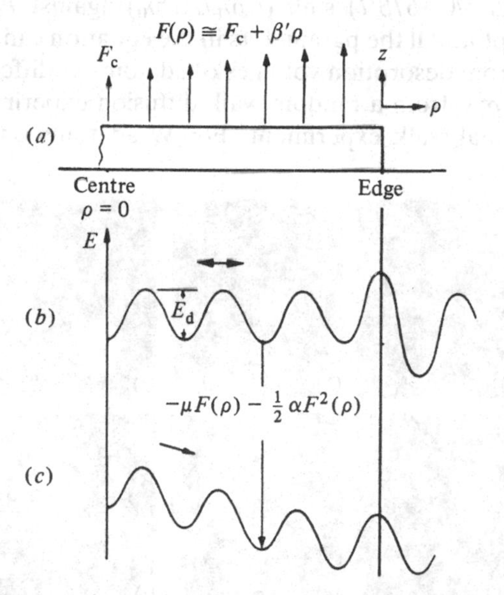

Fig. 4.41 (*a*) Cross-sectional view showing field variation across a plane. (*b*) In the absence of an applied field, the surface potential of an adatom, except at the plane edges, is symmetric. Atomic jumps are symmetric. (*c*) In an applied field, the surface potential of an adatom becomes inclined owing to the additional polarization binding. Atomic jumps are now asymmetric.

With a few rearrangements, one has

$$\mu_0 + \alpha F_c = \frac{2kT}{\beta' l} \sinh^{-1}\left(\frac{l\langle\rho\rangle_F}{2\langle\rho^2\rangle_0}\right), \tag{4.103}$$

where $\langle\rho^2\rangle_0$ is the mean square displacement of the adatom in the absence of the applied field in the same time period at the same temperature. In deriving the equation we have assumed that the polarizability of the diffusing atom does not change along the diffusion path, or the polarizability of the adatom has the same value at the rest sites and at the saddle points of the diffusion path. This assumption should be quite reasonable for smooth planes like bcc {110} and fcc {111}. For rough surfaces like bcc {111}, a correction accounting for this change has to be made. The one-dimensional analysis discussed above may be diectly applied to two-dimensional planes with a minor modification. Fortunately the directional walk of an adatom is very much one-dimensional even on a two-dimensional plane, as can be seen from Fig. 4.42. The effect of restricting the adatom to jump along the surface channels of the W (110), for an example, can be accounted for simply by replacing β' with $\sqrt{3}\beta'/2 = 0.867\beta'$.

One can obtain values of both μ_0 and α from the intercept and the slope of a plot of $(2kT/0.867\beta' l)$ sinh $(l\langle\rho\rangle_F/2\langle\rho^2\rangle_0)$ against F_c, for brevity referred as a τ-plot. All the parameters in the equation can be measured: field gradient from desorption voltages of adatoms at different locations on the plane, $\langle\rho^2\rangle_0$ from a random walk diffusion experiment, and $\langle\rho\rangle_F$ from a directional walk experiment. For W adatoms on the W (110)

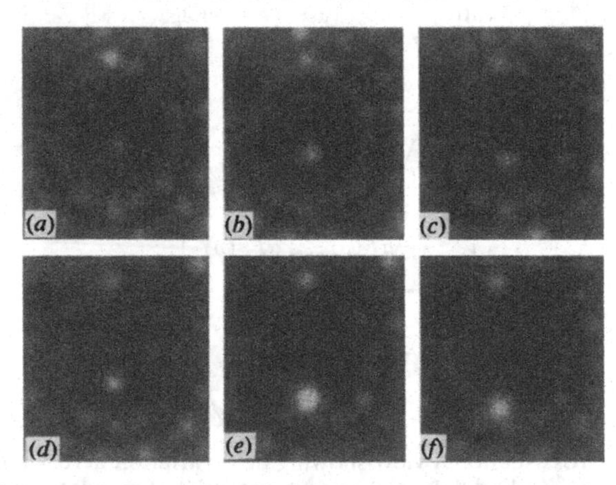

Fig. 4.42 Room temperature, argon promoted helium field ion images showing the directional walk of an adatom from the center of the plane toward the edge of the plane. The adatom drifts along the direction of the maximum field gradient, or in the radial direction of the plane.

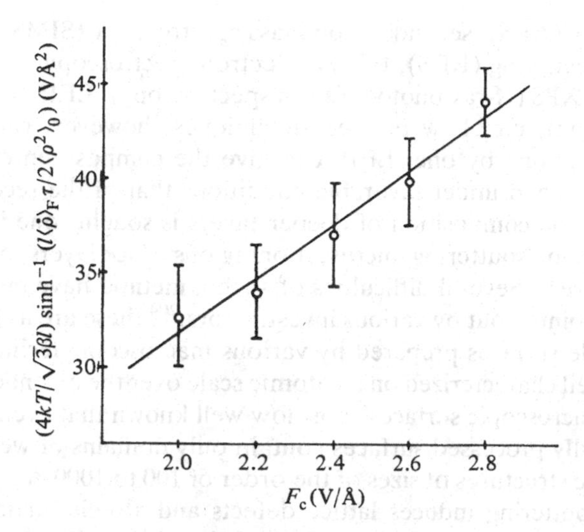

Fig. 4.43 τ-plot for the direction walk of a W adatom on a W (110) surface.

surface, such an experiment has been done as shown in Fig. 4.43.[203] From this plot, the dipole moment of W adatoms is found to be about 0.97 Debye, and the effective polarizability is about 14.4 Å^3. As the experiment involves many measurements of single atom events which usually have large statistical uncertainties, the accuracy of the method is still very limited, and values of μ_0 and α obtained by different measurements vary over a wide range. Nevertheless, the applicability of eq. (4.103) and the chemical specificities of adatom dipole moment and polarizability have been demonstrated in these preliminary studies. The effective polarizabilities of 5-d transition metal adatoms on the W (110) surface are found to decrease monotonically from Ta to Pt, in great contrast to the activation energies of surface diffusion of these atoms on the same surface shown in Fig. 4.27.

4.4 Compositional analysis of surface layers

Basic information needed to understand the physical and chemical properties of solid surfaces and thin solid films include the atomic structures and the compositional variations across the surface and interface layers. The atomic structures can be studied with microscopies and with surface sensitive diffraction and particle scattering techniques. Compositions of surfaces and thin films can be studied with the atom-probe FIM. In general, however, compositional analyses are mostly done with surface sensitive macroscopic techniques, such as auger electron spectroscopy (AES), secondary ion mass spectroscopy (SIMS), Rutherford back scattering (RBS), ESCA (electron spectroscopy for chemical analysis) or XPS (X-ray photoemission spectroscopy), LEIS (low energy ion scattering), etc. Few of these techniques, however, can analyze surface layers one by one. LEIS can give the composition of the top surface layer, and under favorable conditions that of the second layer also. When the composition of deeper layers is sought, one invariably resorts to an ion sputtering microsectioning of surface layers, or through interface layers. Several difficulties of such a method have been recognized and pointed out by various investigators;[204] these are as follows:

1. Sample surfaces prepared by various macroscopic techniques are not well characterized on an atomic scale over the extended area of the macroscopic surfaces. It is now well known that even the most carefully processed surfaces contain only domains of well defined atomic structures of sizes of the order of 100 to 1000 Å.
2. Ion sputtering induces lattice defects and atomic mixing among surface layers. Sputtering yield is different for different chemical species. Thus the composition of a sputtered surface is not necessarily the true composition of that layer.
3. In most surface sensitive spectroscopies, the depth resolution is still much larger than one atomic layer. Thus even under the best conditions a true atomic layer depth resolution cannot be achieved in the compositional analysis.
4. Most surface sensitive spectroscopies are insensitive to some chemical species; for example AES is insensitive to hydrogen, etc.

Most of these difficulties encountered in macroscopic surface analytical techniques are not present in the atom-probe analysis. The composition depth profile of a sample can be derived by a time-of-flight mass analysis of single field evaporated atoms. At low temperatures, field evaporation proceeds atom by atom from the plane edges, and thus also atomic layer by atomic layer. It induces no atomic mixing among surface layers, and true atomic layer depth resolution is achieved. Preferential field evaporation of an alloy species can be avoided in HV pulse atom-probe analysis if the high voltage pulses used for field evaporation are higher than 15 per cent of the total voltage.[205,206] Thus the composition derived for an atomic layer is the true composition of that layer. The capability of the atom-probe in the compositional analysis of single atomic layers was first utilized in a study of the compositions of superlattice layers and fundamental layers of an ordered Pt_3Co,[207] and later in studies of surface and impurity segregation of alloys.[208,209] The atomic layer by atomic layer compositional analysis is now a common practice in atom-probe field ion microscopy. As a composition depth profile derived from an atom-probe measurement has a true atomic layer depth resolution, as compared with the average depth resolution in macroscopic techniques, and the composition is directly calculated from the ion counts of the alloy components, as compared with a measurement of signal strengths of the alloy components in macroscopic techniques, the composition depth profile obtained by atom-probe field ion microscopy is referred by Tsong and co-workers as the 'absolute composition depth profile'.

In general, two different methods can be used to analyze the absolute composition of surface layers in ToF atom-probe field ion microscopy. These methods and volumes sampled are illustrated in Figs 4.44(*a*) and (*b*). The simplest method is to aim the probe-hole near the center of a crystal plane one intends to analyze. The position of the probe-hole is fixed during the entire measurement. As the pulsed-field evaporation proceeds slowly, mass signals from two successive layers are separated by many pulses without detecting an ion between them, i.e. when the plane edge is outside the area covered by the probe-hole. If this method is used to analyze the fundamental planes of a perfectly ordered binary alloy AB, then the sequence of detection of A and B atoms, or ions, will be that shown in (ii). In this method, the number of ions detected per layer is small unless the probe-hole is large. Also, no edges are included in the compositional analysis. Almost all atom-probe microanalyses reported in the literature use this method because of its simplicity: it requires no continuous adjustment of the aiming of the probe-hole.

A much more tedious method, but a better one, is that shown in Fig. 4.44(*b*). The probe-hole is first aimed at the edge of the top surface layer. As the top layer gradually reduces in size, the probe-hole position is readjusted accordingly to match with the plane edge. When the top layer is gone, the probe-hole is then aimed at the edge of the second layer, which is now the top layer. The advantages of this method are that a compositional variation within a surface layer can be detected, edge effects can be studied and also more ions can be detected from one surface layer without having a large probe-hole.

Let us discuss some of the pitfalls one must avoid to achieve a reliable compositional analysis in atom-probe field ion microscopy, some of which have already been discussed in Section 3.2.7. First, one must be sure that there is no preferential d.c. field evaporation of a species by the d.c. holding field. In a HV pulse atom-probe, field evaporation is carried out with nanosecond high voltage pulses, hence the field evaporation rate is at least 10^8 times higher in atom-probe operation than that in d.c. field evaporation. The field, or tip voltage, required is about 15% higher in ns pulsed-field evaporation than in d.c. field evaporation, as has already been discussed in Section 2.2.4(*a*). Thus unless the height of the HV pulses is at least ~15% higher than the d.c. holding voltage, a slow d.c.

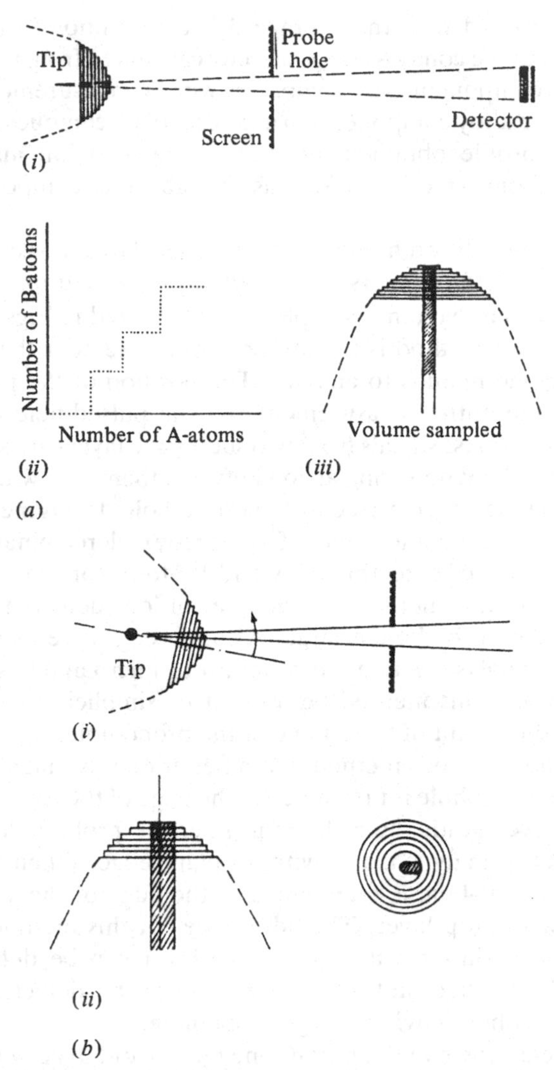

Fig. 4.44 (*a*) The simplest method in atom-probe analysis is to fix the probe-hole to a spot near the center of the plane one intends to analyze. For an ordered AB alloy, the cumulative number of A and B ions detected will look like (*ii*) if the plane is a fundamental plane. The volume sampled is shown in (*iii*). (*b*) Another method is having the probe-hole always aimed at the edge of the top layer. The volume sampled will be that shown in (*ii*).

field evaporation of the surface will occur concurrently with the pulsed field evaporation, and preferential d.c. field evaporation of an alloy species from the edges of a plane can occur. This will of course result in a false analysis of the composition of the sample. Indeed experiments find that a pulse fraction of about 20% is needed to avoid preferential d.c. field evaporation of a species from the plane edges of an alloy sample.[205,206] Second, the field evaporation rate has to be low, i.e. an ion is detected every ten or so pulses, to avoid signal overlaps in high resolution atom-probes, or else a statistical correction has to be done as discussed in Section 3.2.7. Third, a statistically reliable amount of data has to be collected regardless of the detection efficiency of the ion detector. The expected statistical uncertainty of a set of data can be calculated with eq. (3.16). As a rule of thumb, the statistical uncertainty is of the order of $100 \times \sqrt{N}\%$ where N is the number of ions collected in a sample set.

There are some other uncertainties in atom-probe compositional analyses, some of which are still poorly understood. Nishikawa *et al.*[210] and Sakurai *et al.*[211] find that when compound semiconductors are analyzed with a flight time focused HV pulse atom-probe, false compositions are almost always obtained. Nishikawa *et al.*[212] explain that the false compositions are resulted in part from a large energy deficit of ions in HV pulse field evaporation of semiconductors and in part from preferential field evaporation of a chemical species. Mass signals can spread over a large mass range, or mass lines are widely dispersed. If one carefully includes ions in all these tails of mass lines, reliable compositions can still be obtained for GaAs, but not for GaP. The explanation given by Nishikawa *et al.* is quite reasonable. Correct compositions for semiconductors can be obtained when laser pulses are used for the atom-probe operation.[213] In pulsed-laser stimulated field evaporation, if the laser power is not excessively high then ion energy spread does not occur and mass lines are sharp. For metallic alloys, on the other hand, the situation seems to be reversed. While correct compositions of alloys can be obtained with HV pulse atom-probes simply by using a pulse fraction of 15% or higher, when pulsed-laser atom-probes are used for the same purpose the compositions derived for alloys become unreliable. There are two possible explanations. First, the laser power used in these studies may be too high and the emitter surface is thus excessively heated by the laser pulses. Atoms of one of the species may diffuse by a directional walk to a particular region of the emitter surface which is not covered by the probe-hole, and be field evaporated from there. Second, if one tries to use very low laser power, the d.c. holding voltage will have to be properly raised, otherwise pulsed field evaporation will not occur. Then one of the species may be preferentially field evaporated from plane edges by the d.c. holding field. At the moment, for the compositional analysis of poor electrical conductors, the pulsed-laser operation is recommended. For the compositional analysis of metallic alloys, the high voltage pulse operation is recommended. Whether or not the atom-probe can give the correct composition of a sample under the operating conditions can of course be checked by simply analyzing samples of known compositions. In connection with a study of surface segregation and impurity cosegregation of Pt–Rh alloys, Fig. 4.45 shows correct compositions obtained from an atom-probe compositional analysis, provided of course a proper operation condition of the atom-probe is first determined.

4.4.1 Surface segregation and impurity segregation

When an alloy is in thermodynamic equilibrium, the composition of the near surface layers may be quite different from that of the bulk, owing to segregation of one of the alloy components. Surface segregation is one of several interface segregation phenomena which can profoundly affect the physical and chemical properties of materials.[214] These phenomena are important from both a scientific and a technological point of view. Surface segregation and impurity segregation to the surface are particularly important for a basic understanding of interfacial segregation phenomena, as the solid–vacuum interface should be one of the simplest interfaces for theoretical treatments. Surface segregations will also change completely the catalytic properties of alloy catalysts.[215]

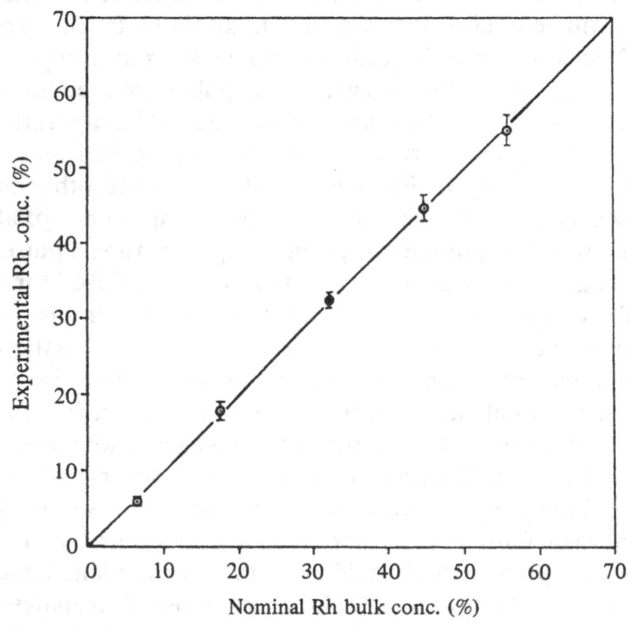

Fig. 4.45 The compositions derived for several Pt–Rh alloys are compared with the nominal compositions of these alloys. They agree very well, indicating that there is no preferential d.c. field evaporation of a species in these analyses.

Surface segregation in binary alloys, such as NiCu alloys, can be studied by changes in catalytic activities[216] and chemisorption properties[217] of the surfaces of binary alloy catalysts. These methods are, however, not really suited for studying surface segregation *per se* as the chemisorption process itself can change the segregation behavior of the alloys. Since then many surface analytical techniques have been used to study surface segregations. These include work function measurements,[218] XAPS,[219] AES,[220] LEIS,[221] UPS and XPS,[222] etc. One must remember that while surface composition changes can be detected by these surface sensitive macroscopic techniques, none of them can give a composition depth profile with true atomic layer depth resolution as already discussed in the last section. Atom-probe FIM, with its capability to analyze compositions of samples atomic layer by atomic layer, should be ideally suited for studying surface segregations.

Atom-probe analyses always start from the surface layers irrespective of whether or not the composition of surface layers is sought. Krishnaswamy *et al.*[223] applied the atom-probe to study the composition of alloy tips of stainless steel 410 after the tips were heated to high temperature between 500 and 900°C. They found a depletion of Cr in tips as a result of preferential thermal evaporation of Cr atoms from the tip surfaces. The kinetic was limited by bulk diffusion, thus the degree of depletion was found to depend on the heating temperature. They further argue that because of the small size of the tip, Cr is depleted from the tip by thermal evaporation at a much faster rate than macroscopic samples used in a study by Park *et al.*[219] Krishnaswamy *et al.* did not consider the depletion of Cr to be a surface segregation phenomenon but rather a preferential thermal evaporation of Cr from the bulk of the sample. Neither did they attempt to determine an equilibrium compositional depth profile, nor to collect ions only from the near surface layer. In fact, they calculated the composition from data collected in a volume of depth about 800 Å. Atomic layer by atomic layer atom-probe compositional analysis was initially reported by Tsong *et al.*[207] in a compositional and image analysis of superlattice layers and fundamental layers of an ordered Pt_3Co alloy, but again this experiment was not intended for studying surface segregations. Studies of equilibrium composition depth profiles of alloys in surface segregation were reported by Tsong and co-workers.[208,209] In their first study, compositions of near surface layers of alloys were analyzed for each surface layer with true single atomic layer depth resolution. However, owing to the insufficient annealing time and insufficient amount of data, no clear evidence of surface segregation was presented in their first attempt. Only some months later were they able to find enrichment of Cu in the (111) planes of microcrystals in Ni–5% Cu tips.[209] They[224] also found an enrichment of Cr in the top surface layer of stainless 410 as a result of Cr segregation to the surface, rather than a depletion in the near surface region of 800 Å depth found by Krishnaswamy *et al.*,[223] and in the study of Park *et al.*[219] Evidence of surface segregation in polycrystalline Ni–Cu alloys had already been reported earlier using macroscopic techniques, as mentioned before. These atom-probe data, however, are derived from specified single crystal facets even though the tips themselves are still polycrystals. Since then, atom-probes have been actively employed to study surface segregation and impurity co-segregation by Tsong and co-workers for a large number of alloys, to study surface segregation of Ni–Cu alloys by Sakurai and co-workers, and by others for different alloys. Recent macroscopic studies also use mostly single crystals. Before we discuss experimental results of these studies, we will consider why the composition of surface layers should be different from that of the bulk.

From a thermodynamic consideration by Gibbs,[225] one can expect surface segregation to occur if the total free energy of a system can be lowered by segregation of a species to the surface. For a dilute solution of B in A, this condition is described as

$$\Gamma_B = -\frac{1}{kT}\left(\frac{\partial\sigma}{\partial \ln x_B}\right), \tag{4.104}$$

where Γ_B is the surface excess concentration of B, x_B is the concentration of B in the surface and σ is the surface free energy. This indicates that if the surface free energy is lowered by an addition of solute B, Γ_B will be positive, and B atoms will segregate to the surface. Unfortunately this equation is not very helpful for predicting which of an alloy species will segregate to the surface since experimental data on compositional dependent surface free energy are largely unavailable. Many theories have been developed to either predict or explain segregating species in a large number of alloys, and many of these theories are very successful in predicting segregation species or some other general aspects of surface segregation such as a non-monotonic depth dependence of composition of near surface layers. Few theories however can explain details of equilibrium composition depth profiles of specific alloys.

In a broken bond model,[226] it is assumed that bond energies are pairwise additive. To form a surface, bonds are broken. The energy of the system is reduced if the weakest bonds are broken. This is accomplished if atoms of a weaker binding energy are placed on the surface. Thus in a binary alloy, the component with the lower sublimation energy should segregate to the surface. In the elastic strain theory, a strain is induced in the solid if the size of solute atoms differs considerably from the solvent atoms. To reduce the strain energy of the system, the solute atoms should segregate to the surface. These simple theories are found to be able to explain correctly available data only about half of the time. An atomic model and a modified atomic model have been developed which combine the two factors, bond strength and atomic size differences, in a self-consistent way.[227] The success rate for the modified atomic model is found to be about 85%. An electronic model in the tight binding approximation has been developed.[228] In this model, the segregation energy is deduced from the electronic densities of states. The driving force for segregation is the lowering of the internal energy ΔE of the system when a component is segregated to the surface. Despite many approximations used in the calculation, this theory is found to predict correctly the segregation species of 29 out of 31 alloys.

For our purpose here, the surface concentration of B, x_s, in a binary alloy $A_{1-x}B_x$ can be assumed to be given by

$$\frac{x_s}{1-x_s} = \frac{x}{1-x}\exp\left(\frac{Q_{seg}}{kT}\right), \tag{4.105}$$

where Q_{seg} is called the energy of segregation. Theories of surface segregation, basically, try to calculate Q_{seg} with different approaches and approximations.[229,230] The segregation behavior of an alloy surface obviously can be affected by chemisorption of a species on the surface.[231] One can expect the species which interacts more strongly with the chemisorbed species to be drawn to the surface. In the monolayer model, eq. (4.105) can be modified slightly to account for the effect of chemisorption:

$$\frac{x_s}{1-x_s} = \frac{x}{1-x}\exp\left[\frac{Q_{seg} + (E_b - E_a)\Theta}{kT}\right], \tag{4.106}$$

where E_a and E_b represent, respectively, the chemisorption energy of adsorbate on A and B metals, and Θ is the degree of coverage of the adsorbate.

(a) *Atom-probe studies of surface segregations*

In a surface segregation study a tip is first carefully degassed, then field evaporated to produce an end form. This surface is next examined in helium field ion images to make sure that there are no lattice defects in the sample, and a proper identification of crystal planes is made. The system is baked to achieve vacuum below 2×10^{-10} Torr. The tip is degassed and field evaporated to end form again. This tip is now ready for surface segregation study. Depending on what alloy the tip is made of, the tip is annealed to a desired temperature, usually between 550 and 750°C, for several minutes to equilibrate the distribution of alloy species in the near surface layers, and then quenched.As the tip volume is very small, the quenching rate is estimated to be over 10^3 °C/s. The temperature and the annealing time are chosen such that within an annealing period, the diffusion length of the alloy components in the bulk should be at least 20 to 100 Å, so that a near-equilibrium distribution of alloy components can be achieved. The diffusion length during the quenching should be less than, say, 0.2 Å, so that no redistribution of the alloy components can occur during the quenching process. Whether or not a composition depth profile obtained represents an equilibrium depth profile or not can easily be checked by changing either the annealing temperature or the annealing time.

In the first successful atom-probe study of surface segregation by Ng *et al.*,[209] a Ni–5% Cu was annealed to 550 ± 50°C for five minutes, and then quenched. After the annealing, a thermal end form of the tip was observed. The {111} poles developed into large facets. The images of the top {111} layers are found to be much dimmer. Atom-probe signals were obtained by slow field evaporation with the probe-hole always aiming at the very edge of the top surface layer, a procedure already described earlier. When the number of Ni ions are plotted against the number of Cu ions according to the sequence of detection of these ions, a lateral concentration profile (shown in Fig. 4.46) was obtained. The slope in each surface layer of this plot gives the composition of that layer. In addition, the first few ions detected come from the plane edges, and thus an edge effect can be investigated. While one can readily derive a composition depth profile from a set of data by simply plotting the slopes of these near surface layers against the depth, the number of ions collected from each layer is usually much too small to make such a depth profile quantitatively reliable. The measurement has to be repeated from several to about 20 times under identical experimental conditions. All the data are then combined. Two absolute composition depth profiles, one for the (111) plane of Ni–5% Cu and one for the (001) plane of Pt–5% Au, are shown in Fig. 4.47(*a*) and (*b*). It is clear that the segregant in this Ni–Cu alloy is Cu and that in this Pt–Au alloy is Au.

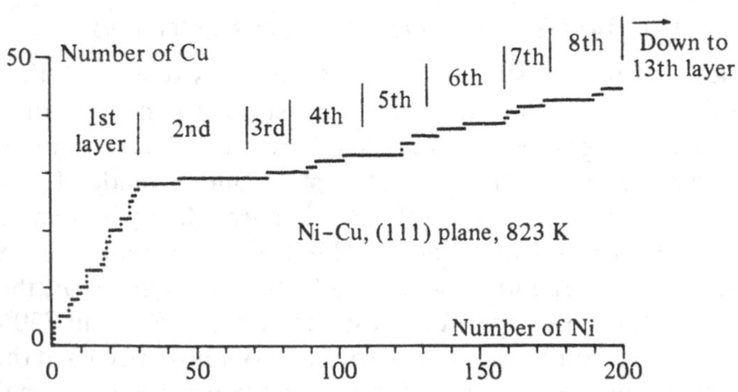

Fig. 4.46 Lateral composition profile obtained for a Ni–5% Cu annealed at 550°C. The first few ions detected are Cu. This may indicate an edge effect, or simply preferential field evaporation of Cu from plane edges of Cu atoms.

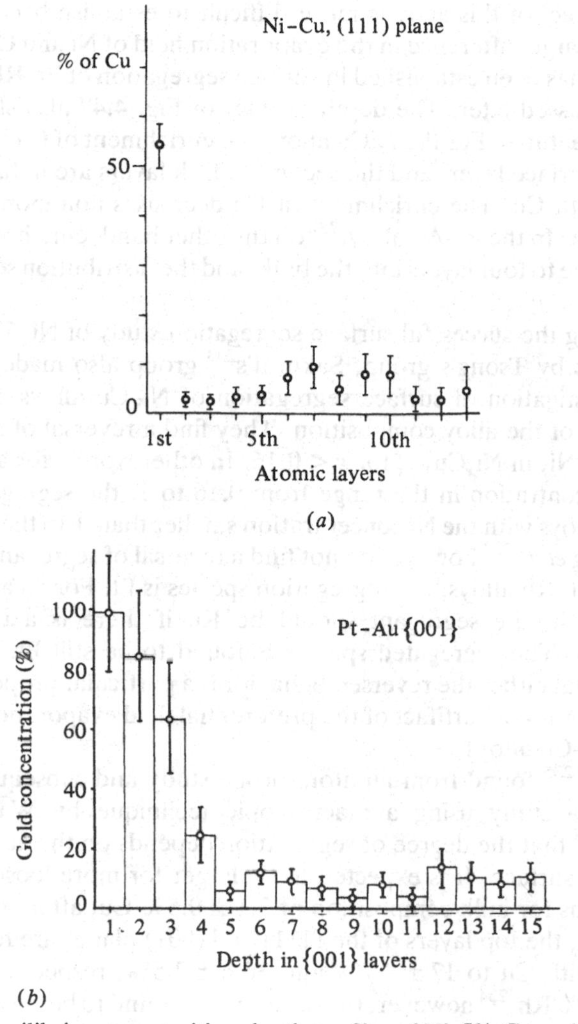

Fig. 4.47 Equilibrium composition depth profiles of Ni–5% Cu and Pt–5% Au. For Ni—5% Cu the enrichment of Cu is confined to the top layer. The enrichment of Au in Pt–5% Au extends to 3 to 4 layers in depth and shows an oscillatory tail.

Some interesting features can be noticed in these lateral concentration profiles and composition depth profiles. For the Ni–5% Cu alloy, the first few ions detected are Cu ions, even though the composition within the top layer is rather uniform. This can be due to an edge effect, in other words the edges of the (111) layers are much more enriched with Cu atoms than inside the top (111) layer. This can also be simply produced by a preferential field evaporation of Cu atoms from the layer edges. While the edge effect of this alloy is more difficult to establish because of an expectedly large difference in the evaporation field of Ni and Cu, a small edge effect has been established in surface segregation of Pt–Rh alloys as will be discussed later. The depth profiles of Fig. 4.47 also show some important features. For the NiCu alloy, the enrichment of Cu is confined to the top surface layer, and the second to fifth layers are in fact slightly depleted with Cu. The enrichment of Cu decreases non-monotonically into the bulk. In the Pt–Au alloy,[232] on the other hand, enrichment of Au extends three to four layers into the bulk, and the distribution seems to be oscillatory.

Following the successful surface segregation study of Ni–5% Cu and other alloys by Tsong's group, Sakurai's[233] group also made an atom-probe investigation of surface segregation of Ni–Cu alloys covering a large range of the alloy composition. They find a reversal of segregant, from Cu to Ni, in Ni_xCu_{1-x} for $x < 0.16$. In other words, for alloys with the Ni concentration in the range from 0.16 to 1, the segregant is Cu, while for alloys with the Ni concentration smaller than 0.16 the segregant is Ni. Tsong *et al.*[234] however do not find a reversal of segregant in Pt–Rh alloys. In Pt–Rh alloys, the segregation species is Pt. For an alloy of Pt–6.4 at.% Rh, the segregant should be Rh if there is a reversal of segregation. The segregated species is found to be still Pt. One must conclude that either the reversed behavior is a particular property of Ni–Cu alloys, or it is an artifact of the preferential field evaporation of Cu in Cu-rich Ni–Cu alloys.

Ng *et al.*[235] found from an atom-probe study and subsequently also found by a study using a macroscopic technique by Wandelt and Brundle,[236] that the degree of segregation depends on the atomic structure of the surface. It is expected to be larger for more loosely packed planes. Thus for bulk composition of 3.1 ± 0.5% Cu, after annealing at 650 ± 20°C, the top layers of the (111) and (001) planes are found to be enriched with Cu to 17 ± 2.7% and 36.6 ± 7.5%, respectively.[235] For Pt–44.9 at.% Rh,[234] however, the difference is found to be very small, if it exists at all. For the (111) and (001) planes the Pt concentrations are found to be 73.2 ± 5.0% and 71.1 ± 6.0%, respectively, or well within the statistical uncertainties of the experiments.

Another important question is whether surface segregation is enhanced at the edge of a plane. In fact a few people outside field ion microscopy continue to question the validity of comparing atom-probe data with data obtained by macroscopic techniques. They argue that the plane edges will greatly enhance the surface segregation and thus will make the atom-probe data unreliable. The advantage of the atom-probe in surface segregation study is that an edge effect, if it exists, can be found and studied. This, in fact, has been done for the (111) and (001) surfaces of a Pt–44.8 at.% Rh alloy.[241] For this purpose, the Pt concentration of the (111) plane edges is calculated from the first three ions detected from 20 sets of data. It is found to be 78.3 ± 5.3 at.%. This is only very slightly higher than the 73.2 ± 5.0 at.% average Pt concentration of the first layers if one considers the statistical uncertainties of these data. For the (001) plane, from 23 sets of data, Pt concentration at plane edges is found to be 82.6 ± 4.6 at.%, which is also only slightly larger than the 71.1 ± 6.0 at.% of the average Pt concentration of the first layers. A plane edge effect is clearly seen even though it is very small. As three ions constitute only 5 to 10 per cent of the total ions collected from a top surface layer and the plane edge effect observed is very small, we must conclude that the overall inaccuracy in the analysis of the composition of the top layer by ignoring the plane edge effect is negligibly small, and the atom-probe data are reliable and should represent well with those of macroscopic surfaces.

Surface segregation can be changed by chemisorption of a gas species. Segregation behavior derived under UHV condition, while needed for understanding surface segregation *per se*, is not really suited for interpreting surface reactivity of an alloy surface in promoting chemical reaction. One must consider the effect of chemisorption. This has been done for Ni–5% Cu.[237] It is found that when the alloy is annealed in 10^{-3} Torr of nitrogen, the plane edges are greatly enriched with Cu. On the other hand when it is annealed in oxygen then oxide layers of Ni are formed, and within these layers no Cu atoms are detected.

Table 4.8 summarizes composition depth profiles in atom-probe surface segregation studies and a study of impurity cosegregation, to be discussed in the next section. The segregants found in atom-probe experiments are in excellent agreement with other experimental and theoretical studies. However, atom-probe studies provide greater details. For an example, atom-probe experiments found that the enrichment of segregant in some alloys extends to only the top surface layer, while for others the enrichment extends to a depth of a few atomic layers. The nature of convergence can be monotonic, or non-monotonic, possibly oscillatory, depending on the alloy systems. The non-monotonic behavior has been explained in theoretical studies[238] and a Monte Carlo simulation.[240] Interested readers should consult the original papers.

(*b*) *Atom-probe studies of impurity segregation*

In the original study of Ahmad & Tsong[239] of surface segregation of Pt–Rh alloys, the segregation species was found to be Rh, in disagreement with other studies.[240] However, in this study, Ahmad & Tsong also noticed the formation of an adsorption layer of sulfur upon annealing the alloy surface. Further studies by Ren & Tsong[241] with the same samples,

Table 4.8 *ToF atom-probe composition depth profiles in alloy segregation*

Alloy system	Segregated species		Depth of segregation in atomic layer	Nature of convergence	References
	Experimental	Theoretical			
Surface segregation					
Ni–Cu(5%)	Cu	Cu	1	Non-monotonic	209
Ni–Cu(Ni > 0.84)	Cu	Cu	—	Non-monotonic	233
Ni–Cu(Ni < 0.16)	Ni	Cu	—	—	233
Pt–Au(5%)	Au	Au	4	Oscillatory	232
Pt–Rh	Pt	Pt	1	Non-monotonic	234
Pt–Ru	Pt	Pt	1	Non-monotonic	234
Impurity cosegregation					
Pt–Ir(S)	Pt		2	Monotonic	239
Pt–Rh(S)	Rh		1	Non-monotonic, & Reversed segregation	239
Pt–Ru(S)	Pt		1	Non-monotonic	234
Segregation with chemisorption					
Ni–Cu(N_2)	Cu		1	Plane edge effect	237
Ni–Cu(O_2)	Ni		Many	Ni oxide formation	237

but with the impurity sulfur in the samples now carefully removed by annealing the Pt–Rh alloy wires at about 800°C in one atmospheric pressure of high purity oxygen, find that the segregation species is Pt, in agreement with other macroscopic studies of this alloy system.[240] We must conclude that the segregation behavior found by Ahmad & Tsong represents that of surface segregation of S impurity and co-segregation of Rh with S in the Pt–Rh alloys. Impurity segregation is a well known phenomenon in interface science. Impurity segregations to the surfaces have been studied by Ichimura *et al.*[242] and by Shek,[243] but none of these studies obtain composition depth profiles, neither do they find reversed segregation of alloys as a result of co-segregation with impurities.

In the work of Ahmad & Tsong, the Pt–Rh alloys used contained a very small amount of sulfur. On the basis of an atom-probe analysis of the bulk composition, the sulfur content is found to be less than 50 parts per million. When a tip is annealed to 700°C for 5 minutes, field ion images reveal the formation of an overlayer of impurities. The image spots of these impurities are sharper than those of metal atoms. The overlayer is slightly field evaporation resistant, and field evaporates together with the top surface layer. Thus in an atom-probe analysis of top surface layer one obtains simultaneously the degree of coverage of the overlayer and the composition of the top surface layer. Composition depth profiles of these alloys containing about 50 p.p.m. of sulfur show an enrichment of Rh for the top surface layer, a slight depletion of Rh for the second layer, and the bulk Rh concentration for the third and the subsequent layers, as shown in Fig. 4.48. It is found that the degrees of coverage of the overlayers are linearly proportional to the Rh concentration in the top surface layer. The impurity overlayer has a $c(2 \times 2)$ structure for the Pt–5.9 at.% Rh alloy. For other compositions of the alloy samples, the overlayer structures are not identified. Since in high purity Pt–Rh alloys the segregation species is Pt, whereas in Pt–Rh(S) alloys the segregation is reversed to Rh by the segregated overlayer of sulfur, and in addition the Rh concentration of the top surface layer is proportional to the coverage of the sulfur overlayer, we must conclude that Rh atoms are drawn to the surface by the impurity S atoms. It is most interesting that as a result of co-segregation with sulfur impurity the compositions of both the top and the second surface layers are reversed, as shown in Fig. 4.49. In Fig. 4.50, we summarize the top layer compositions obtained from different techniques for Pt–Rh alloys of different bulk compositions. Before we go to other topics, let us explain briefly how the impurity atoms detected in the Pt–Rh experiment were identified as sulfur atoms, not oxygen atoms. Sulfur has three isotopes of masses 32 (95.0%), 33 (0.8%) and 34 (4.2%), whereas oxygen has only one major isotope of mass 16 (99.8%). From the isotope abundance of the detected ions, the ion species can be unequivo-

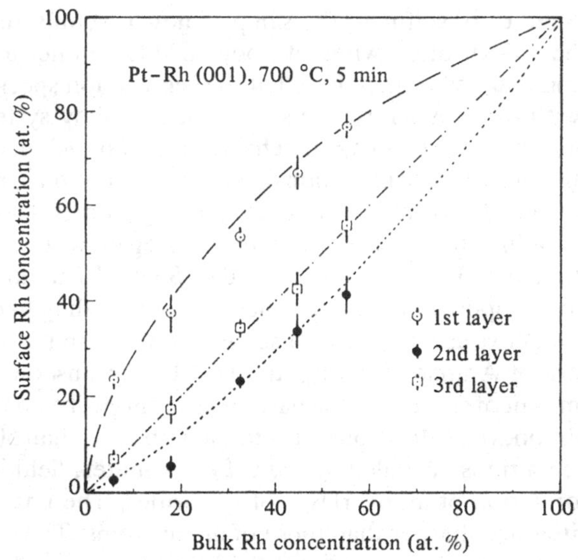

Fig. 4.48 Composition of the first three atomic layers of Pt–Rh alloys, thermally equilibrated at 700°C, which contain about 50 p.p.m. of sulfur impurity.

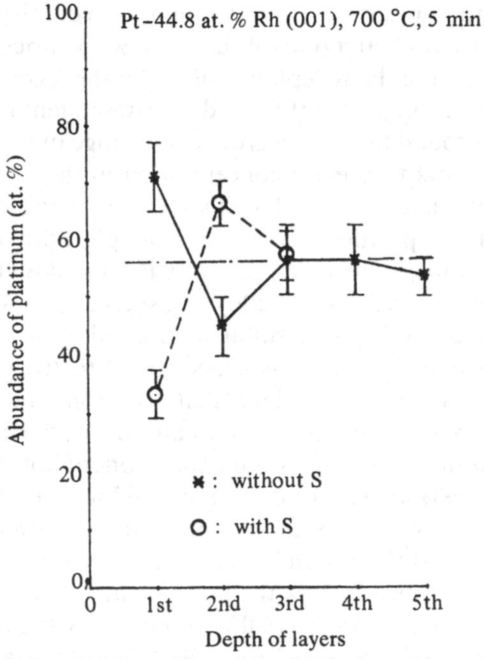

Fig. 4.49 Composition depth profiles of the (001) plane of two Pt–44.8 at.% Rh alloys, one containing sulfur impurity and one not. Note the reversal of the enriched species, in both the top and the second layers.

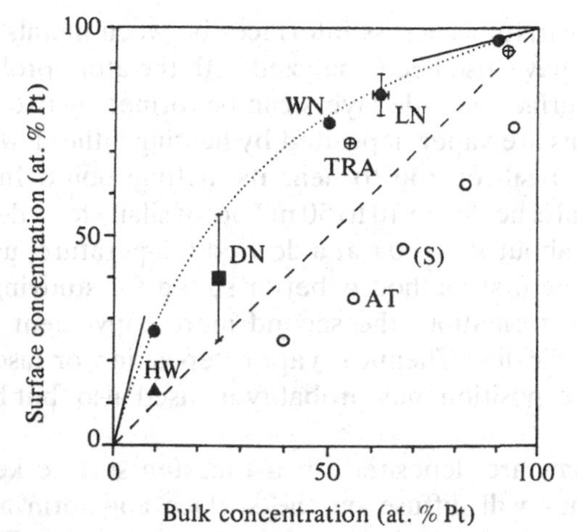

Fig. 4.50 Concentration of the top surface layer in surface segregation of Pt–Rh alloys obtained by different surface analytical techniques. Data points indicated with TRA are from Tsong, Ren & Ahmad, LN is from Langenveld & Niemantsverdriet, WN are from Williams & Nason, DN is from van Deft & Nieuwenhuys and HW is from Holloway & Williams. Data points AT are from Ahmad & Tsong, obtained with samples containing ~50 p.p.m. of sulfur impurities. See references 234 and 240 for the origin of the data.

cally identified. In addition, sulfur forms 1+, 2+ and 3+ atomic ions, whereas oxygen usually forms O^+ and O_2^+ ions. Thus they can also be easily distinguished from one another.

4.4.2 Metal–semiconductor interface formation

The importance of silicon technology in modern society is well known. Interface atomic structures and compositional variations across interfaces can affect the electronic properties of interfaces, or the performance of ohmic and Schottky contacts and gates in electronic devices, and thus have attracted the considerable interest of many scientists.[244] Most studies of these interfaces use either transmission electron microscopy, or macroscopic surface analytical techniques. Only recently have field ion microscope studies started to make useful contributions in this research. In field ion microscopy an early study of silicide formation on tungsten surfaces lacked good-quality images and no meaningful results were obtained.[245] The situation was changed in 1983 when Nishikawa *et al.*[246] and Tsong *et al.*[247] reported successful studies of silicide formation on tungsten surfaces. In subsequent studies, formation of one to a few atomic layers of silicide on various metal and silicon emitters have been reported and atomically resolved images of these silicide layers obtained.

Compositional variations across interfaces between metals and various semiconductors have also been analyzed with the atom-probe.

On a metal surface, silicide layers can be formed by two methods. In the first, Si atoms are vapor deposited by heating either a well degassed silicon wafer or a silicon rod to near its melting point. In the second method the metal is heated in 10 to 50 mTorr of silane for a desired length of time, usually about 10 to 60 s at a desired temperature, usually about 300 to 700°C. The first method is better suited for studying very early stages of silicide formation, the second more convenient for growing thick layers of silicides. Chemical vapor deposition or laser enhanced chemical vapor deposition may probably be used also, but have not yet been explored.

When Si atoms are deposited on a tungsten surface kept at room temperature, they will diffuse on the surface and form an overlayer structure of minimum free energy and maximum entropy. The diffusion parameters of silicon adatoms on the W (110) surface have been measured by Casanova and Tsong[156] to be $E_d = 0.7$ eV, and $D_0 \simeq 3.1 \times 10^{-4}$ cm^2/s. Silicon adatoms interact with a non-monotonic distant dependent potential. The adatom cohesive energy, i.e. the magnitude of potential energy at the site of absolute minimum, is less than ~0.05 eV, thus Si adatoms do not interact with one another on the surface by forming covalent bonds. When the coverage is less than 0.25, islands of a p(2 × 1) structure are formed. Atom transport between two islands is found to be always achieved by random walk diffusion of single Si adatoms between the islands, not by diffusion of Si clusters. Silicon atoms are continuously dissociated from these islands and reabsorbed to them again, or two-dimensional vaporization and condensation occur continuously. Si adatoms at terraces also repel with the W (110) plane ledge atoms at the nearest neighbor distance of the substrate. They tend to keep at a distance corresponding to the third nearest neighbor distance of the substrate. At room temperature, many atomic processes take place for Si adatoms, but W atoms are not displaced from their lattice sites by the presence of silicon adatoms, and no reaction between Si atoms and W atoms occurs. Thus no silicide layers are formed.

If a tungsten emitter surface, which has a very thin layer of Si vapor deposited on it, is heated to ~1000 K for a few minutes, a thin silicide layer will start to grow epitaxially from the {001} plane. The {110} facets of the tungsten emitter remain intact, and the W {112} facets are covered with one atomic layer of 'silicide'. The composition of this silicide layer has not yet been analyzed, but the image exhibits a (2 × 1) structure. This is the earliest stage, also called the first stage, of silicide growth on the W emitter surface.[247] Further deposition of Si atoms will of course increase the thickness of silicide layers grown on the {001} facets, but the W {110} facets remain bare of silicide. The {112} facets are now covered with about three atomic layers of silicide. These facets show characteristically bright images of atomic rows at a very wide row spacing. This second stage of growth is very stable. Even if a much larger amount of Si atoms is deposited, these Si atoms will be consumed in forming thicker layers of silicide on the {001}, but the number of layers formed on the {112} will not change. However, if hundreds of layers of Si atoms are vapor deposited on the W emitter surface, and a short, 1 to 3 min, annealing is made, then the entire W emitter surface will be covered with thin silicide layers. The thin silicide crystals grow nearly epitaxially from the {001} facets. On other regions of the emitter surface, mismatches are perhaps accommodated by localized small defects in the interface layers. These silicide crystals, grown from the {001} facets, will meet at the [110]-zone lines, forming boundaries with mismatches and defects since the lattice structure of WSi_2 does not match that of the W substrate. But at the same time the basal plane of WSi_2 ($a = b = 3.21$ Å) matches well with the {001} of W ($a = b = 3.16$ Å).[246] This stage, the third stage, is metastable. If one heats the tip further then silicon atoms diffuse to the {001} facets along the shank region, and a structure corresponding to the second stage is again formed in the tip region. Thus this stage is metastable. The final stage is characterized by the growth of many well ordered tungsten silicide polycrystallines over the entire tungsten emitter surface, as shown in Fig. 4.51. Since it is very difficult to vapor deposit a sufficient amount of silicon atoms to grow a thick layer of polycrystalline silicide on the surface of the entire emitter, i.e. including the shank of the tip, this final stage of silicide formation can be more easily produced by heating the tip in silane. The composition of the tungsten silicide has been analyzed in the atom-probe. Figure 4.52 shows that the stoichiometry of this silicide is WSi_2, the same as that identified in Rutherford backscattering experiments. A comparison of the atomic images of these silicide surfaces and the atomic structures of WSi_2 surfaces reveals that silicon atoms do not give rise to field ion images, a case similar to that of ordered alloys where one of the species may not be seen. It is found that the interface between W and WSi_2 is very sharp everywhere, as can be seen in Fig. 4.53. However, atomic structures at the interface are very difficult to figure out from these micrographs.

Silicide layers can also be grown on Pt surfaces.[247] The preferential facets of epitaxial growth are the {111}. Atom-probe data reveal that the stoichiometry of the silicide phase is Pt_2Si, and the Pt–Pt_2Si interface is also very sharp. However, a small fraction of silicon atoms can diffuse into the Pt matrix. Formation of silicide layers on a nickel emitter surface is much more complicated where silicide layers of varying stoichiometries are formed.[246,247] Owing to the statistical nature of the atom-probe data, identification of all the silicide phases in a nickel silicide layer is at best uncertain.

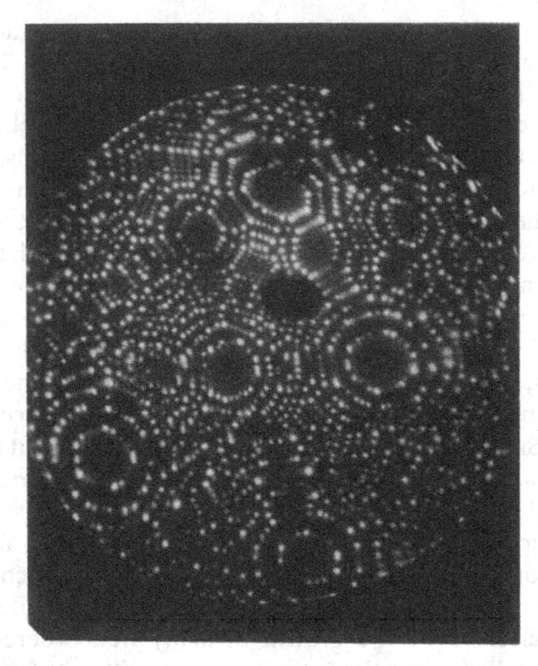

Fig. 4.51 35 K helium field ion image of polycrystalline tungsten silicide layers grown on a W emitter surface. Atom-probe analysis shows that the stoichiometry is WSi_2. Images of a few crystal planes of WSi_2 have also already been shown in Fig. 4.14. The dark area is the hole of the 45° mirror of the atom-probe.

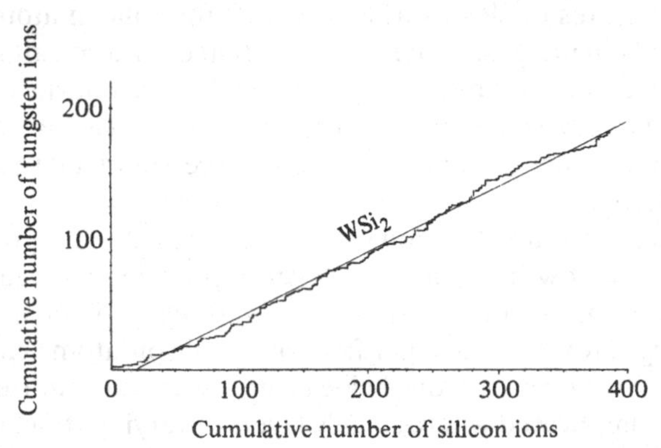

Fig. 4.52 Atom-probe compositional analysis of the polycrystalline silicide layers shown in Fig. 4.51. The slope gives the composition to be WSi_2.

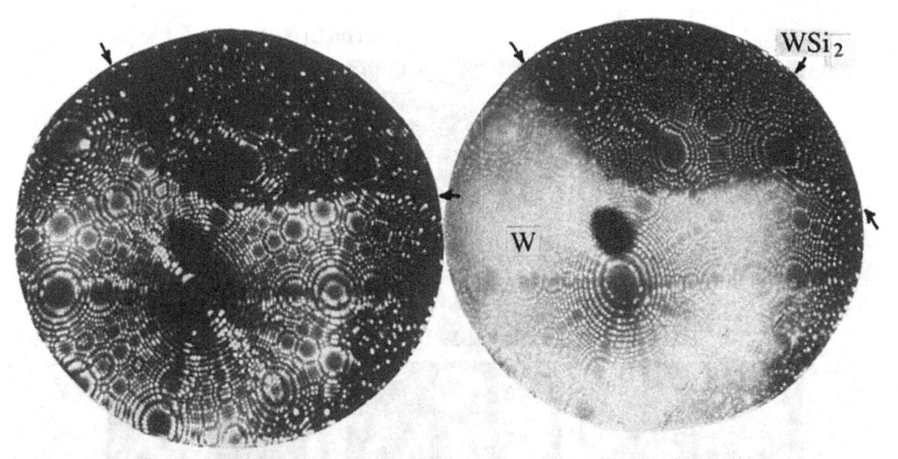

Fig. 4.53 Helium field ion images showing the sharpness of the boundary between a W crystal and a WSi_2 crystal. The two images are from the same surface, one taken at the best image voltage of the WSi_2 surface, and one the best image voltage of the W surface.

Formation of one to a large number of atomic layers of silicide and germanium compound on various faces of fcc metals such as Pt, Ir and Rh has also been studied by Liu *et al.*[248] When the silicide or germanium compound is only one atomic layer in thickness, atomically resolved and well ordered atomic structures can be seen on {110} and {001} surfaces. For an example, two atomic structures have been found for an atomic layer of IrSi grown on an Ir (001) surface. One shows a c(2 × 2) structure of the substrate. This is a highly strained structure; thus when the layer is gradually reduced in size by field evaporation, this unit cell relaxes from the square structure to a rhombic one. The other shows a rectangular unit cell. The exact atomic structural relation of this silicide layer with the substrate is not established, but the size of the unit cell is very large, about 4.5 ± 1.5 Å by 6.0 ± 2.0 Å. This structure is not strained. Thus when it is reduced in size by field evaporation to only four seen atoms, the structure retains its original rectangular form. Similarly two structures, one strained and one unstrained, are also found for an atomic layer of germanium compound grown on the Ir (001) surface. Figure 4.54 shows images of these strained and unstrained silicide layers.

Formation of silicide layers on silicon surfaces is much more difficult to study, as good silicon tips are very difficult to prepare; also atomically resolved and well ordered images of silicon have been obtained only very recently. Originally Liu & Tsong obtained a large number of well ordered images of thin silicide layers of Ir, Rh and Pt grown on silicon emitter surfaces.[249] However, these images resemble too closely the images of pure silicon surfaces. Thus it is as yet very uncertain whether these silicide

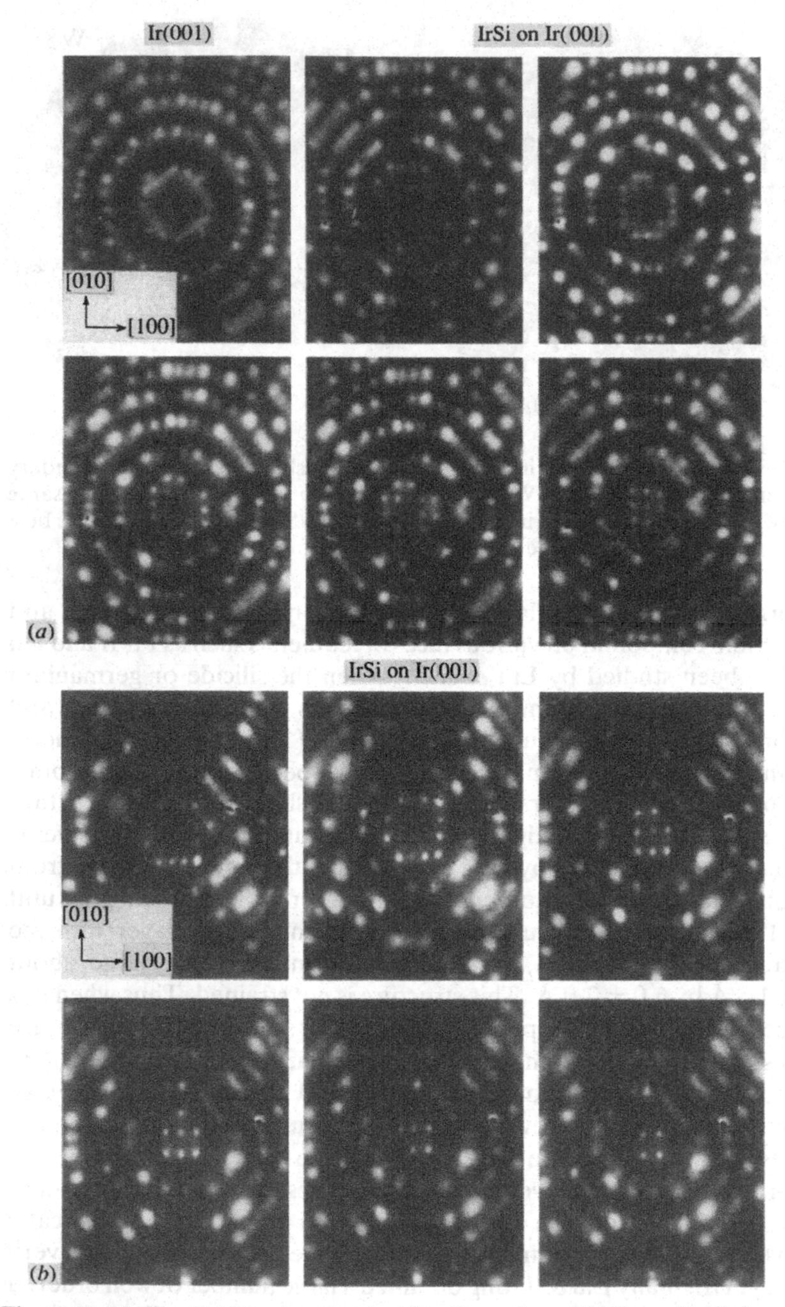

Fig. 4.54 Helium field ion images showing formation of iridium silicide layers of two different atomic structures on the Ir (001) surface. The first micrograph in (*a*) shows the Ir surface before formation of an IrSi layer.

images are really images of silicides, or just images of pure silicon. Further investigation is needed to establish the atomic structures of thin silicide layers on silicon emitter surfaces. Liu *et al.* have studied formation of thick layers of nickel and cobalt silicide on silicon surfaces.[250] $NiSi_2$ and $CoSi_2$ have a similar structure (CaF_2) to silicon, and the lattice misfit is very small, only about 1%. Transmission electron microscope studies find that two kinds of interfaces can be formed between $NiSi_2$ or $CoSi_2$ and the Si (111) surface; these are called A-type and B-type interfaces.[251] The structures of the silicide phase in these two types of interfaces are oriented 180° from one another. In field ion microscope, if the interface has the A-type structure, the image of the silicide grown on a (111) oriented silicon tip will not change its symmetry from the original symmetry of the Si tip. On the other hand, if a B-type interface is formed, the symmetry will change by 180°. Indeed, two types of interfaces have been observed, one without changing the image symmetry and one with a 180° change in the image symmetry. In UHV, the B-type interface is the preferred type for the $NiSi_2$–Si (111) interface.

Interfaces of metals and compound semiconductors have also been analyzed with the atom-probe, particularly by the groups of Nishikawa, Nakamura, and Sakurai. Nishikawa *et al.*[252] reported a study of Al–Ga exchange reaction at the Al–GaAs interface. An atomically abrupt interface is observed for the (110) surface of a Si-doped GaAs when Al is deposited at 200 K. Subsequent heating to 300 K to 500 K results in the formation of an Al–As mixed layer with increasing As concentration, but without Ga in the layer. At 900 K heating, all the Al atoms are consumed to form AlAs phase. Therefore in this Si-doped GaAs sample an Al–As layer is formed by substitutional replacement of Ga atoms by Al atoms. On the other hand, a mixed layer of Ga, As and Al can be formed at 200 K for a Zn-doped GaAs, the formation of which is attributed to a promoted reaction by Zn atoms. A decrease in the thickness of the unreacted Al layer and an increase in the As concentration in the mixed layer with temperature, as well as the abrupt AlAs–GaAs interface, all indicate Al diffusion into GaAs and stable Al–As bonding. Other systems studied include Pd/Si, Pd/GaAs, Pd/Ti/GaAs, Ti/Si, and Ti/SiC etc.[253,254] Ti is known to be inert to GaAs and can be used as a marker in the study of the interface reactions. These are systems of considerable practical technological interest. Interested readers should consult original papers for details.

4.5 Gas–surface interactions

Field ionization mass spectroscopy of chemically reactive molecular gases found often dissociated species and associated species of these molecules.[255,256] They are mostly produced by field dissociation and field association, and probably have only a little to do with gas–surface interactions since field ionization takes place a few Å above the emitter surface. Most of these studies were intended to find the suitability of field ionization as an ion source for mass spectroscopy and field effects in field ion emission. In any case, few attempts were made to identify the ionic species observed in field ionization with those expected from surface promoted, or catalyzed, simple chemical reactions. Early atom-probe studies by Müller and co-workers[257] on field evaporation of metals in chemically reactive gases found formation of metal–gas complex ions, and ionic species of mostly fragmented gas molecules. These reactions were field induced, and again no correlations with surface promoted chemical reactions were attempted or made. Early studies of formation of H_3^+ and $(H_2O)_nH^+$ ions in field ionization and metal–gas complex ions in field evaporation again emphasized field ion emission *per se*.[258] Only recently have field desorption studies of hydrogen and exchange reactions with deuterium to form HD_2, H_2D, and HD and so on, and formation of NH_3 from H_2 and N_2, been concerned with field and surface enhanced chemical reactions, and tried to correlate with the chemical properties of the emitter surfaces of different metals, such as the dissociative and non-dissociative chemisorption behavior of these surfaces.

There are some difficulties of studying non-field-enhanced surface reactions with field ion emission techniques. Aside from the uncertain effects of the very high applied field, there is also the disturbing effect of field dissocation, which will complicate a correct identification of the true desorbed species. However, some of these difficulties are not insurmountable. A few precautionary procedures which can be followed are listed below:

1. Field desorption rather than field ionization has to be used, so that all the ions come directly from the surface; the coverage of the reactants adsorbed on the surface also has to be low.

2. The disturbing effects of the applied field can be reduced by using pulsed-laser stimulated field desorption which needs a field strength only about half that needed in regular field desorption and the effects of the field should be reduced to about one-quarter of the original ones since an energy level shift by the polarization effect is proportional to F^2; electric field, in fact, is believed to be one of the factors which can be responsible for the enhanced chemical reactivity of the surface.

3. Whether or not an ionic species is desorbed directly from the surface or produced by field dissociation above the emitter surface can easily be determined with a precision measurement of the ion kinetic energy distribution.

4. One can always check the validity of chemical reaction by its

temperature dependence and compare it with other studies where there is no strong applied field. Thus even though the field desorption technique is not without drawbacks, there is hope of using it to study site specific chemical reactions with an atom-probe where ions can be collected from well specified atomic sites. In any case, there is no other technique which can provide directly such information. We discuss here a study of H_3 formation and a preliminary study of NH_3 formation based on the above mentioned procedures.

The reactivity of a surface depends on many factors. These include the adsorption energies of chemical species and their dissociation behavior, their diffusion on the surface, the adatom–adatom interactions, the active sites where a chemical reaction can occur, and the desorption behavior of a new chemical species formed on the surface. The site specificity depends on at least three factors: the atomic configuration of the surface, the electronic structures of the surface, and the localized surface field. In atom-probe experiments, the desorption sites can be revealed by a time-gated image of an imaging atom-probe as well as by an aiming study with a probe-hole atom-probe, the electronic structure effect of a chemical reaction can be investigated by the emitter material specificity, and the surface field can be modified by the applied field.

On a metal surface, the surface field is produced by an electronic charge smoothing effect[259] and a gradual decaying of the electronic charge density beyond the positive ion core background. The surface field should depend on the atomic structure of the surface in question, and the field strength can be estimated from a jellium model calculation.[260] It is found that for transition metals a typical surface field varies from ~3 to 1 V/Å at a distance 0.5 to 2.5 Å from the surface. This is comparable in strength to the applied field used in field ion emission experiments. The applied field needed in field ion emission experiments is therefore not expected to change drastically the reactivity of a metal surface, since an intrinsic surface field of comparable strength also exists on the surface. The applied field however extends far above the surface and thus desorbed molecules can be field ionized and can be detected. The applied field can also enhance the gas supply to the surface by a field enhancement factor of $\alpha F_0^2/2kT$, where α is the polarizability of the molecules. This gas supply enhancement factor is in general greater than 100. The effect of gas supply enhancement is equivalent to an increase in the gas pressure of the reactants.

4.5.1 Surface reactivity in the formation of H_3^+

In a time-of-flight atom-probe, if the vacuum condition is not good enough, many hydride ions of the tip material can be found in the mass spectra. The presence of a small partial pressure of residual hydrogen can also result in field induced chemical etching of the tip surface at a much reduced field. These are in general considered nuisances in the chemical analysis of the tip material and in a study of the atomic structure of the surface. On the other hand, as hydrogen can promote field evaporation of the tip of silicon, oxide compounds, and some low melting point metals such as iron and steel, at a much lower field, this process can be utilized to reduce the field stress of the tip during field evaporation and can thus improve the surviving rate of the tip. In addition, hydrogen etching tends to occur from plane edges where atoms have many more dangling bonds to react with hydrogen atoms. Thus the atomic structures of the surfaces, particularly Si and oxide compounds, can be better developed. For some metals such as W and Mo, atoms are not reactive with hydrogen and its field evaporation behavior as well as surface diffusion behavior (both the substrate and the adatoms have to be non-reactive to hydrogen) are not significantly affected by the presence of a small amount of residual hydrogen.

In field ionization mass spectroscopy of hydrogen from W surfaces, Clements & Müller[261] find that most of the ions formed are H^+ and H_2^+, but a small fraction of H_3^+ can be found in the field range from 2.0 to 2.5 V/Å, coming mostly from protruding sites of the W emitter surface. This atomic site dependence can be directly seen in the time-gated images of H_3^+ in pulsed-laser stimulated field desorption. Molecular H_3 is not stable in free space although its optical spectrum has been obtained by Herzberg[262] in a plasma gas discharge tube. On the other hand, H_3^+ ions have been known to be stable since J. J. Thomson's mass spectroscopy study in 1912. In hydrogen plasma, H_3^+ is the most abundant ionic species. We now discuss briefly field adsorption and desorption of hydrogen from metal surfaces and reactivities of surfaces in the formation of H_3^+ ions in field ion emission.

In field ionization, hydrogen molecules near the tip region are attracted to the tip surface. They either hop around the tip surface or are field adsorbed on it. As the hopping motion and the field adsorption are dynamical phenomena, some of the ionic species detected may also come from field adsorbed states, not necessarily just from the gas phase. On the other hand, in pulsed-laser stimulated field desorption, where gas pressure is very low, of only $\sim 1 \times 10^{-8}$ Torr, gas molecules are thermally desorbed by laser pulses from their field adsorbed and chemisorbed states. When they pass across the field ionization zone some of them are field ionized. The critical ion energy deficit in pulsed-laser stimulated field desorption of a gas is therefore found to be identical to that found in field ionization. In both pulsed-laser stimulated field desorption and field ionization of hydrogen, the majority of ions detected are H_2^+ and H^+. H_2^+

is more abundant at low field and H^+ is more abundant at high field. When there is no formation of H_3^+, the field dependence of the ratio of the abundances of H_2^+ and H^+ is found to be dependent neither on the tip material nor on the surface plane; it depends only on the field strength. These observations can be understood by postulating that field adsorption occurs in molecular form of H_2. In low laser power pulsed-laser stimulated field desorption, chemisorbed H atoms on the surface cannot be desorbed because of the large binding energy of H atom; about 2 eV, with the surface. H_2^+ comes from field ionization of pulsed-laser thermally desorbed H_2, whereas H^+ is the field dissociation product of H_2^+. Thus both the two ionic species, H_2^+ and H^+, observed are the products of the applied field and have nothing to do with the chemisorption properties or the reactivity of the tip surface. Chemisorbed H atoms can be desorbed only when the substrate atoms are also field evaporated. They usually come out as hydride ions of the tip material. In Fig. 4.55(*a*), we show the universal behavior of the field dependence of the relative abundances of H_2^+ and H^+; i.e. this behavior is independent of tip materials and surface planes.

The formation of H_3^+, on the other hand, is found to be very sensitive not only to the atomic structure of the surface, but also to the chemical properties of the tip material. Figure 4.55(*b*) is a good example. Ai & Tsong[263] have studied formation of H_3^+ in pulsed-laser stimulated field desorption from different emitter surfaces of over 20 metals. Although no systematics of its formation with the chemical properties of these metals in the periodic table are found, there are some general trends in the formation of H_3^+ on the surfaces of these metals. For some hcp metals like Be, Ti, Hf and Re, the maximum abundance of H_3^+, when the field strength is about 2.5 V/Å, can be as high as 40%. For bcc metals like W, Mo, Ta and Fe, it is about 10 to 20%. For fcc metals, the maximum abundance of H_3^+ is the lowest, generally less than 10%, with the exception of Au which is over 20%. This material specificity in the formation of H_3^+ clearly indicates that H_3^+ observed in field ion emission experiments is a true product of the surface reactivity and not just an artifact of the applied field, although its formation may have been greatly enhanced by the applied field.

An interesting question is how H_3^+ is formed on the emitter surface and whether H_3 molecules can exist on the surface. This question can be investigated with a measurement of the appearance energy of H_3^+ ions. Jason *et al.*[264] find H_3^+ in field ionization of condensed layers of hydrogen, and measure the appearance energy to be 12.7 eV. This value is 2.9 eV smaller than that of H_2^+. Ernst & Block conclude[265] from a similar measurement in field ionization mass spectrometry of hydrogen that an H_3^+ ion is formed at the moment when a chemisorbed H atom combines

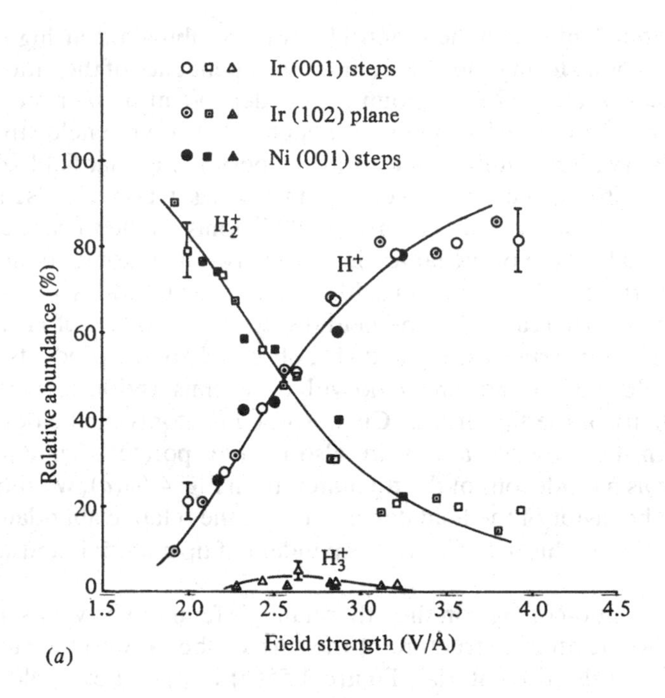

Ir (001) steps
Ir (102) plane
Ni (001) steps
H₂⁺
H⁺
H₃⁺
Relative abundance (%)
Field strength (V/Å)
0
20
40
60
80
100
1.5
2.0
2.5
3.0
3.5
4.0
4.5
(a)

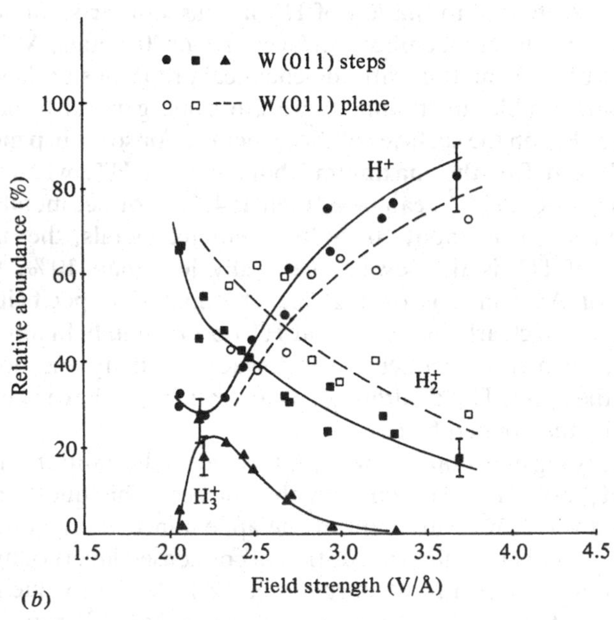

W (011) steps
W (011) plane
H⁺
H₂⁺
H₃⁺
Relative abundance (%)
Field strength (V/Å)
0
20
40
60
80
100
1.5
2.0
2.5
3.0
3.5
4.0
4.5
(b)

with a field adsorbed H_2 molecule via

$$H_2(\text{field ad.}) + H(\text{chem. ad.}) \rightarrow H_3^+(g) + e^-(\text{metal}).$$

In these 'steady state' experiments, hydrogen molecules are supplied to the emitter surface continuously, and H^+, H_2^+ and H_3^+ ions are steadily emitted from the surface. The role of the substrate is much less transparent. Tsong and his co-workers[266] also measured the appearance energy of H_3^+, but formed in pulsed-laser stimulated field desorption. It is 13.2 eV, or within experimental uncertainties the same as those obtained by Jason *et al.* and Ernst & Block. Since H_3^+ ions can be formed on surfaces at temperatures where no dissociative chemisorption of hydrogen can occur, the mechanism suggested by Ernst & Block cannot explain the formation of H_3^+ on these metal surfaces at low temperatures. Tsong and co-workers therefore favor the existence of neutral H_3 molecules in field adsorption states. These molecules are produced by the effect of both the reactivity of the surface (thus showing material specificity) and the applied field and have a life time at least of the order of ms, about the time it takes to have one gas atom field adsorbed on an initially bare surface atom, in the field adsorption states in the strong applied field.

Another question is how H_3 molecules are formed on the emitter surface. At least two distinctive mechanisms can be envisioned. They are dissociative and associative mechanisms as represented by:

$$2^* + H_2(\text{gas}) \rightarrow 2H(\text{ad.})$$
$$H(\text{ad.}) + H_2 \rightarrow H_3(\text{field ad.}), \qquad (4.107)$$

or

$$* + H_2(\text{gas}) \rightarrow H_2(\text{field ad.})$$
$$3H_2(\text{field ad.}) \rightarrow 2H_3(\text{field ad.}). \qquad (4.108)$$

An asterisk (*) here represents an adsorption site. There are several observations which provide strong support for the association mechanism.[266] First, H_3^+ is observed from some metals such as Au, which is known not to show dissociative chemisorption of H_2 at low temperatures. Second, for some metals such as Ir, even though chemisorption of H_2 is dissociative, few H_3^+ ions are observed. Third, when a H_2–D_2 mixed gas is used, atomic exchanges always occur regardless of whether hydro-

Fig. 4.55 When few H_3 ions are formed, the field dependence of the abundances of H_2^+ and H^+ is dependent neither on the atomic structure of the surface nor on the chemical species of the emitter surface, indicating that formation of these two ionic species depends only on the applied field. (*b*) Formation of H_3^+, on the other hand, depends on both the surface atomic structure and the chemical species of the emitter tip, indicating that H_3^+ is a surface and field catalyzed chemical reaction product.

gen adsorption on the surface is dissociative or not. Of course some other mechanisms cannot be ruled out. An interesting point in these studies is that H_3^+ can be produced by field adsorption–desorption processes. It is well known that in interstellar space H_3^+ is one of the most abundant ionic species. These ions can be produced by field adsorption–desorption processes from interstellar dust particles. A strong electric field can be produced on the surface of these particles by bombardment of energetic particles such as photons and electrons and thus the particles may lose their neutrality. High electric field and surface promoted chemical reactions should therefore also be of basic scientific interest beyond field ion emission and field ion microscopy.

4.5.2 Atomic steps and reaction intermediates in chemical reactions

Identification of reaction intermediates is generally considered to be a key to understanding heterogeneous catalytic reactions on the atomic and molecular level and also of the selectivity of catalysts in catalytic reactions.[267] There have been a few ToF and imaging atom-probe studies of gas surface interactions, such as formation of NH_3 and CH_4 in pulsed-laser stimulated field desorption of co-adsorbed N_2 and H_2 and of co-adsorbed CO and H_2, done specifically for this purpose.[263] In practical catalysis, gas pressures and temperatures are usually very high. These conditions are chosen for the purpose of improving the reaction rates and the turnover numbers. Also catalysts are almost always supported alloy or compound particles, and are not single crystal surfaces. In surface science studies of surface promoted chemical reactions, gas pressure is usually very low and the experiments are almost always performed on single crystal faces at low temperatures. It is not certain that conclusions derived from surface science experiments can be directly used to interpret practical catalytic reactions. However, a fundamental understanding of surface catalyzed chemical reactions can come only from simple systems performed under best defined physical conditions. Field ion emission experiments belong to this class of studies.

Early field ion emission studies of gas–surface interactions use field ionization mass spectrometry. Gas molecules are supplied continuously to the tip surface by a polarization force and by the hopping motion of the molecules on the tip surface and along the tip shank. These molecules are subsequently field ionized. The role of the emitter surface in chemical reactions is not transparent and has not been investigated in detail. Only in recent pulsed-laser stimulated field desorption studies with atom-probes are these questions addressed in detail. We now discuss briefly a preliminary study of reaction intermediates in NH_3 formation in pulsed-laser stimulated field desorption of co-adsorbed hydrogen and nitrogen, reported by Ai & Tsong[263] and Liu & Tsong.[268] In their experiments, an emitter surface is first developed by low temperature field evaporation in neon or helium. The image gas is removed from the system and a N_2–H_2 mixed gas is introduced into it. The tip temperature is adjusted to a desired value, and pulsed-laser stimulated field desorption is then carried out with the desorption species identified from their flight times. Temperature, gas pressure and field dependences of the relative abundances of different ionic species are studied from which conclusions are drawn. Before we discuss experimental results of these experiments, let us consider two proposed mechanisms of ammonia formation on solid surfaces, namely the dissociative and associative mechanisms.[269] It is generally accepted that heterogeneous catalysis involves the active surface site, which will be represented by an asterisk (*). The detailed atomic steps and the reaction intermediates of the two possible mechanisms are listed in Table 4.9. From this table it is clear that the two mechanisms will give rise to two different sets of reaction intermediates. Thus by finding the reaction intermediates the atomic steps of the catalytic reaction can be identified.

In pulsed-laser stimulated field desorption, if the field is high enough, the adsorbed species can be thermally field desorbed, most probably within one to a few atomic vibrations. If the activation barrier of evaporation has been reduced by the applied field to much less than the surface diffusion barrier, then the adsorbed species will be desorbed before they have any chance of interacting with other atoms or molecules on the surface. Thus the desorbed species should represent well the

Table 4.9 *Dissociative and associative mechanisms in ammonia synthesis on solid surfaces*[a]

Dissociative	Associative
$2^* + N_2 \rightleftarrows 2^*N$	$^* + N_2 \rightleftarrows {}^*N_2$
$2^* + H_2 \rightleftarrows 2^*H$	$^*N_2 + H_2 \rightleftarrows {}^*N_2H_2$
$^*N + {}^*H \rightleftarrows {}^*NH + {}^*$	$^*N_2H_2 + H_2 \rightleftarrows {}^*N_2H_4$
$^*NH + {}^*H \rightleftarrows {}^*NH_2 + {}^*$	$^*N_2H_4 + H_2 \rightleftarrows 2NH_3 + {}^*$
$^*NH_2 + {}^*H \rightleftarrows {}^*NH_3 + {}^*$	$N_2 + 3H_2 \rightleftarrows 2NH_3$
$^*NH_3 \rightleftarrows NH_3 + {}^*$	
$N_2 + 3H_2 \rightleftarrows 2NH_3$	
Reaction intermediates	
N, H, NH, NH_2, NH_3	N_2, N_2H_2, N_2H_4

[a] An asterisk (*) represents an active site on the surface, which may well be a site of a displaced surface atom or a kink atom.

reaction intermediates on the surface. Figure 4.56(*a*) and (*b*) show two pulsed-laser stimulated field desorption ToF spectra, one taken at ~140 K and one at ~90 K, each at a gas pressure of 4×10^{-7} Torr, of a $H_2 : N_2 = 3:1$ mixed gas. The probe-hole is aimed at a lattice step near the (001) pole of the Pt tip. At ~140 K, the major ion species detected are NH_2^+, NH_3^+ and NH^+. Some N^+, N_2^+, N_2H^+ and $N_2H_2^+$ are also detected, but the amount is much smaller. When the temperature is lowered by ~50 K to ~90 K, NH_2^+, NH_3^+ and NH^+ disappear rapidly and the most abundant ion species are now N_2^+ and N_2H^+. Whenever NH_3^+ is observed, so are N^+, NH^+ and NH_2^+. One must conclude that the reaction intermediates under the low temperature and gas pressure conditions are N^+, NH^+, NH_2^+ and NH_3^+. The dissociative mechanism is therefore the correct one. At very low temperatures when dissociative chemisorption of nitrogen becomes improbable, the reaction intermediates disappear quickly, and only field adsorbed species N_2 and N_2H are observed. There is another piece of information in the mass line of NH^+ which further confirms the conclusions drawn above. The line shape of this ion species shows two peaks: one sharp primary peak and one diffusive secondary peak. Ions in the sharp primary peak originate from the surface and are the true reaction intermediates. Ions in the secondary peak come from field dissociation of NH_2^+ and NH_3^+. The mass peaks of NH_2^+ and NH_3^+ are sharp and thus ions of these species are desorbed directly from the surface.

Ai & Tsong, using a pulsed-laser imaging atom-probe, find that an atomically perfect iron surface, prepared by low temperature field evaporation, is quickly corroded by chemisorption of nitrogen. Iron atoms are displaced from lattice sites to some adsorption sites. Gated images show that it is from these sites that reaction intermediates are desorbed. Thus a study of the reactivity of an atomically perfect single crystal surface using a macroscopic technique may not represent the true reactivity of the perfect surface. This result indicates clearly that the reactivity of a surface is determined by the small number of displaced atoms. It can also explain why a small fraction of a monolayer of impurity atoms such as S on the surface can either poison or promote the entire surface in a surface catalyzed chemical reaction. Gated images explaining these features are shown in Fig. 4.57.

The experiments of Ai & Tsong and Liu & Tsong have been repeated by Li *et al.*[270] They report no formation of NH_2^+ or NH_3^+. They argue that the ion species observed by the former authors are most probably O^+ and OH^+, etc., in field desorption of residual gas molecules. However, in one experiment, Ai & Tsong show a continuous growth of $NH_2^+ + NH_3^+$ mass peak when hydrogen is gradually leaked into the system which contains nitrogen. A reversed behavior is observed when too much N_2 is leaked

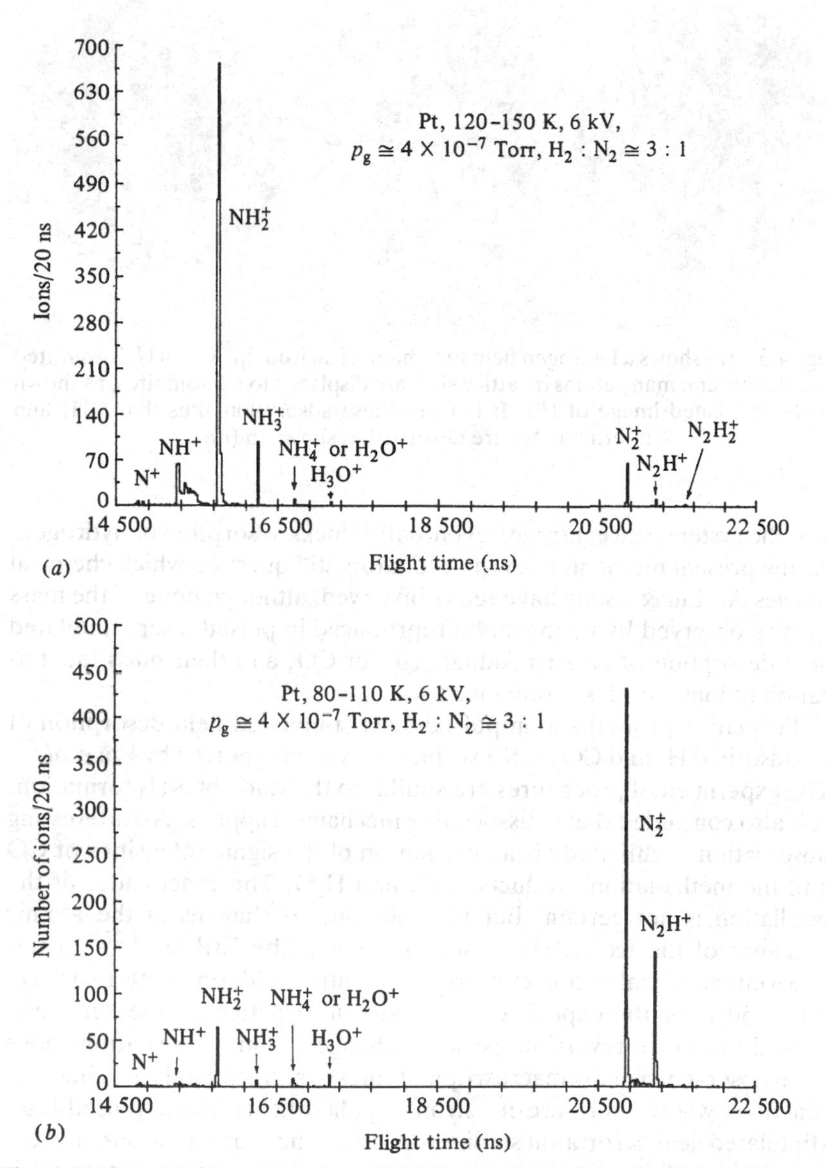

Fig. 4.56 In pulsed-laser stimulated field desorption with coadsorption of N_2 and H_2, if the temperature of the Pt tip is between 120 and 150 K, then reaction intermediates of NH_3 can be detected as shown in (*a*). If the surface temperature of the tip is only ~50 K lower, no reaction intermediates of NH_3 can be detected. Instead, only field adsorption products, N_2^+ and N_2H^+ are detected.

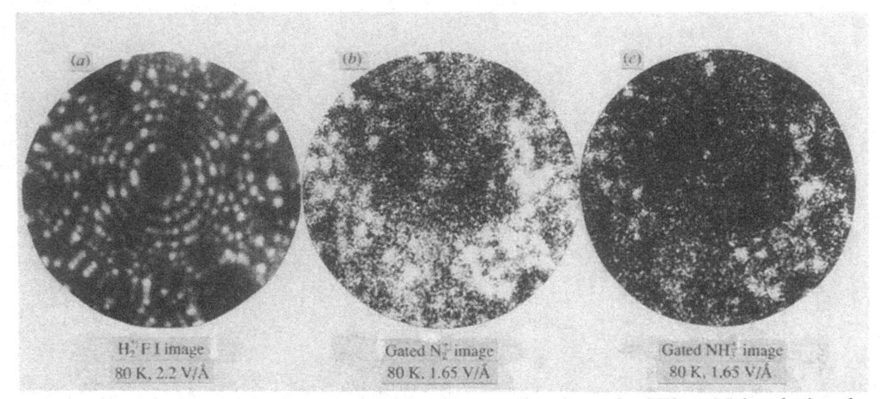

Fig. 4.57 (*a*) shows a hydrogen field ion image of an iron tip. When N_2 is admitted into the system, many atoms in lattice sites are displaced to adatom sites, as shown in the N_2^+ gated image of (*b*). It is from these adsorption sites that NH_3 and intermediates are desorbed as shown in (*c*).

into the system, since nitrogen eventually blocks adsorption of hydrogen. At the present moment, a few investigators still question which chemical species Ai, Liu & Tsong have really observed, although none of the mass spectra observed by them can be reproduced in pulsed-laser stimulated field desorption of either residual H_2O or CO, and their mass identification of ionic species is reliable.

Formation of methane in pulsed-laser stimulated field desorption of co-adsorbed H_2 and CO on Rh surfaces has been reported by Liu *et al.*[271] The experimental procedures are similar to the study of NH_3 formation. It is also concluded that a dissociative mechanism applies. An interesting observation of this study is an oscillation of the signal intensities of CO and the methanation products, CH_4 and H_2O. The exact cause of the oscillation is not certain, but it is not due to changes in the atomic structure of the fcc (001) surface as reported by Ertl *et al.*[272] In this experiment, signal is collected from the entire field ion emitter surface. At ~150 K of the experiment, the atomic structure of the surface is unlikely to make reversible periodic changes. Similar oscillations have been observed in many macroscopic studies of surface catalyzed chemical reactions where there are no strong applied fields. Since pulsed-laser stimulated field desorption studies reproduce these observations, and the reactions studied are also very sensitive to surface temperature, it is unlikely that the results derived are entirely artifacts of the applied field. Nevertheless, the role of the applied field has to be more carefully investigated in these studies. Of course, the effect of an electric field in promoting a chemical reaction is itself also a subject of basic scientific interest, and worthy of pursuing.

4.5.3 Some field induced effects

(a) Formation of metal helide ions

The field ion emission phenomena discussed in Chapter 2 are effects produced by high electric fields. We discuss here an additional effect, i.e. formation of noble compound ions by the catalytic effect of the applied field and the solid surface. As already mentioned earlier, metal helide ions[273] and metal neide ions[274] can be formed in low temperature field evaporation of metals. While little is known about neide formation, a considerable amount of data has been accumulated on metal helide formation. It is found[275] that all the metals whose low temperature evaporation fields are greater than about 4.5 V/Å form metal helide ions of certain charge states, while metals with low temperature evaporation fields lower than this value do not form helide ions; thus there is a critical field strength in the formation of helide ions in low temperature field evaporation of metals. For tungsten and platinum, dihelide and a very small number of trihelide ions can also be formed.[275] The abundance of dihelide ions is a few per cents of helide ions, and the abundance of helide ions can be as large as several tens per cent of the total field evaporated ions. In low temperature field evaporation of a metal in helium, metal ions of different charge states usually coexist, but helide ions exist only in one of the charge states, either the 2+ or 3+ state, but never the 1+ state. In Table 4.10 we summarize ion species observed in low temperature field evaporation of refractory metals in helium. Other novel ions such as ArH^+ have also been observed, but few details are available.

The critical ion energy deficits of metal helide ions are found to be only ~0.2 to 0.5 eV higher than the corresponding metal ions. The differences, presumably representing the binding energies of these helide ions, are so small that a reliable measurement has not yet been reported, although they should be within the reach of a high resolution pulsed-laser

Table 4.10 *Ion species in low temperature field evaporation in helium*

Element	Ion species	d.c. evaporation field (V/Å)
Mo	Mo^{2+}, Mo^{3+}, Mo^{4+}; $MoHe^{3+}$	~4.8
Rh	Rh^{+}, Rh^{2+}; $RhHe^{2+}$	~5.0
Ta	Ta^{2+}, Ta^{3+}; $TaHe^{3+}$	~4.8
W	W^{2+}, W^{3+}, W^{4+}; WHe^{3+}, WHe_2^{3+}, WHe_3^{3+}	~5.7
Re	Re^{+}, Re^{2+}, Re^{3+}; $ReHe^{3+}$	~5.0
Ir	Ir^{2+}, Ir^{3+}; $IrHe^{2+}$	~5.3
Pt	Pt^{+}, Pt^{2+}; $PtHe^{2+}$, $PtHe_2^{2+}$	~4.6

atom-probe developed in the author's laboratory. Metal helide ions, while formed by the interaction between metals and high electric fields, are of some fundamental interest since their stability can be explained only if relativistic corrections are made in Hartree–Fock calculations[276] of compound ions. The binding energies of metal helide ions are calculated to be in the 0.1–0.5 eV range. There have been no other techniques available to measure these energies. Measurement of ion energy distribution of helide ions in field evaporation may be a good way to derive such data for comparing with theoretical calculations.

(*b*) *Atomic structure of field adsorbed atoms*

In Section 4.1.1 we have already discussed that the atomic structure of a surface thermally equilibrated in an applied field may be very different from the structure equilibrated in the absence of the applied field. This is also true for adsorbed atoms. While it is generally very difficult to image adsorbed atoms of volatile species with field ion microscope, Faulian & Bauer[277] successfully imaged the atomic structures of halogen atoms adsorbed on the tungsten {111} surface. They find that halogen atoms are adsorbed on apex sites and form a $(\sqrt{3} \times \sqrt{3})R45°$ structure on the surface; one such is shown in Fig. 4.58. As this structure has not been observed by other surface techniques, it is attributed to a field induced effect. It is also doubtful whether in the absence of the strong image field

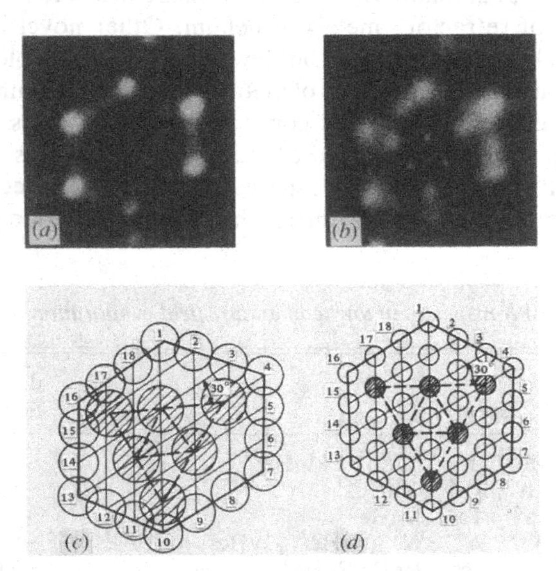

Fig. 4.58 Field ion images of Br atoms adsorbed on the W (111) surface obtained by Faulian & Bauer, and diagrams showing the adsorption sites of Br atoms in the image field.

the chemisorption sites will be the apex sites. Obviously in this case the field adsorption energy in the image field exceeds the chemisorption energy and the adsorption sites observed correspond to those in field adsorption. The apex site field adsorption proposed by Tsong & Müller, as discussed in Chapter 2, thus finds a most direct confirmation from this observation. This experiment also demonstrates convincingly that in interpreting an atomic structure observed in field ion microscope one must carefully consider whether the structure represents that of lowest surface free energy in the absence of the applied field as discussed in Section 4.1.1.

References

1 T. T. Tsong & E. M. Müller, *Phys. Rev.*, **181**, 530 (1969).
2 C. F. Ai & T. T. Tsong, *Surface Sci.*, **127**, L165 (1983).
3 S. Nishigaki & S. Nakamura, *J. Appl. Phys.*, **14**, 769 (1975).
4 T. T. Tsong, *Surface Sci.*, **81**, 28 (1979); **85**, 1 (1979).
5 A. J. Melmed & R. J. Stein, *Surface Sci.*, **49**, 645 (1975).
6 T. Sakurai, R. J. Culbertson & A. J. Melmed, *Surface Sci.*, **78**, L221 (1978).
7 Others like G. L. Kellogg, O. Nishikawa and T. T. Tsong and their co-workers have also obtained silicon images of similar quality, i.e. with ring structures visible but not atomic structures.
8 T. Sakurai, T. T. Tsong & R. J. Culbertson, *J. Vac. Sci. Technol.*, **15**, 647 (1978).
9 H. M. Liu, T. T. Tsong & Y. Liou, *Phys. Rev. Lett.*, **58**, 1535 (1987).
10 See for examples M. A. Van Hove and S. Y. Tong, *Surface Crystallography by LEED*, Springer Ser. Chem. Phys., Vol. 2, Springer, Berlin–Heidelberg, 1979; P. J. Estrup & E. G. McRea, *Surface Sci.*, **25**, 1 (1971).
11 See for example H. Niehus & G. Comsa, *Surface Sci.*, **151**, L171 (1985); **152/3**, 93 (1985).
12 G. Binnig, H. Rohrer, Ch. Gerber & E. Weibel, *Phys. Rev. Lett.*, **50**, 120 (1983); R. S. Becker, J. Golovchenko & B. S. Swartzentruber, *Phys. Rev. Lett.*, **54**, 2678 (1985); R. M. Tromp, R. J. Hammer & J. Demuth, *Phys. Rev. Lett.*, **55** 1303 (1985).
13 L. D. Mark & D. J. Smith, *Nature* (*London*), **303**, 316 (1983); L. D. Mark, *Phys. Rev. Lett.*, **51**, 1000 (1983).
14 G. L. Kellogg, *Phys. Rev. Lett.*, **55**, 2168 (1985).
15 Q. J. Gao & T. T. Tsong, *Phys. Rev. Lett.*, **57**, 452 (1986); *Phys. Lett.*, **A117**, 132 (1986).
16 Most STM micrographs published in the literature show such features.
17 T. T. Tsong, H. M. Liu & D. L. Feng, *Phys. Rev.*, **B36**, 4446 (1987).
18 K. Takayanagi, Y. Tanishiro & M. Takahashi, *Surface Sci.*, **164**, 367 (1985); K. Takayanagi and Y. Tanishiro, *Phys. Rev.*, **B34**, 1034 (1986).
19 G. Binnig, H. Rohrer, Chr. Gerber & E. Weibel, *Surface Sci.*, **131**, L379 (1983).
20 G. K. Binnig, H. Rohrer, Ch. Gerber & E. Stall, *Surface Sci.*, **144**, 321 (1984).
21 T. Hasegawa, K. Kobayashi, K. Takayanagi & Y. Yagi, *Jap. J. Appl. Phys.*, **25**, L366 (1986).

22 Q. J. Gao & T. T. Tsong, *Phys. Rev.*, **36**, 2547 (1987).
23 K. Yonehara & L. D. Schmidt, *Surface Sci.*, **25**, 238 (1971).
24 T. E. Felter, R. A. Barker & P. J. Estrup, *Phys. Rev. Lett.*, **38**, 1138 (1977).
25 M. K. Debe & D. A. King, *Phys. Rev. Lett.*, **39**, 708 (977).
26 M. K. Debe & D. A. King, *Surface Sci.*, **81**, 193 (1979).
27 T. T. Tsong & J. Sweeney, *Solid State Commun.*, **30**, 767 (1979).
28 A. J. Melmed, R. T. Tung, W. R. Graham & G. D. W. Smith, *Phys. Rev. Lett.*, **43**, 1521 (1979).
29 R. T. Tung, W. R. Graham & A. J. Melmed, *Surface Sci.*, **115**, 576 (1982).
30 See for examples D. P. Jackson, T. E. Jackman, J. A. Davies, W. N. Unertl & P. R. Norton, *Surface Sci.*, **126**, 226 (1983); C. M. Chan, M. A. van Hove, W. H. Weiberg & E. D. Williams, *Surface Sci.*, **91**, 440 (1980).
31 For example see C. M. Chan, K. L. Luke, M. A. van Hove, W. H. Weiberg & E. D. Williams, *J. Vac. Sci. Technolog.*, **16**, 642 (1979).
32 A. J. Melmed, *J. Appl. Phys.*, **38**, 1885 (1967).
33 H. P. Bonzel, in *Surface Physics of Materials*, Vol. II, ed. J. M. Blakely, Academic Press, New York, 1975, 279.
34 H. P. Bonzel & S. Ferrer, *Surface Sci.*, **118**, L263 (1982).
35 H. F. Liu, H. M. Liu & T. T. Tsong, *J. Appl. Phys.*, **59**, 1334 (1986).
36 T. T. Tsong & Q. J. Gao, *Surface Sci.*, **182**, L257 (1987).
37 See for examples T. T. Tsong, *Rep. Prog. Phys.*, **51**, 759 (1988); G. Erlich & K. Stolt, *Ann. Rev. Phys. Chem.*, **31**, 603 (1980); D. W. Bassett, in *Proc. NATO Advanced Summer Institute on Surface Mobility of Solid Materials*, Vol. B86, Plenum, New York, 1983.
38 P. J. Estrup, in *Chemistry and Physics of Solid Surfaces*, eds. R. Vanselow & R. Howe, Springer, Berlin, 1984, **5**, 205.
39 T. E. Jackman, K. Griffiths, W. N. Unertl, J. A. Davies, K. H. Gurtler, D. A. Harrington & P. R. Norton, *Surface Sci.*, **179**, 297 (1987); G. Kleinle, V. Penka, R. J. Behm, G. Ertl & W. Moritz, *Phys. Rev. Lett.*, **58**, 148 (1987).
40 G. J. R. Jones, J. H. Onuferko, D. P. Woodruff & W. Holland, *Surface Sci.*, **147**, 1 (1984).
41 G. L. Kellogg, *Phys. Rev.*, **B37**, 4288 (1988).
42 J. T. Grant, *Surface Sci.*, **18**, 228 (1969).
43 S. Hagstrom, H. B. Lyon & G. A. Somorjai, *Phys. Rev. Lett.*, **15**, 491 (1965).
44 D. G. Fedak & N. A. Gjostein, *Phys. Rev. Lett.*, **16**, 171 (1966).
45 M. A. van Hove, R. J. Koestner, P. C. Stair, J. P. Biberian, L. L. Kesmodel, I. Bartos & G. A. Somorjai, *Surface Sci.*, **103**, 189, 218 (1981).
46 J. Witt & K. Müller, *Phys. Rev. Lett.*, **57**, 1153 (1986).
47 T. T. Tsong & Q. J. Gao, *Phys. Rev.*, **B35**, 7764 (1987); Q. J. Gao & T. T. Tsong, *Phys. Rev.*, **B36**, 2547 (1987).
48 M. Ahmad & T. T. Tsong, *J. Chem. Phys.*, **83**, 388 (1985); R. Vanselow & M. Munschau, *Surface Sci.*, **176**, 701 (1986).
49 E. Land, K. Müller, K. Heinz, M. A. Van Hove, R. J. Koestner & G. A. Somorjai, *Surface Sci.*, **127**, 347 (1983); N. Bickel & K. Heiz, *Surface Sci.*, **163**, 435 (1985).
50. K. Heinz, E. Lang, K. Strauss & K. Müller, *Surface Sci.*, **120**, L401 (1982).
51 D. G. Brandon, *Surface Sci.*, **5**, 137 (1966).
52 T. T. Tsong & E. W. Müller, *Appl. Phys. Lett.*, **9**, 7 (1986).
53 H. N. Southworth & B. Ralph, *Phil. Mag.*, **14**, 383 (1966); T. F. Page & B. Ralph, *Proc. Roy. Soc. (London)*, **A339**, 223 (1974).
54 T. T. Tsong & E. W. Müller, *J. Appl. Phys.*, **38**, 545 (1967); 3531 (1967).

55 For a review of early works, see B. G. Lefervre, *Surface Sci.*, **23**, 144 (1970).
56 R. W. Newman & J. J. Hren, *Phil. Mag.*, **16**, 211 (1967); *Surface Sci.*, **8**, 373 (1967).
57 H. Berg, T. T. Tsong & J. B. Cohen, *Acta Met.*, **21**, 1589 (1973).
58 T. Kingetsu, Y. Yamamoto & S. Nenno, *Surface Sci.*, **103**, 13 (1981), and references therein.
59 Q. J. Gao & T. T. Tsong, unpublished data (1986).
60 H. F. Liu, H. M. Liu & T. T. Tsong, *Phys. Rev. Lett.*, **56**, 65 (1986).
61 For example see D. Haneman, *Rep. Prog. Phys.*, **50**, 1045 (1987).
62 H. M. Liu & T. T. Tsong, *J. Appl. Phys.*, **59**, 1532 (1987); T. T. Tsong, D. L. Feng & H. M. Liu, *Surface Sci.*, **199**, 421 (1988).
63 A. J. W. Moore, *J. Chem. Phys. Solids*, **23**, 907 (962).
64 B. Z. Olshanetski & V. I. Mashanov, *Surface Sci.*, **111**, 414 (1981), and references therein.
65 E. Bauer, *Ultramicroscopy*, **17**, 51 (1985); W. Telieps & E. Bauer, *Ultramicroscopy*, **17**, 57 (1985).
66 N. Osakabe, Y. Tanishiro, K. Yagi & G. Honjio, *Surface Sci.*, **109**, 353 (1981).
67 I. T. Inal, L. Keller & F. G. Yost, *J. Mat. Sci.*, **15**, 1947 (1980); P. Jacobaeus, J. U. Madsen, F. Krapg, R. M. J. Cotterill, *Phil. Mag.*, **B41**, 11 (1980); J. Piller, *Proc. 27th Intern. Field Emission Symp.*, eds Y. Yashiro and N. Igata, Tokyo, 1980, 285; see also Fig. 31a in R. Wagner, *Field Ion Microscopy in Materials Science*, Springer, Berlin–Heidelberg–New York, 1982; J. Piller & P. Hassen, *Acta Metall.*, **30**, 1 (1982).
68 S. Nakamura, Y. S. Ng, T. T. Tsong & S. B. McLane, *Surface Sci.*, **87**, 656 (1979).
69 M. A. Fortes & B. Ralph, *Proc. Roy. Soc. (London)*, **A307**, 431 (1968); B. Ralph, S. A. Hill, M. J. Southon, M. P. Thomas & A. R. Waugh, *Ultramicroscopy*, **8**, 361 (1982).
70 M. K. Wu, J. R. Ashburn, C. J. Thorn, P. H. Hor, R. L. Meng, L. Gao, Z. J. Huang, Y. Q. Wang & C. W. Chu, *Phys. Rev. Lett.*, **58**, 908 (1987).
71 G. L. Kellogg & S. S. Brenner, *Appl. Phys. Lett.*, **51**, 1851 (1987).
72 A. J. Melmed, R. D. Shull, C. K. Chiang & H. A. Fowler, *Science*, **239**, 176 (1988).
73 W. S. Williams, *J. Appl. Phys.*, **39**, 2131 (1968); C. L. Chen & T. T. Tsong, *MRS Symp. Proc.*, **139**, 181 (1989); R. S. Khairnar, C. V. Dhamadhikari & D. S. Joag, *J. Appl. Phys.*, **65**, 4935 (1989).
74 O. Nishikawa & M. Nagai, *Phys. Rev.*, **B37**, 3685 (1988).
75 M. K. Miller, in Workshop on Statistical Analysis of Atom-Probe Data, *35th Intern. Field Emission Symp.*, Oak Ridge, TN, 1988, unpublished.
76 J. Liu & T. T. Tsong, *Phys. Rev.*, **B38**, 8490, (1988).
77 T. T. Tsong, *Appl. Phys. Lett.*, **45**, 1149 (1984); *Phys. Rev.*, **B30**, 4946 (1984).
78 A. J. Melmed & R. Klein, *Phys. Rev. Lett.*, **56**, 1478 (1986).
79 H. B. Elswijk, P. M. Bronsveld & J. Th. M. De Hossen, *J. de Physique*, **48**, C6-305 (1987); H. B. Elswijk, PhD Thesis, Univ. Groningen (1988).
80 J. A. Thornton, *Annu. Rev. Mater. Sci.*, **7**, 239 (1977); R. A. Roy & R. Messier, *J. Vac. Sci. Technol.*, **A2**, 312 (1984).
81 S. V. Krishnaswamy, R. Messier, Y. S. Ng & T. T. Tsong, *Appl. Phys. Lett.*, **35**, 870 (1979); *J. Vac. Sci. Technol.*, **18**, 309 (1981); *Thin Solid Films*, **79**, 21 (1981).
82 A. V. Crewe, J. Wall & J. Langmore, *Science*, **168**, 1338 (1970). Also see

Proc. Nobel Symp. 47: *Direct Imaging of Atoms in Crystal and Molecules*, ed. L. Kihlborg, Chemica Scripta, Vol. 14 (1979).
83 G. Binnig, C. F. Quate, and Ch. Gerber, *Phys. Rev. Lett.*, **56**, 930 (1986).
84 T. T. Tsong, *Phys. Rev.*, **B6**, 417 (1972).
85 E. W. Müller, *Z. Electrochem.*, **61**, 43 (1957).
86 G. Ehrlich & F. G. Hudda, *J. Chem. Phys.*, **44**, 1039 (1966).
87 G. Ehrlich & C. F. Kirk, *J. Chem. Phys.*, **48**, 1465 (1968).
88 E. W. Plummer & T. N. Rhodin, *J. Chem. Phys.*, **49**, 3474 (1969).
89 T. T. Tsong, *J. Chem. Phys.*, **54**, 4205 (1971).
90 D. W. Bassett & M. J. Parsely, *Nature (London)*, **221**, 1046 (1969).
91 T. T. Tsong, *Phys. Rev. Lett.*, **31**, 1207 (1973).
92 T. T. Tsong & R. J. Walko, *Phys. Status Solidi*, **12**, 111 (1972).
93 P. L. Cowan & T. T. Tsong, *Surface Sci.*, **67**, 158 (1977).
94 T. T. Tsong & R. Casanova, *Phys. Rev. Lett.*, **47**, 113 (1981).
95 H. W. Fink & G. Ehrlich, *Surface Sci.*, **143**, 125 (1984).
96 M. Volmer & J. Easterman, *Z. Phys.*, **7**, 13 (1921).
97 See for example J. A. Venable, G. D. T. Spiller & M. Hanbücken, *Rept. Progr. Phys.*, **47**, 399 (1984).
98 See for examples: J. M. Blakely, *Progr. Mater. Sci.*, **10**, 395 (1963); A. G. Naumovets & Yu S. Vedula, *Surface Sci. Rept.*, **4**, 365 (1985); H. P. Bonzel, *CRC Crit. Rev. Solid State Sci.*, **6**, 171 (1976); G. Ehrlich & K. Stolt, *Ann. Rev. Phys. Chem.*, **31**, 603 (1980); R. Gomer, *Scientific American*, August, 1982; T. T. Tsong, *Rept. Progr. Phys.*, **51**, 759 (1988); articles by D. W. Bassett, by R. Gomer and by others in *Proc. NATO Adv. Summer Inst. on Surface Mobilities of Solid Surfaces*, Vol. B86, ed. V. T. Binh, Plenum, New York, 1983.
99 M. Drachsler, *Z. Electrochem.*, **58**, 340 (1954).
100 W. P. Dyke & W. W. Dolan, in *Adv. Elect. Elect. Phys.*, Vol. 8, ed. L. Marton, Academic Press, New York, 1956.
101 H. Utsugi & R. Gomer, *J. Chem. Phys.*, **37**, 1706 (1962); A. J. Melmed & R. Gomer, *J. Chem. Phys.*, **34**, 1802 (1961).
102 G. Ehrlich & F. G. Hudda, *J. Chem. Phys.*, **35**, 1421 (1961).
103 A. J. Melmed, *J. Appl. Phys.*, **37**, 275 (1966).
104 L. W. Swanson, R. W. Strayer & L. E. Davis, *Surface Sci.*, **9**, 165 (1968).
105 Ch. Kleint, *Surface Sci.*, **25**, 394, 411 (1971).
106 G. A. Odisharia & V. N. Shrednik, *Trudy I Vsesojuznogo soveshtshania po avtoionnoj mikroskopii*, Kcharkov, s.81, 1976.
107 R. Gomer, *Surface Sci.*, **38**, 373 (1973).
108 J. R. Chen & R. Gomer, *Surface Sci.*, **79**, 413, **81**, 589 (1979).
109 R. DiFoggio & R. Gomer, *Phys. Rev. Lett.*, **46**, 1258 (1981).
110 A. G. Naumovets, *4th Intern. Conf. Solid Films and Surfaces*, Hamamatsu, Japan, 1987, unpublished.
111 This is similar to an argument given for field desorption by atomic and ionic tunneling. See R. Gomer & L. W. Swanson, *J. Chem. Phys.*, 38, 1613 (1963); T. T. Tsong, *Surface Sci.*, **10**, 102 (1968).
112 K. Lakatos-Lindenberg & K. Shuler, *J. Math. Phys.*, **12**, 633 (1971).
113 S. Chandrasekhar, *Rev. Mod. Phys.*, **15**, 1 (1943).
114 G. Ehrlich, *J. Chem. Phys.*, **44**, 1050 (1966).
115 T. T. Tsong, P. Cowan & G. L. Kellogg, *Thin Solid Films*, **25**, 97 (1975).
116 D. A. Reed & G. Ehrlich, *J. Chem. Phys.*, **64**, 4616 (1976).
117 P. Cowan, PhD Thesis, Pennsylvania State University, 1976.

118 T. T. Tsong, in *Proc. NATO ASI on Surface Mobilities of Solid Materials, Les Arcs, France*, NATO ASI Series, Vol. 86, Plenum, New York, 1980.
119 See for example J. R. Manning, *Diffusion Kinetics for Atoms in Crystals*, Van Nostrand, Princeton, NJ, 1968.
120 T. T. Tsong, *Phys. Rev.*, **B7**, 4018 (1973).
121 G. Ayrault & G. Ehrlich, *J. Chem. Phys.*, **124**, 41 (1961).
122 D. W. Bassett & P. R. Webber, *Surface Sci.*, **70**, 520 (1978).
123 P. G. Flahive & W. R. Graham, *Surface Sci.*, **91**, 449 (1980).
124 D. W. Bassett & M. J. Parsely, *Brit. J. Appl. Phys.*, **2**, 13 (1969).
125 T. T. Tsong & G. L. Kellogg, *Phys. Rev.*, **B12**, 1343 (1975).
126 T. T. Tsong, *Progr. Surface Sci.*, **10**, 165 (1980); *Rep. Prog. Phys.*, **51**, 759 (1988).
127 S. C. Wang & G. Ehrlich, 35th International Field Emission Symp., Oak ridge, TN, 1988. *J. de Physique, Coll.*, **49**, C6–263 (1988).
128 G. Mazenko, J. R. Banavar & R. Gomer, *Surface Sci.*, **107**, 459 (1981).
129 H. P. Bonzel & E. E. Latta, *Surface Sci.*, **76**, 275 (1978).
130 V. T. Binh & P. Melinon, *Surface Sci.*, **161**, 234 (1985).
131 M. N. Barber & B. W. Ninham, *Random and Restricted Walks*, Gordon & Breach, New York, 1970.
132 E. W. Montroll, *J. Math. Phys.*, **6**, 167 (1965).
133 G. Ehrlich, *J. Chem. Phys.*, **44**, 1050 (1966).
134 K. Stolt, W. R. Graham & G. Ehrlich, *J. Chem. Phys.*, **65**, 3206 (1975).
135 G. Ehrlich, *J. Vac. Sci. Technol.*, **17**, 1 (1980).
136 T. T. Tsong & R. Casanova, *Phys. Rev.*, **B22**, 4632 (1980).
137 D. R. Bowman, R. Gomer, K. Muttalib & M. Tringides, *Surface Sci.*, **138**, 581 (1984).
138 M. Tringides & R. Gomer, *Surface Sci.*, **155**, 254 (1985).
139 See ref. 122. Also R. T. Tung & W. R. Graham, *Surface Sci.*, **97**, 73 (1980).
140 J. Wrigley & G. Ehrlich, *Phys. Rev. Lett.*, **44**, 661 (1980).
141 M. R. Mruzik & G. M. Pound, *J. Phys. F.*, **11**, 1403 (1981); S. H. Garofalini & T. Halicioglu, *Surface Sci.*, **104**, 199 (1981); De Lorenzi, G. Jacucci & V. Pontikis, *Proc. NATO ASI Series*, **86**, 556 (1981).
142 D. W. Bassett, C. K. Chung & D. R. Tice, *Le Vide*, **176**, 39 (1975).
143 S. C. Wang & T. T. Tsong, *Surface Sci.*, **121**, 85 (1982).
144 P. L. Cowan & T. T. Tsong, *Phys. Lett.*, **53A**, 383 (1975).
145 D. W. Bassett, *J. Phys. C*, **9**, 2491 (1976).
146 W. R. Graham & G. Ehrlich, *Phys. Rev. Lett.*, **31**, 1407 (1973); T. Sakata & S. Nakamura, *Surface Sci.*, **51**, 313 (1975).
147 C. H. Bennett, *Thin Solid Films*, **25**, 65 (1975).
148 W. R. Graham & G. Ehrlich, *Thin Solid Films*, **25**, 85 (1975).
149 T. T. Tsong, P. L. Cowan & G. L. Kellogg, *Thin Solid Films*, **25**, 97 (1975).
150 T. T. Tsong & R. J. Walko, 19*th Field Emission Symp.*, Urbana-Champain, IL, 1972; also *physica status solidi*, **a12**, 111 (1972).
151 D. W. Bassett, *J. Phys. C*, **9**, 2491 (1976).
152 P. L. Cowan, PhD Thesis Pennsylvania State University, 1977.
153 T. T. Tsong & R. Casanova, *Phys. Rev.*, **B21**, 4564 (1980); T. T. Tsong, *Phys. Rev.*, **B25**, 5234 (1982).
154 See for example C. Kittel, *Solid State Physics*, John Wiley, New York, 1953; G. L. Kellogg, T. T. Tsong & P. L. Cowan, *Surface Sci.*, **70**, 485 (1978).
155 O. Nishikawa, M. Wada & M. Konishi, *Surface Sci.*, **97**, 16 (1980).
156 R. Casanova & T. T. Tsong, *Thin Solid Films*, **93**, 41 (1982).

157 T. T. Tsong, W. A. Schmidt & O. Frank, *Surface Sci.*, **60**, 109 (1977).
158 T. T. Tsong, Y. Liou & J. Liu, *J. Vac. Sci. Technol.*, **A7**, 1758 (1989).
159 For theories of chemisorption, see J. P. Muscat & D. M. Newns, *Prog. Surface Sci.*, **9**, 1 (1978).
160 T. M. Lu, G. C. Wang & M. G. Lagally, *Phys. Rev. Lett.*, **39**, 411 (1977).
161 L. D. Roelofs, A. R. Kortan, T. L. Einstein & R. L. Park, *J. Vac. Sci. Technol.*, **18**, 492 (1981).
162 D. A. King, *CRC Critical Rev. Solid State Mater. Sci.*, **7**, 167 (1978).
163 V. N. Shrednik; see references cited in *Proc. 7th Intern. Vacuum Congr. and 3rd Intern. Conf. Solid Surfaces*, Vienna, 1977, 2455.
164 W. Kohn & K. H. Lau, *Solid State Commun.*, **18**, 533 (1976).
165 T. L. Einstein, *CRC Crit. Rev. Solid State Mater. Sci.*, **7**, 261 (1978).
166 T. B. Grimley, *Proc. Phys. Soc.*, **B90**, 751; **B92**, 776 (1967).
167 T. L. Einstein & J. R. Schrieffer, *Phys. Rev.*, **B7**, 3629 (1929); T. L. Einstein, *Phys. Rev.*, **B16**, 3411 (1977).
168 N. R. Burke, *Surface Sci.*, **58**, 49 (1976); A. M. Gabovich, L. G. Il'Chenko, E. Pashitaskii & Yu. A. Romanov, *Surface Sci.*, **94**, 179 (1980).
169 W. Ho, S. L. Cunningham & W. H. Weinberg, *Surface Sci.*, **62**, 662 (1977).
170 D. W. Bassett & D. R. Tice, *Surface Sci.*, **40**, 499 (1973).
171 T. Sakata & S. Nakamura, *Jap. J. Appl. Phys.*, **14**, 943 (1975).
172 D. W. Bassett, *Surface Sci.*, **23**, 240 (1970).
173 T. T. Tsong, *Phys. Rev. Lett.*, **31**, 1207 (1973).
174 T. T. Tsong & R. Casanova, *Phys. Rev.*, **B24**, 3063 (1981).
175 P. J. Feibelman, *Phys. Rev. Lett.*, **58**, 2766 (1987).
176 F. Watanabe & G. Ehrlich, Abst. 35th Intern. Field Emission Symp., Oak Ridge, TN, 1988. *J. de Physique, Coll.*, **49**, C6–267 (1988).
177 T. T. Tsong, *J. Chem. Phys.*, **55**, 4658 (1971).
178 D. A. Read & G. Ehrlich, *J. Chem. Phys.*, **64**, 4616 (1976).
179 K. Stolt, W. R. Graham & G. Ehrlich, *J. Chem. Phys.*, **65**, 3206 (1976); K. Stolt, J. D. Wrigley & G. Ehrlich, *J. Chem. Phys.*, **69**, 1151 (1978).
180 P. L. Cowan & T. T. Tsong, *22nd Intern. Field Emission Symp.*, Atlanta, GA (1975); *Proc. 7th Int. Vacuum Congr. and 3rd Int. Conf. Solid Surfaces*, Vienna, 1977, 2475; see also ref. b of 154.
181 See for example, U. Landman, *Israel J. Chem.*, **22**, 339 (1982).
182 P. L. Cowan & T. T. Tsong, *Surface Sci.*, **67**, 158 (1977).
183 D. W. Bassett, *Thin Solid Films*, **48**, 237 (1978).
184 P. R. Schwoebel & G. L. Kellogg, *Phys. Rev. Lett.*, **61**, 578 (1988).
185 T. T. Tsong, *Surface Sci.*, **122**, 99 (1982).
186 J. Amar, S. Katz & J. D. Gunton, *Surface Sci.*, **155**, 667 (1985).
187 H. W. Fink & G. Ehrlich, *Surface Sci.*, **110**, L611 (1981).
188 B. A. Boiko, D. A. Gorodetskii & A. A. Yas'ko, *Fiz. Tverd. Tela*, **15**, 3145 (1973); *Sov. Phys.-Solid State*, **15**, 2601 (1974).
189 T. L. Einstein, *Surface Sci.*, **84**, L497 (1979).
190 K. Takayanagi, Y. Tanishiro & K. Murooka, *J. de Physique, Coll.*, **C6**, 525 (1987).
191 A. V. Crewe, *Chem. Scr.*, **14**, 17 (1978).
192 M. Isaacson, D. Kopf, M. Utlaut, M. Parker & A. V. Crewe, *Proc. Natl. Acad. Sci. USA*, **74**, 1802 (1977).
193 M. Utlaut, *Phys.* Rev., **B22**, 4605 (1980).
194 R. C. Jaklevic & L. Elie, *Phys. Rev. Lett.*, **60**, 120 (1988).

195 T. T. Tsong, X. Li, S. D. Chi, Y. Liou & R. Messier, *Thin Solid Films*, **172**, L91 (1988).
196 E. Ganz, K. Sattler & J. Clarke, *Phys. Rev. Lett.*, **60**, 1856 (1988).
197 J. W. Gadzuk, J. K. Hartman & T. N. Rhodin, *Phys. Rev.*, **B4**, 241 (1971).
198 T. T. Tsong & G. L. Kellogg, *Phys. Rev.*, **B12**, 1343 (1975).
199 G. L. Kellogg & T. T. Tsong, *Surface Sci.*, **62**, 343 (1977).
200 E. W. Plummer & T. N. Rhodin, *Appl. Phys. Lett.*, **11**, 194 (1967).
201 K. Besocke & H. Wagner, *Phys. Rev.*, **B8**, 4597 (1973).
202 T. T. Tsong, *J. Chem. Phys.*, **54**, 4205 (1971).
203 S. C. Wang & T. T. Tsong, *Phys. Rev.*, **B26**, 6470 (1982).
204 J. W. Coburn, *J. Vac. Sci. Technol.*, **13**, 519 (1976); Z. L. Liau, B. Y. Tsaur & J. W. Mayer, *J. Vac. Sci. Technol.*, **16**, 121 (1979).
205 M. K. Miller, P. A. Beaven & G. D. W. Smith, *Surface and Interface Analysis*, **1**, 149 (197).
206 M. Yamamoto & D. N. Seidman, *Surface Sci.*, **118**, 535 (1982).
207 T. T. Tsong, S. V. Krishnaswamy, S. B. McLane & E. W. Müller, *Appl. Phys. Lett.*, **23**, 1 (1973).
208 T. T. Tsong, Y. S. Ng & S. V. Krishnaswamy, *Appl. Phys. Lett.*, **32**, 778 (1978).
209 Y. S. Ng, T. T. Tsong & S. B. McLane, *Phys. Rev. Lett.*, **42**, 588 (1978).
210 O. Nishikawa, H. Kawada, Y. Nagai & E. Nomura, *J. de Physique, Coll.*, **C9**, 465 (1984).
211 T. Sakurai, T. Hashizume, A. Jimbo & T. Sakata, *J. de Physique, Coll.*, **C9**, 453 (1984).
212 O. Nishikawa, E. Nomura, H. Kawada & K. Oida, *J. de Physique, Coll.*, **C2**, 297 (1986).
213 A. Cerezo, C. R. M. Grovenor & G. D. W. Smith, *Appl. Phys. Lett.*, **46**, 567 (1985).
214 D. McLean, *Grain Boundaries in Metals*, Oxford University Press, Oxford 1957.
215 J. H. Sinfelt & J. A. Cusumano, in *Advanced Materials in Catalysis*, eds J. J. Burton & J. L. Garten, Academic Press, New York, 1977.
216 P. van der Plank & W. M. H. Sachtler, *J. Catal.*, **12**, 35 (1968).
217 J. H. Sinfelt, J. L. Carter & D. J. C. Yates, *J. Catal.*, **24**, 283 (1972).
218 W. M. H. Sachtler & G. J. H. Doregelo, *J. Catal.*, **4**, 654 (1965).
219 R. L. Park, J. E. Houston & D. G. Schreiner, *J. Vac. Sci. Technol.*, **9**, 1023 (1971).
220 C. R. Helms & K. Y. Yu, *J. Vac. Sci. Technol.*, **12**, 276 (1975); K. Watanabe, M. Hashiba & T. Yamashina, *Surface Sci.*, **61**, 483 (1976).
221 H. H. Brongersma & T. M. Buck, *Surface Sci.*, **53**, 649 (1975); H. H. Brongersma, M. J. Sparnaay & T. M. Buck, *Surface Sci.*, **71**, 657 (1978).
222 D. T. Ling, J. N. Miller, I. Lindau, W. E. Spicer & P. M. Stefan, *Surface Sci.*, **74**, 612 (1978).
223 S. V. Krishnaswamy, S. B. McLane & E. W. Müller, *J. Vac. Sci. Technol.*, **11**, 899 (1974).
224 Y. S. Ng, T. T. Tsong & S. B. McLane, *Surface Sci.*, **84**, 31 (1979).
225 J. W. Gibbs, *The Collected Works of J. Willard Gibbs*, Vol. 1, Yale Univ. Press, New Haven, 1948.
226 F. L. Williams & D. Nason, *Surface Sci.*, **45**, 377 (1974); P. Wynblatt & R. C. Ku, *Surface Sci.*, **65**, 511 (1977).

227 F. F. Abraham, N. H. Tsai & G. M. Pound, *Surface Sci.*, **83**, 406 (1979); F. F. Abraham, *Phys. Rev. Lett.*, **46**, 546 (1981).
228 Ph. Lambin & J. P. Gaspard, *J. Phys. F.*, **10**, 2413 (1980).
229 See for examples V. Kumar, *Phys. Rev.*, **B23**, 3756 (1981); G. Kerker, J. L. Moran Lopez & K. H. Bennemann, *Phys. Rev.*, **B15**, 638 (1977).
230 T. S. King & R. G. Donnelly, *Surface Sci.*, **141**, 417 (1984).
231 D. Tomanek, S. Mukherjee, V. Kumar & K. H. Bennemann, *Surface Sci.*, **114**, 11 (1982).
232 T. T. Tsong, Y. S. Ng & S. B. McLane, *J. Chem. Phys.*, **73**, 1464 (1980).
233 T. Sakurai, T. Hashizume, A. Jimbo, A. Sakai & S. Hyodo, *Phys. Rev. Lett.*, **55**, 514 (1985).
234 T. T. Tsong, D. M. Ren & M. Ahmad, *Phys. Rev.*, **B38**, 7428 (1988).
235 Y. S. Ng, S. B. McLane & T. T. Tsong, *J. Vac. Sci. Technol.*, **17**, 154 (1980).
236 K. Wandelt & C. R. Brundle, *Phys. Rev. Lett.*, **46**, 1529 (1981).
237 T. T. Tsong, Y. S. Ng & S. B. McLane, *J. Appl. Phys.*, **51**, 6189 (1980).
238 R. N. Bennett, R. G. Barrera, C. L. Cleveland & U. Landman, *Phys. Rev.*, **B28**, 1667 (1983); R. N. Bennett, U. Landman & C. L. Cleveland, *Phys. Rev.*, **B28**, 6647 (1983).
239 M. Ahmad & T. T. Tsong, *J. Chem. Phys.*, **83**, 388 (1985).
240 F. L. Williams & G. C. Nelson, *Appl. Surface Sci.*, **3**, 409 (1979); P. H. Holloway & F. L. Williams, *Appl. Surface Sci.*, **10**, 1 (1982); F. C. M. J. M. van Deft & B. E. Nieuwenhuys, *Surface Sci.*, **162**, 538 (1985); A. D. Van Langeveld & J. W. Niementsverdriet, *Surface Sci.*, **178**, 880 (1986).
241 D. M. Ren & T. T. Tsong, *Surface Sci.*, **184**, L439 (1987).
242 S. Ichimura, M. Shikata & R. Shimizu, *Surface Sci.*, **108**, L393 (1981).
243 M. L. Shek, *Surface Sci.*, **149**, L39 (1985).
244 See for examples: P. S. Ho and K. N. Tu, eds, *Thin Films and Interfaces, Mat. Res. Soc. Symp. Proc.*, Vol. 10, 1982; L. J. Brillson, *Surface Sci. Rept.*, **2**, 123 (1982); P. S. Ho, *J. Vac. Sci. Technol.*, **A1**, 745 (1983).
245 A. P. Janssen & J. P. Jones, *Surface Sci.*, **41**, 257 (1974).
246 O. Nishikawa, Y. Tsunashimas, E. Nomura, S. Horie, S. Wada, M. Shibata, T. Yoshimura & R. Uemori, *J. Vac. Sci. Technol.*, **B1**, 6 (1983).
247 T. T. Tsong, S. C. Wang, H. F. Liu, H. Cheng & M. Ahmad, *J. Vac. Sci. Technol.*, **B1**, 915 (1983).
248 H. F. Liu, H. M. Liu & T. T. Tsong, *Phys. Rev. Lett.*, **56**, 65 (1985); *Appl. Phys. Lett.*, **48**, 1661 (1986).
249 H. M. Liu, H. F. Liu & T. T. Tsong, *Surface Sci.*, **179**, L71 (1987).
250 H. F. Liu, H. M. Liu & T. T. Tsong, *Appl. Phys. Lett.*, **47**, 524 (1985).
251 R. T. Tung, J. M. Gibson & J. M. Poate, *Phys. Rev. Lett.*, **50**, 429 (1983).
252 O. Nishikawa, M. Kaneda, M. Shibata & E. Nomura, *Phys. Rev. Lett.*, **53**, 1252 (1984).
253 A. Kobayashi, T. Sakurai, T. Hashizume & T. Sakata, *J. Appl. Phys.*, **57**, 3448 (1986).
254 S. Nakamura, T. Hashizume, Y. Hasegawa & T. Sakurai, *Surface Sci.*, **172**, L551 (1986).
255 H. D. Beckey, *Field Ionization Mass Spectrometry*, Pergamon Press, Oxford (1971).
256 H. D. Beckey, K. Levesen, F. W. Röllgen & H. R. Schulten, *Surface Sci.*, **70**, 325 (1978); F. W. Röllgen & H. D. Beckey, *Surface Sci.*, **26**, 100 (1971).
257 See E. W. Müller & T. T. Tsong, *Prog. Surface Sci.*, **4**, 1 (1973).

258 A. R. Anway, *J. Chem. Phys.*, **50**, 2012 (1969); W. A. Schmidt, O. Frank & J. H. Block, *Surface Sci.*, **44**, 185 (1974); D. L. Cocke & J. H. Block, *Surface Sci.*, **70**, 363 (1978).
259 R. Smoluchowski, *Phys. Rev.*, **60**, 661 (1941).
260 See N. D. Lang, *Solid State Phys.*, **28**, 225 (1973).
261 T. C. Clements & E. W. Müller, *J. Chem. Phys.*, **37**, 2684 (1962).
262 G. Hertzberg, *J. Chem. Phys.*, **70**, 4806 (1979).
263 C. F. Ai & T. T. Tsong, *Surface Sci.*, **138**, 339 (1984); *J. Chem. Phys.*, **81**, 2845 (1984).
264 A. J. Jason, B. Halpern, M. G. Inghram & R. Gomer, *J. Chem. Phys.*, **52**, 2227 (1970).
265 N. Ernst & J. H. Block, *Surface Sci.*, **126**, 357 (1983).
266 T. T. Tsong, T. J. Kinkus & C. F. Ai, *J. Chem. Phys.*, **78**, 4763 (1983); T. T. Tsong & T. J. Kinkus, *Phys. Rev.*, **29**, 529 (1984).
267 M. Boudart, in *Physical Chemistry*: *An Advanced Treatise*, eds H. Eyring, W. Jost & D. Henderson, Academic Press, New York, 1975.
268 W. Liu & T. T. Tsong, *Surface Sci.*, **156**, L26 (1986).
269 For surface science studies of ammonia synthesis, see also G. Ertl, *J. Vac. Sci. Technol.*, **A1**, 1247 (1983); *CRC Crit. Rev. Solid State Mater. Sci.*, **10**, 349 (1982).
270 Y. Li, W. Drachsel, J. H. Block & F. Okuyama, *J. de Physique*, *Coll.*, **C7**, 413 (1986).
271 W. Liu, C. L. Bao, D. M. Ren & T. T. Tsong, *Surface Sci.*, **180**, 153 (1987).
272 G. Ertl, *Surface Sci.*, **152**, 328 (1985).
273 E. W. Müller, S. B. McLane & J. A. Pantiz, *Surface Sci.*, **17**, 430 (1969); E. W. Müller & S. V. Krishnaswamy, *Phys. Rev. Lett.*, **31**, 1282 (1984).
274 E. W. Müller & T. Sakurai, *J. Vac. Sci. Technol.*, **11**, 878 (1974).
275 T. T. Tsong & T. J. Kinkus, *Phys. Rev.*, **B29**, 529 (1984); T. T. Tsong, *Phys. Rev.*, **B30**, 4946 (1984); T. T. Tsong & Y. Liou, *Phys. Rev. Lett.*, **55**, 2180 (1985).
276 P. Pykko, *J. Chem. Soc. Faraday Trans. II*, **75**, 1256 (1978); M. Hotokka, T. Kindsted, P. Pykko & B. O. Roos, *Mol. Phys.*, **52**, 23 (1984).
277 K. Faulian & E. Bauer, *Surface Sci.*, **70**, 71 (1978).

5

Selected topics in applications

5.1 Observation and analysis of lattice defects

5.1.1 Point defects

The field ion microscope, with its spatial resolution sufficient to resolve atoms in most high index planes of an emitter surface and the capability of reaching into the bulk of the sample by removing surface layers with field evaporation, can be used to observe and analyze intrinsic and extrinsic lattice defects in the sample. The atom-probe, on the other hand can analyze the chemistry of a solid with comparable spatial resolution, and has been actively used to study compositional variations across phase boundaries, in precipitates, and across a grain boundary in impurity segregations, etc. In fact, atom-probe and field ion microscopy is currently the only technique capable of imaging lattice defects and simultaneously analyzing the chemistry of these defects in a solid on an atomic scale. Some of the most actively pursued and important applications in field ion microscopy have been in materials science and physical metallurgy. At least half the papers published in field ion microscopy are concerned with these subjects. As comprehensive reviews of these studies by experts in this field already exist,[1] only selected topics will be briefly discussed here.

The simplest lattice defects as far as FIM observations are concerned are point defects, such as vacancies, self-interstitials and substitutional as well as interstitial impurity atoms. Vacancies invariably show up as dark spots in the field ion images. Other point defects may appear as either bright image spots or 'vacancies' in the image. Thus these defects can be identified from field ion images of high index planes where all the atoms in a plane are fully resolved.

(a) *Thermal vacancies*

FIM was applied by Müller[2] to study vacancy concentration in a Pt wire annealed and subsequently quenched from near the melting point. By removing 71 successive {012} planes, 5 monovacancies were found out of a total of 8500 atomic sites. This gives a vacancy concentration of 5.9×10^{-4} and thus a formation energy of ~1.15 eV according to the equation $n_v/n_0 = \exp(-E_v/kT)$, where n_v/n_0 is the atomic fraction of vacancies, and E_v is the formation energy. Unfortunately, it is soon recognized that a vacancy concentration as high as 10^{-2} can be found on this surface in samples made from low temperature annealed high purity Pt wires.[3] Obviously, these 'vacancies' are artifacts of the imaging process, and are not real thermal vacancies of the metal. Most such 'vacancies' are produced by hydrogen, or another chemically reactive gas, induced field evaporation, also known as field etching, of Pt atoms from the surface,[4] while a small fraction of them may represent non-imaged or preferentially field evaporated impurity atoms also.[5] Figure 5.1 shows the image contrast of vacancies and also the unusually high concentration in the Pt {012} planes due to a field induced chemical etching effect.

Berger *et al.*[6] report a comprehensive study of the concentrations of monovacancy and divacancy in brine quenched high purity Pt wires from 1700°C, where artifact vacancies due to field induced chemical etching are carefully avoided. They also find unusually large concentrations of mono- and divacancies in the {012} and {137} surfaces, thus these surfaces cannot be used in such a study. The artifact vacancies, however, can be largely avoided if one uses images of other surfaces such as the {135}, {023}, {124}, {123} and {167}. Out of a total of 593 800 sites examined, 157 monovacancies and 9 divacancies are found. The monovacancy concentration in this sample is therefore derived from these FIM images to be $(2.64 \pm 0.14) \times 10^{-4}$ at. fr., while the ratio of divacancies and monovacancies is 0.06 ± 0.02. This monovacancy concentration is much lower than the calculated monovacancy concentration at 1700°C, indicating the loss of vacancies during quenching of the wire, or more probably during polishing of the tips in the molten salt even though they try to avoid this loss by using a low temperature method for the initial polishing of the wires. They assume that at a critical temperature the mobility of monovacancies becomes so slow that they are effectively

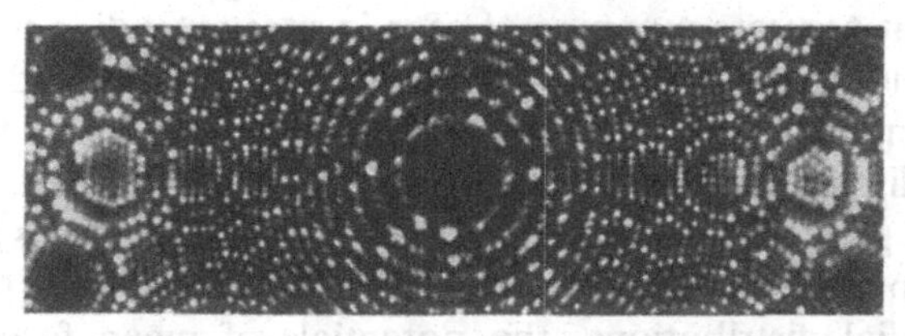

Fig. 5.1 Field ion image of a Pt surface showing the image contrast of vacancies on the {012} planes. These vacancies are however produced by a field induced chemical etching effect, not equilibrium thermal vacancies.

frozen in the sample. This temperature is found to be 443°C. On the basis of known monovacancy migration energy and pre-exponential factors, the divacancy binding energy at 443°C is then calculated to be 0.23 eV. It is also found that the number of divacancies decreases much faster than monovacancies upon annealing the sample at 350°C, signifying that the mobility of divacancies is much faster than that of monovacancies. Attempts have also been made to measure vacancy concentrations in an ordered alloy.[7] Owing to the problem of image contrast where one of the species may not give rise to a field ion image, so that 'vacancies' can also be produced by misplaced atoms if the sample is not perfectly ordered or if the stoichiometry is not perfect, the conclusions drawn are unlikely to be correct. In a study of the concentration of lattice defects at a given equilibration temperature, it is also important that tips are prepared by a low temperature method so that no annealing effects occur during the tip polishing process.

(b) Clustering and interactions of solute atoms in dilute alloys

Field evaporated surfaces of solid solution alloys appear in general rather disordered in the field ion microscope. While ring structures around low index planes can be seen, atomic arrangements appear to be disordered. For dilute alloys, however, surfaces developed by field evaporation can be quite regular, except for some 'vacancies' and bright image spots. As discussed in Section 4.1.3, owing to selective field ionization and selective field evaporation effects, solute atoms in alloys can often be distinguished from matrix atoms. They can appear either as dark spots or bright spots. Unfortunately, dark spots can also be produced by a field induced chemical etching effect, i.e. an atom inside a net plane can be field evaporated at a much lower field by chemisorption of a chemically reactive gas molecule. Preferential field evaporation may also occur for matrix atoms surrounding a solute atom. It is possible that selective field ionization is not confined to a solute atom, but also extends to its surrounding matrix atoms. Unequivocal identification of solute atoms in alloys from field ion images alone is therefore a difficult problem. An additional problem in the study of the short range order of solute atoms in an alloy is that accurate mapping of atomic sites extending to different atomic layers is often difficult. Fortunately for certain high index planes, interlayer atomic site mapping may be accurate enough for finding three-dimensional distributions of solute atoms in dilute alloys where the image quality is still good. Thus radial distributions of solute atoms and short range order parameters can be derived from these site distributions. From the radial distributions, the potentials of mean force, or the effective pair potentials if the solute concentrations are sufficiently small, can be derived. Gold & Machlin[8] and Chen & Balluffi[9] have studied interactions and short range ordering of solute atoms of Au and Ni in Pt base dilute alloys. Their results are summarized in Fig. 5.2. In the work of Gold & Machlin, possible field induced chemical etching effect was not considered, whereas Chen & Balluffi carefully avoided such an effect by not using fcc {012} and {137} planes. Solute atom interactions show a very weak interaction of only ~0.1 eV and a non-monotonic distance dependence, similar to adatom–adatom interactions as already discussed in Chapter 4. In adatom–adatom interaction studies, there is no uncertainty in image identifications and the distance range covered is also much larger.

Atomic clustering or atomic complex formation in materials can also

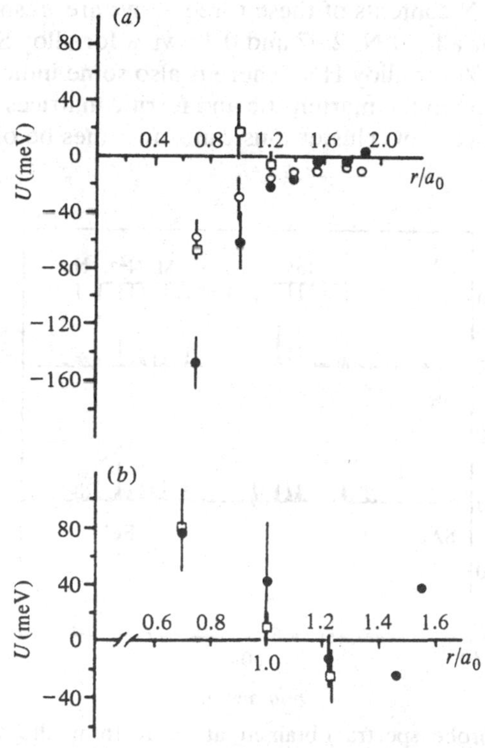

Fig. 5.2 (a) Interaction potential between two gold atoms in Pt–4 at.% Au (open circles), in Pt–0.62 at.% Au (filled circles), and in Pt–3.1 at.% Au (squares). Data are compiled from those of Chen & Bulluffi,[9] and of Machlin.[8]
(b) Interaction potential between two nickel atoms in Pt–0.65 at.% Ni (filled circles) and in Pt–0.31 at.% Ni (squares); from the same sources.

be studied with the atom-probe. In a dilute alloy of matrix atoms M and solute atoms X and Y, if the solute concentrations are very small, the probability of detecting XY complex ions in low temperature field evaporation should be very small, of the order of $c_x c_y$ where c_x and c_y are the fractional abundances of X and Y, unless X atoms and Y atoms already form XY clusters or complexes in the alloy. Andren *et al.*[10] studied the austenitic phase of four nitrogen containing duplex austenitic–ferritic stainless steels by microhardness measurements and atom-probe analysis. They find the hardness increases with the nitrogen content, and atom-probe analysis shows that a large fraction of the nitrogen atoms appears as MoN cluster ions. They argue that the observation of such cluster ions indicates the existence of Mo–N complexes in austenite. Between 70 and 100% of the least abundant Mo and N atoms are observed as MoN cluster ions, as shown in Fig. 5.3. The average Mo and N contents of these three alloys are, respectively, 3.05 and 0.05 wt% for alloy LN, 2.97 and 0.13 wt% for alloy SAF2205, and 3.02 and 0.20 wt% for alloy HN. There is also some indication of MoC complex formation in the martensitic and ferritic matrices of high speed stainless steel since MoC cluster ions can sometimes be observed. This

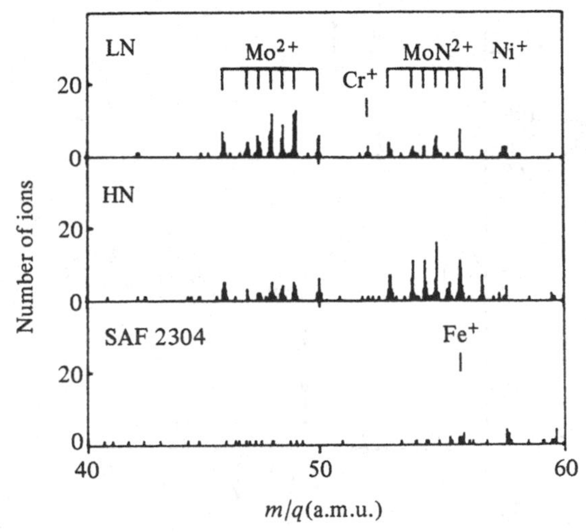

Fig. 5.3 Atom-probe spectra obtained at 90 K from alloys LN, HN and SAF2304. These alloys contain ~0.02 to 0.03 wt% C, ~0.3 to 0.4 wt% Si, ~1.4 to 1.6 wt% Mn, ~22 wt% Cr, ~4.5 to 7.0 wt% Ni, ~3.0 wt% Mo, ~0.05 to 0.20 wt% N, and the balances Fe. Yet MoN^{2+} ions are observed with abundances between 70 and 100%. This can happen only if Mo–N clusters already exist in the alloys. (Courtesy of H. O. Andren.)

experiment thus clearly shows the power of the atom-probe in studies of complex formation in dilute alloys. However, observation of MoN ions does not indicate formation of MoN complexes in the alloy. It merely indicates that Mo_nN_m are formed which then field evaporate in part as MoN^{2+} ions.

5.1.2 Line defects

Field ion microscope image structures due to one- to three-dimensional lattice defects were subjects of intense investigation during the 1960s, and detailed discussions can be found in the book by Bowkett & Smith (reference 13(c) of Chapter 1). Field ion image structures from the emergence of dislocations to the emitter surface were also studied with the facilitation of computer simulations. In earlier FIM works, spiral image structures were interpreted to represent screw dislocations.[11,12] Ranganathan[13] and Pashley[14] show that this interpretation is oversimplified and is not necessarily correct. One can easily understand that if an edge dislocation emerges from an atomically resolved surface and the dislocation core line is also perpendicular to the surface, one should observe a mismatch of the atomic rows at the position of the core, and the Burger's vector can easily be found. Since only a small number of surface atoms are imaged and only a small fraction of these atoms belong to the same top surface layer, the chance of having such a situation is rare. In general, only the long range effect of a dislocation can be observed and a dislocation, regardless of whether it is a screw dislocation or an edge dislocation, will appear as a spiral in the field ion image. The nature of the spiral configuration is determined by the dot product $\mathbf{g} \cdot \mathbf{b}$, where $\mathbf{b}$ is the Burger's vector and $\mathbf{g} = \mathbf{n}/s_{hkl}$, where $\mathbf{n}$ is a unit vector normal to the (hkl) plane and s_{hkl} is the interplanar distance of that plane. When $\mathbf{g} \cdot \mathbf{b} = 0$, the Burger's vector lies in the plane. This is the situation we have just discussed. Unless the dislocation core emerges out of an atomically fully resolved plane, the dislocation may escape from being observed. In general, dislocations will give rise to interleaved helicoid structures. The number of leaves is determined by the value of $\mathbf{g} \cdot \mathbf{b}$, which represents the vertical component of the Burger's vector in units of s_{hkl}. Detailed discussions of both field ion image structures and computer simulation images can be found in the literatures[15] and reviews,[16] and will not be further discussed here. It is, however, well recognized that proper interpretations of field ion images of dislocation networks and lattice distortions around a dislocation core are very difficult owing to the complicated geometry of the field ion emitter surface and the non-uniform image magnification, the small number of surface atoms which are imaged, and the still limited resolution of the field ion microscope. It appears that atomic structures and displacements of atoms around dislocation cores can be more easily studied with TEM lattice images of thin films where geometrical relations are simpler and are easier to interpret. However, certain types of dislocations may not be stable in very thin films.

5.1.3 Planar defects and interfaces

(*a*) *Stacking faults*

Many different kinds of planar defects have been either observed or investigated in detail by field ion microscopy. There are at least three features of interfaces which can give rise to distinctive image contrasts in the field ion microscope. These are a difference in the atomic structures across a planar boundary, in the chemical species across a phase boundary, and in the atomic geometries at two sides of a boundary as a result of a slight difference in their evaporation fields. One of the simplest planar defects is the stacking fault. The bcc lattice can be considered to be built up of {112} layers in an ordered stacking sequence of ABCDEFABCD . . . and so on. The fcc and hcp lattices are built up of closely packed planes in ABCABCAB . . . and ABABAB . . . ordered sequences respectively. Faults in the normal stacking sequence can arise from a dissociation of dislocations which will lower the elastic energy, but this process is balanced by an increase in the misfit energy.[17] In field ion microscopy of cobalt, Nishikawa & Müller[18] observe many band-like image structures which arise from stacking faults. Cobalt is a particularly interesting metal for studying stacking faults because of the expected low stacking fault energy, and also because of the exhibition of an fcc–hcp phase transformation. They find that faults are basically parallel to the basal plane. However, every 20 to 30 Å, a small jog in the same direction can be observed, so that the faults appear to cross the basal plane with an angle of about 10°. Owing to the still limited resolution of the FIM, it is impossible to sort out the exact stacking sequence across these faults. However, the fault density can easily be derived.

(*b*) *Twin boundaries*

Twinning is a mode of deformation favored at low temperatures and high strain rate.[19] In field ion microscopy of fcc metals such as Ir, Pt and Rh, field stress of the image field often can cause the tip to slip and twin boundaries may suddenly appear within the tip cap, or the imaged region of the tip. A field ion image of a twin slice in Ir is already shown in Chapter 3 in connection with a discussion of computer image simulation. Rendulic & Müller[20] observed a three-fold symmetric twin slice structure in a (111) oriented Ir tip. Upon field evaporation of approximately eighty atomic layers, the twin structure disappeared. Since the twin structure is completely symmetric with respect to the tip axis, it is most likely produced by the field stress of the image field rather than originally existing in the sample. The twin boundary is characterized by a complete ring matching across the boundary at some crystal planes.

(c) *Grain boundaries*

One of the earliest contributions of field ion microscopy to physical metallurgy was to help clarify the atomic structure of high angle grain boundaries. The imaged area in a field ion microscope is only of the order of 10^{-10} cm^2 or less. The chance of having a grain boundary passing across the imaged area of a tip is very small unless the grain size of the tip material is made to be very small. Despite such an oddity, grain boundaries are often observed in the field ion microscope. To improve the chance of having a grain boundary running across a field ion emitter surface, the wire can be bent several times and the tip prepared from the straightened out, cold worked section of the wire. But then the tip can easily be ruptured during the field evaporation and imaging processes. The grain boundaries so prepared do not have thermally equilibrated atomic structures either. A better method, developed by Loberg *et al.*,[21] is to prepare a tip from a section of a wire where TEM images already show the existence of a grain boundary. The wire can of course be carefully annealed first. A considerable amount of work on FIM studies of grain boundary structures was reported in the 1960s, and some of these studies were also in conjunction with TEM observations. In grain boundary studies, one is also interested in impurity segregation to the boundary, which can change profoundly the physical properties of a material. Grain boundary segregation and chemistry can, of course, be analyzed in the atom-probe. Figure 5.4 shows a thermally equilibrated structure of a grain boundary running across the tip surface. The boundary plane is nearly parallel to the tip axis.

McLane & Müller[22] reported an early analysis of grain boundary structures in tungsten. Field ion micrographs showed no amorphous region between two crystals as early theoretical models had suggested. The boundary was no wider than two lattice spacings, and in most regions there was a complete fit of seemingly undisturbed lattices. The geometrical relations of the two crystals were traced by controlled field evaporation through 368 successive surface layers. The boundary planes were found to be an (001) for one crystal and a (343) for the other crystal. The misfit of the two planes was found to be fairly small, since the atomic structure of the (343) surface was very close to the $(\sqrt{2} \times 3\sqrt{2})R45°$ superlattice of the (001) surface. The unit cell sizes of the (343) surface

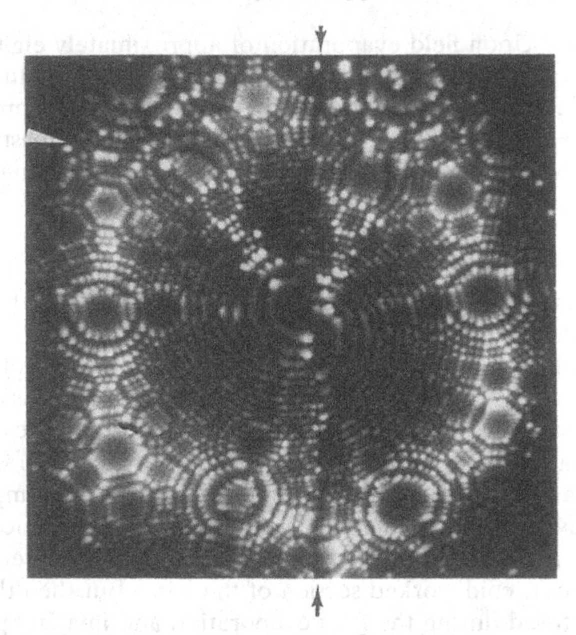

Fig. 5.4 Grain boundaries in a tungsten tip. One of them runs nearly parallel to the tip axis across the central section of the tip, as indicated by black arrows. One is near the upper left-hand corner, which is more difficult to visualize. The lattice match of the crystal grains is excellent. The tip has been carefully annealed, thus the grain boundaries are thermally well equilibrated.

and the $(\sqrt{2} \times 3\sqrt{2})R45°$ superlattice of the (001) surface were 4.47×13.05 Å and 4.47×13.41 Å, respectively. The misfit was therefore only ~2.7% in one direction and 0% in the other. In other words, there was a coincident lattice at the interface of the two crystal grains, a structure similar to that in a model discussed by Kronberg & Wilson.[23] On the basis of FIM studies Brandon *et al.*,[24] and subsequently Ranganathan,[25] Smith and co-workers[26] and many others contributed extensively to the interpretation of field ion images and to the further development of the 'coincident site lattice' model of the atomic structure of high angle grain boundaries which forms the basis for the further development of other models of grain boundary structures.[27] Many methods have been developed for the purpose of studying grain boundaries with the FIM. Bolin *et al.*[28] develop a method to digitize FIM images of a grain boundary so that three-dimensional contours of the boundary plane can be reconstructed from these images. They find that although large sections of the boundary are relatively planar, the boundary also contains protrusions, ledges and serrations, etc. There are a considerable amount of FIM works on grain boundary structures. Details of these analyses can be found in the book by Bowkett & Smith. In general, agreement with the simple coincident site lattice model is fair, but more sophisticated models have to be developed to explain specific boundaries observed. One must keep in mind, however, that unless FIM samples are carefully annealed, the grain boundary structures observed, particularly those protrusions, ledges, and serrations, etc., found by Bolin *et al.*, may not represent structures of lowest interface free energy and comparisons with theoretical models may not be very meaningful. The problem can of course be overcome by using the approach of Loberg *et al.* in which specimens are prepared from a section of a wire in which TEM images already show the existence of a grain boundary. The sample wire can be carefully annealed before being used for preparing tip samples.

Segregation of either alloy components or impurities to grain boundaries is a subject of great importance in physical metallurgy since such segregation may change profoundly the physical, or mechanical, and chemical properties of the material.[29] Many techniques have been used to study interface segregation phenomena. These include indirect methods such as measurements of microhardness or grain boundary energy and direct methods such as auger electron spectroscopy, analytical electron microscopy, and secondary ion mass spectroscopy and microscopy. These, while having their advantages, all have a severe drawback, i.e. the lack of spatial resolution required for an atomistic understanding of interface segregation phenomena. Field ion microscopy, particularly atom-probe field ion microscopy, with its atomic resolution in both structural and chemical analyses, is ideally suited for such studies and has been making important contributions in this area of research.

Fortes & Ralph[30] report observation of bright image spots in iridium samples containing oxygen solute atoms. Solute enrichments in the vicinities of grain boundaries are measured by counting the number of bright image spots near the boundaries. They find that oxygen segregates to the poorly fitted parts of the boundaries with a concentration about 6 times higher than the average concentration for a sample containing 470 p.p.m. of oxygen. The width of the segregation zone is as wide as 450 Å. Howell *et al.*,[31] on the other hand, find that the width of the segregation zone of Cr at tungsten grain boundaries and of Nb at stacking fault interfaces in Co is no more than 5 Å. The Cr at the tungsten boundaries is found to distribute randomly while Nb at the Co stacking fault planes appears to be clustered.

In his early attempt to study diffusion of foreign atoms along the grain boundary, Tsong[32] observed diffusion of Co atoms into a grain boundary of tungsten when the surface was heated, as shown in Fig. 5.5. In this preliminary experiment, about 20 layers of Co atoms were vapor deposited on a tungsten surface with a grain boundary. Upon heating the surface to 700°C for 60 seconds, Co atoms diffused into the grain

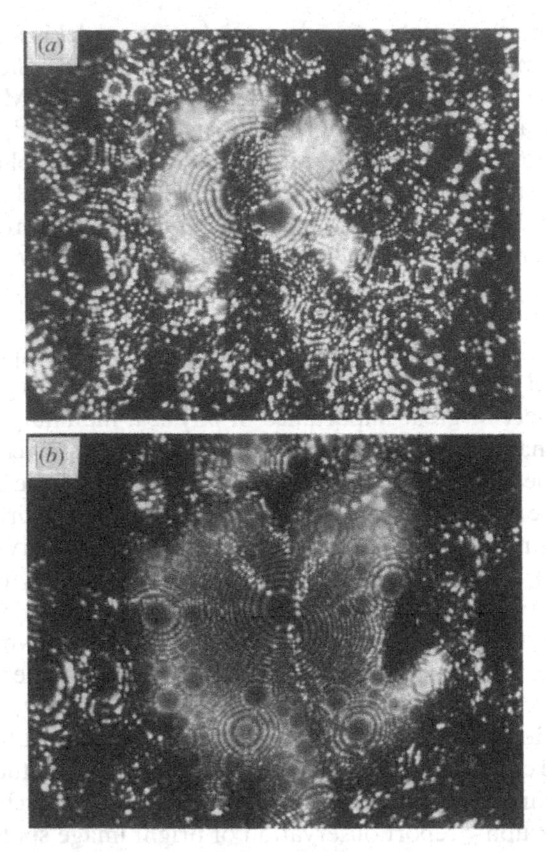

Fig. 5.5 (*a*) About 20 atomic layers of Co are vapor deposited on the tungsten tip surface shown in Fig. 5.4, and then annealed to 700°C for 60 s. It appears that an ordered W–Co alloy is formed on the tip surface and also in the grain boundaries. The grain boundaries are widened by diffusion of Co atoms into the boundaries. (*b*) About 20 (110) layers have been field evaporated. The grain boundaries are now decorated with bright image spots due to Co atoms within these grain boundaries.

boundary. Contrary to bulk Co metal, which did not give a stable helium field ion image at 78 K, the polycrystalline Co formed on the tungsten surface gave stable helium ion images at that temperature. As a result of the diffusion, the grain boundary near the surface region was widened considerably. However, when about 20 (110) layers of tungsten were field evaporated, the grain boundary returned to its normal width, but now it was decorated with bright image spots. These bright image spots could only be due to the presence of Co atoms. He did not pursue this subject further because of other involvements. Although in principle quantitative data on grain boundary diffusion can be studied with this method as was also pointed out in the book by Müller & Tsong[32], no study in this direction has so far been reported. From the much better resolved and very well ordered images of the 'polycrystalline' Co obtained, an ordered alloy phase may have formed on the tungsten surface and inside the grain boundary. It appears that this is an interesting subject where both the formation of a thin layer of ordered alloy on the surface and the surface diffusion of foreign atoms along the grain boundary can be studied, and may be worth further pursuit.

An FIM study of grain boundary segregation is possible only if solute atoms give distinctive image contrast. It is perhaps not terribly important whether it is the solute atoms which give rise to bright image spots, or whether their presence causes the images of the surrounding atoms to become brighter, unless of course an atomic map of the solute atoms is sought. Image contrast may also depend on the crystallographic region. Another difficulty of using a simple FIM for studying grain boundary segregations is that when there are more than two components in the sample it is impossible to tell which of the species segregate to the boundary. An atom-probe, with its single atom chemical analysis capability and a spatial resolution nearly comparable to an FIM, is of course a more ideal tool for studying interface segregations, which include surface segregation, segregation to grain boundaries and phase boundaries, etc. Atom-probe studies of surface segregation and impurity co-segregation to the surface, or the solid–vacuum interface, have already been discussed in Chapter 4. Turner & Parazian[33] appear to have been the earliest workers to apply the atom-probe to study grain boundary segregation, in this case segregation of carbon to grain boundaries in an Fe–0.05 at.% C alloy annealed at 723 K. They find a broad segregation zone of about 500 Å in width where the carbon concentration is over 0.4 at.%, but the peak value is still much less than one would expect from a monolayer of carbon at the boundary plane. However, the amount of data collected seems too small to be statistically reliable, and the probe-hole used seems too large to be able to measure the peak carbon concentration reliably. Grain boundary and phase boundary segregations constitute one of the most actively pursued subjects for nearly all the atom-probe metallurgy groups in the world. For a greater detailed discussion of these works, we refer to reviews by Wagner, by Brenner & Miller,[34] by Ralph *et al.*,[35] by Karlsson, Norden & Andren,[36] by Sakurai *et al.*,[1] and by Miller & Smith.[1] Here we will describe briefly a few of these works.

A careful, quantitative study on grain boundary segregation, boron segregation in austenitic stainless steels of types 316L (with 206 at. p.p.m. of boron, and with <5 at. p.p.m. of boron) and 'Mo-free 316L' (75 at. p.p.m. boron), has been reported by Karlsson, Norden, and Andren[36] using a combination of TEM, FIM, AP and IAP techniques. Concen-

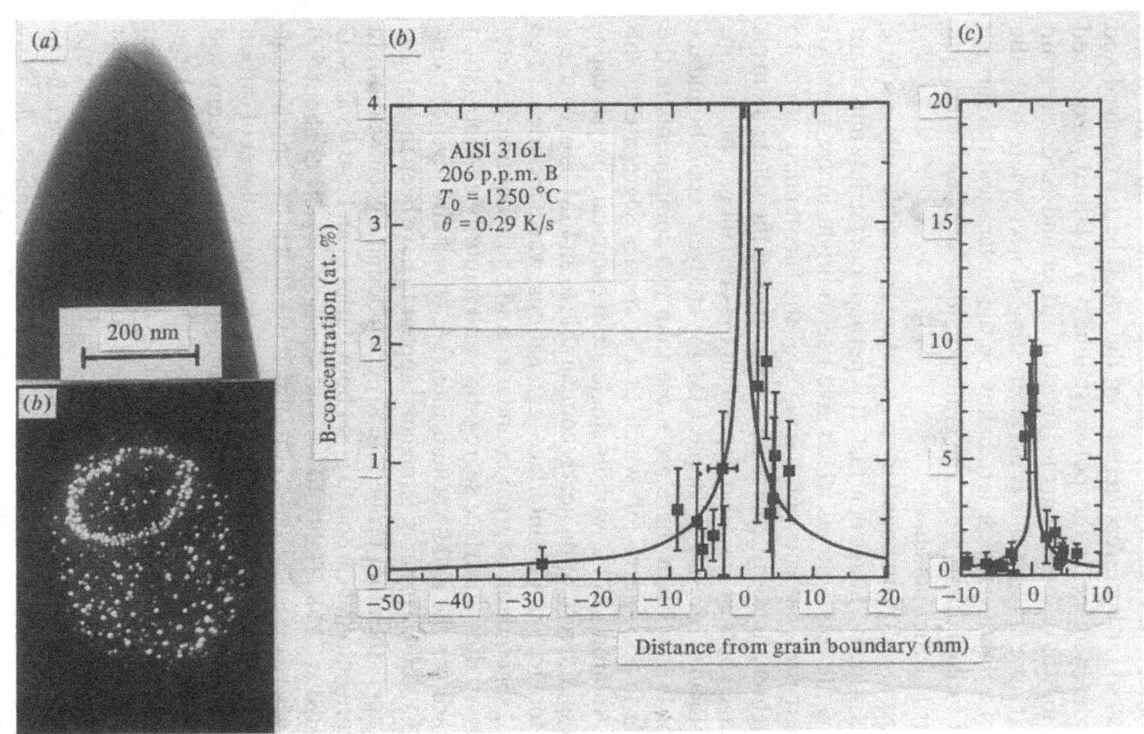

Fig. 5.6 Concentration variations of boron in grain boundary segregation of AISI 316L austenitic stainless steel. (*a*) TEM micrograph showing the existence of a grain boundary near the cap region of a tip. (*b*) Field ion image of the tip. Note that the grain boundary is decorated with bright image spots of about two atomic layers width. (*c*) Concentration profiles of boron near the grain boundary; by Karlsson and Norden. (Courtesy of H. Norden.)

tration profiles across grain boundaries are derived for samples annealed at different temperatures, 800, 1075 and 1250°C, and then quenched at different quenching rates, from 0.29 to 530°C/s. It is found that the boron concentration profiles peak at the boundaries and decay smoothly away from the boundaries. The width of the profiles is narrower for a faster quenching rate and a lower annealing temperature. The profiles are mainly non-equilibrium ones, except those obtained at 800°C annealing and subsequently rapidly quenched. The binding energy of boron to austenite grain boundaries is determined to be 0.65 ± 0.04 eV for the 'high-B' 316L and 'Mo-free' 316L stainless steel. By comparing non-equilibrium concentration profiles obtained at different quenching rates and a computer model, the diffusion coefficients for boron atoms (D_B) and vacancy–boron complexes (D_{vB}) are found, respectively, to be $D_B = 2 \times 10^{-7} \exp\{-1.15\,(\text{eV})/kT\}$ and $D_{vB} = 2 \times 10^{-6} \exp\{-1.15\,(\text{eV})/kT\}$ cm^2/s. In Fig. 5.6 some of the data given by these authors are presented.

Iron and steel are materials most actively investigated in field ion microscope studies of metallurgical problems. The partitioning of alloy elements in commercial steels has been studied by Goodman, Brenner & Low,[37] Miller, Smith, Williams[38,39] and many others.[40] In a pearlitic steel of 0.8C–0.7Mn–0.3 wt% Si, manganese is found to segregate to a region of about 10 to 15 nm of the carbide–matrix interface. For the thicker carbide, the Mn level in the central part of the carbide is comparable to that in the ferrite matrix. Williams *et al.* study distributions of Cr, Mn and Si in a 0.6C–0.86Cr–0.66Mn–0.26 wt% Si steel aged for 2 min at 596°C. They find an enrichment of Cr, Mn and Si at the interface of the cementite and Ferrite phases. Cr enrichment is confined to the cementite, Mn enrichment extends to both the cementite and the ferrite and Si enrichment is confined to the ferrite. In the cementite phase, the Cr concentration is about 3 at.%, whereas at the interface it is about 5 at.%; on the other hand, Si is depleted in the cementite. An interesting finding of these studies is that they all report a significant redistribution of alloy components across the interfaces, and the distribution is confined to a width of about 10 nm or less. The exact width of the redistributed zone is difficult to establish because of the probe-hole size, the limited number of ions which can be collected from the area covered by a very small probe-hole, the ion trajectories, and the statistical fluctuations of the atom-probe data. All the segregation zone widths reported in these studies probably represent only the spatial resolution of the data. The real widths should be much smaller.

Interface segregations in other materials have also been investigated. Horton & Miller[41] studied segregation of boron to the grain boundary and antiphase boundary in rapidly solidified ordered Ni_3Al alloy doped with boron. This alloy exhibits high temperature properties superior to other corrosion resistant materials.[42] Unfortunately the polycrystalline stoichiometric material has poor mechanical properties and fails intergranularly. The addition of boron produces a ductile fabricable material that has good high temperature properties. The alloy studied by them has the stoichiometry of Ni–24.0 at.% Al containing 0.24 at.% B. Atom-probe analysis found that boron distribution in the matrix is non-uniform, with local variations ranging from 0.07 to 0.2 at.%. At the grain boundary, boron concentration may rise to 2 to 5 at.% The Al level is also higher than the matrix to about 32 at.%. It is proposed that a thin boron-containing grain boundary phase is formed which may be more accommodating to slip and thereby to making the alloy more ductile. Another type of interface segregation is impurity segregation to the phase boundaries of precipitates. Brenner & Walck[43] found that Sb segregated to nitride interfaces in an Fe–3 at.% Mo alloy. The segregation coefficient is about 50. The trapped Sb can reside as much as 20 Å from the interface, although again the effect of ion trajectory in atom-probe analysis cannot be isolated. On the other hand, P does not accumulate at the interface but is absorbed into the nitride up to a concentration of 4 to 6 at.%.

Interface segregation can also be studied with the imaging atom-probe (IAP), particularly with an IAP equipped with a position sensitive particle detector. An advantage of IAP is that segregation of different alloy species to an interface and their distributions near and along the interface can be easily displayed. An example demonstrating the powerfulness of IAP in interface segregation studies is given by Waugh & Southon.[44] in their study of oxygen segregation to a grain boundary in molybdenum as illustrated in Fig. 5.7. In Fig. 5.7(*a*) a helium field ion image of a Mo tip with a grain boundary is shown. In (*b*) a flight-time gated field desorption image of Mo^{3+} ions is shown. While the grain boundary can still be traced because of a slight focusing–defocusing effect of the ion trajectories, there is nothing unusual about this image. When the gated time is adjusted to the expected flight time of O^+ ions, a desorption image as shown in (*c*) is obtained. This image clearly shows segregation of a considerable amount of oxygen to the grain boundary. The width of the segregated zone appears to be less than 1 to 2 nm. This width is probably produced by the ion trajectories rather than the true width of the oxygen distribution near the grain boundary. There are very few oxygen atoms in the matrix, in great contrast to an earlier FIM study of grain boundary segregation of oxygen in iridium. It is of course much more difficult to quantify IAP data than those from probe-hole type atom-probes. It should be possible to alleviate this difficulty with the use of a position sensitive particle detector. Another problem with IAP is the limited mass resolution, thus a clever choice of systems has to be made to avoid overlapping of mass signals.

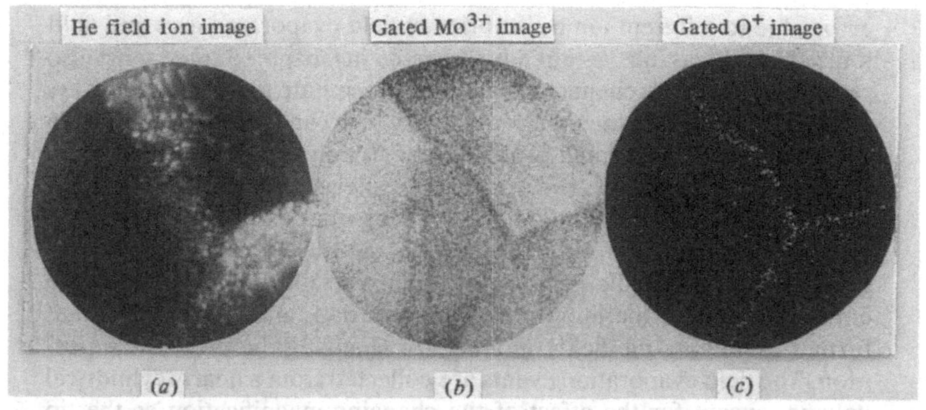

Fig. 5.7 (*a*) Helium field ion image of a Mo tip with intersecting grain boundaries.
(*b*) Time-gated image of Mo^{3+} ions. Except for the characteristic image brightness variations, there is nothing unusual about this image.
(*c*) Time-gated image of O^{+} ions which clearly shows segregation of oxygen atoms to the grain boundaries. (Courtesy of A. R. Waugh.)

5.2 Chemical analysis of alloy phases

The formation mechanisms of different phases in alloys and the effects of precipitate morphology, distributions of chemical species in different phases and across phase boundaries, etc., constitute another area where field ion microscopy has made significant contributions. Field ion microscopy is particularly suited for studying small precipitates whose sizes range from a few Å to ~100 Å where from a few tens to just a few precipitates can be present in the imaged area of a field ion tip, and where other techniques, such as electron microprobe or STEM combined with an energy-dispersive X-ray analyzer, find their limitations because of insufficient spatial resolution, especially in chemical analysis. In other words, in principle, field ion microscopy is capable of studying the very early stages of decomposition in supersaturated alloys, the mechanisms of phase transformations in alloys, and the kinetics and dynamics of these phase transformations.

Before we discuss some field ion microscope and atom-probe studies of partitioning in alloys, let us touch briefly here upon some of the methods commonly used for analyzing atom-probe data. Obviously if a precipitate or an alloy phase can be distinguished from the matrix from the field ion image, the size can easily be found. For the composition the probe-hole may be aimed at the precipitate, or the alloy phase, and analyze the composition of this volume. If the precipitate is very small and the probe-hole is large, matrix atoms will be detected concurrently regardless of where the probe-hole is aimed. Another problem is aiming errors due to different ion trajectories of field evaporated ions and field ionized image gas ions, as already discussed in Chapter 3. One must also recognize that in the chemical analysis of precipitate formation and alloy phase decomposition the number of ions which can be collected within a small volume using a small probe-hole is usually much too small to be statistically meaningful, and a repeated measurement with identical precipitates has to be done. With a few exceptions, most atom-probe studies are not carried out in such a detailed but tedious manner. In general, the probe-hole is aimed at a plane of the tip, and during the entire duration of the atom-probe analysis data are collected without further examining the field ion image or changing the probe-hole position. The field evaporation events are collected from a nearly cylindrical volume, except for the effect of the changing magnification as the tip becomes duller by the field evaporation. These events form a Markov chain of field evaporation events.[45] This chain is then analyzed using different statistical techniques to find the average particle size and the average inter-particle distance. Such a simple method actually has the advantage of being able to collect a large amount of data from a large volume to satisfy the needed statistics without undue effort. More importantly, there is no need for the material to be able to stand the imaging field and to give rise to a good field ion image. In principle, any material which does not rupture or field evaporate in chunks can be analyzed.

The method most commonly used is an autocorrelation analysis.[46] A set of data collected are divided into n subsets of the same size. The correlation coefficient $R(k)$ is calculated according to

$$R(k) = \frac{n \sum_{i=0}^{n-k} (c_i - c_0)(c_{i+k} - c_0)}{(n-k) \sum_{i=1}^{n} (c_i - c_0)^2},$$

where k is the correlation length, c_i and c_{i+k} denote, respectively, the solute concentration in the i-th and $(i+k)$-th subsets, c_0 is the average solute concentration of the sample, and n is the total number of subsets in the data set. If one chooses a subset to be approximately the amount of data which can be collected in the area covered by the probe-hole from one atomic layer of the (hkl) plane, then c_i represents the solute concentration in the i-th plane of this surface and k is the correlation length measured in units of s_{hkl}, the interlayer spacing of the (hkl) plane. The first minimum in the correlation coefficient then represents the average particle size in units of the inter-planar distance, and the first maximum represents the average inter-particle distance in the same units. Another method often used is a Fourier spectral analysis,[47] which gives the Fourier components in reciprocal space of the periodically modulated structure, assumed to be present in the sample. The accuracy achievable by different statistical methods in finding the sizes of periodically modulated alloy phases, such as precipitates or compositional modulations in spinodal decomposition, etc., in samples by blinded atom-probe analyses has been a subject of great interest, some concern and intense discussions.[48]

Ralph and co-workers reported some of the earliest FIM studies of precpitate formation in steels. Schwartz & Ralph[49] studied the shapes and growth rates of carbide precipitates in Fe–2% V–0.2% C steel by isothermal transformation between 600 and 700°C. The V_4C_3 precipitates formed have a slightly higher field evaporation voltage and therefore appear brighter in the FI images. Thus the shapes and sizes of precipitates can be measured quite accurately through field evaporation, and from these the size distributions and the mean sizes can be derived. They found that the mean size of vanadium carbide particles grew upon annealing following the $t^{1/2}$ law, rather than the $t^{1/3}$ law expected from a diffusion controlled growth. They therefore concluded that the growth was controlled by the interface. Brenner and co-workers performed one of the earliest atom-probe studies of precipitate formation in Fe–Cu systems. This system exhibits the classical age hardening response, but electron microscopy could detect structure changes only past the age hardening peak when precipitate particles reached the size of about 50 Å. In field ion microscopy, particles as small as 10 Å can be detected. Goodman *et al.*[37] studied precipitation in an Fe–1.4 at.% Cu alloy isothermally aged at 500°C. Upon ageing, Cu rich precipitates are formed and are visible in the field ion images as dark areas if the particle diameters are less than 50 Å. Particles of larger sizes can show some atomic structure details of the precipitates and can therefore be distinguished from the matrix also. Particle sizes and number density as functions of the ageing time are studied up to 120 hours of ageing. While the number density decreases continuously with time, the size and size dispersion increase with time. Atom-probe chemical analysis shows that the distance over which the composition changes from that of the precipitate to that of the matrix is less than three lattice spacings. The copper content of the precipitates increases with ageing time from about 40% up to ~100 Å when it becomes 100% Cu. As in the early stages the precipitate sizes are very small, it is impossible to avoid detecting matrix atoms in the atom-probe analysis. Thus the less than 100% composition of small copper precipitates may be an artifact of the atom-probe analysis. From a small angle neutron scattering experiment, Kampman *et al.*[50] suggest that the Cu content of the precipitates is 100% from the very beginning of their

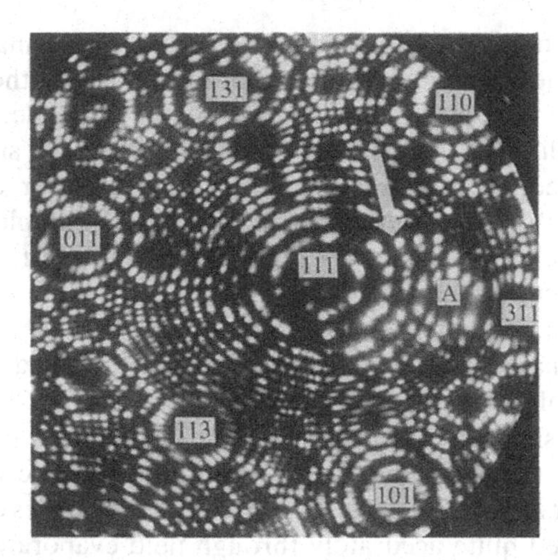

Fig. 5.8 Ne field ion image of a γ-Fe particle (pointed to by the arrow), taken at 4.9 kV, in a Cu–1.5% Fe alloy which has been aged at 790 K for 3.6×10^5 s. (Courtesy of M. Wada.)

formation. An analysis with an imaging atom-probe equipped with a position sensitive particle detector seems to confirm this suggestion.[51] Youle & Ralph[52] and Wendt & Wagner[53] have also made a similar study for Fe–1.5 at.% Cu and Cu–1 at.% Fe, respectively. At lower ageing temperature, the number densities of precipitates are found initially to increase with the ageing time, to reach a maximum, and then to decrease continuously. The mean size, on the other hand, remains constant until the number density starts to decrease, then it increases steadily. Wada *et al.*[54] also report a study of precipitate formation in Fe–1.5% Cu alloy. Their atomically resolved neon ion images show excellent coherence of the fcc γ-Fe particles with the Cu matrix in a well-aged sample. An example is shown in Fig. 5.8. Atom-probe analysis of γ-Fe particles of greater than 50 Å size shows that these particles are almost 100% iron and their interfaces with the Cu matrix are very sharp. Also, no dislocations are found at these interfaces, which explains why these particles are so mechanically stable, and capable of resisting the large mechanical stress of the image field at ~20 K.

In compositional analysis of very small precipitates, or in interface segregation studies, using a probe-hole type atom-probe, one is always faced with the fact that the probe-hole may cover both the matrix and the precipitate phases, or the interface as well as the matrix. Thus any abrupt compositional changes will be smeared out by the size of the probe-hole and also by the effect of ion trajectories. A similar uncertainty seems to exist in the compositional analysis of nitride platelets formed in nitrided Fe–3 at.% Mo alloy, aged between 450 and 600°C, where Wagner &

Brenner[55] find three different ranges of nitride precipitates having their stoichiometries consistent with $(Fe, Mo)_{16}N_2$, $(Fe, Mo)_6N_2$ and $(Fe, Mo)_{1\ \mathrm{to}\ 2}N$. Huffman and Podgurski,[56] on the basis of a Mossbauer study, conclude that the nitride phase has the stoichiometry of MoN, and if Fe exists it is smaller than 10 at.%. It appears that a more careful evaluation of atom-probe data is warranted to ascertain the atom-probe compositions derived for nitride precipitates. In the field ion microscope, nitride precipitates form platelets of bright images and can easily be distinguished from the matrix, thus the growth rate can be studied quite accurately. The thickness of nitride plates does not change very much, but at first the diameters grow following the $t^{1/3}$ law, then eventually slow down. The study of nitriding is important since it is a well known process for hardening steels.

Many other atom-probe analyses of different phases in different types of steels exist as steels are one of the most important materials. It is possible to investigate how the magnetic properties of alloys are correlated to the microstructures of different phases in the alloys.[57,58,59] The chemical contents, growth process and structures of metallic carbides in different alloy steels have been studied with the field ion microscope and the atom-probe field ion microscope.[60,61,62,63] We refer the reader to some of the original papers published on these subjects.

Other important alloys actively studied by field ion microscopy are aluminum based alloys. The calculated evaporation field of aluminum is about 1.9 V/Å as discussed in Chapter 2. In principle, it is possible to obtain field ion images using either hydrogen or argon as an image gas. Hydrogen tends to reduce the evaporation fields of most chemically reactive metals which include Al, Fe, Co, Pt and many others by forming hydride ions.[64] For imaging aluminum, it is found that either Ar or Ne can be used, and the residual hydrogen partial pressure has to be very low. The tip temperature has to be kept as low as possible. Boyes *et al.*[65] found that the evaporation field for aluminum in neon at 15 K is as high as 3.2 V/Å and that in UHV or helium at 5 to 10 K it is 3.3 V/Å. Thus neon, or even helium, field ion images of aluminum can be obtained. Abe *et al.*[66] also found that if the tip temperature is kept around 10 K and the image gas is carefully cooled first, good helium field ion images of pure aluminum can be obtained even if the field is well below the best image field and slow field evaporation may also occur continuously. It appears that the theoretical evaporation field of Al is on the low side, and the real low temperature evaporation field is at least somewhere between 3.0 to 3.5 V/Å. Thus good neon field ion images can be obtained for aluminum and aluminum base alloys if the tip temperature is low and the background vacuum is good. Precipitates such as GP zones can therefore be seen and studied with the field ion microscope.

One of the important topics in aluminum alloys is the formation of GP zoners, or solute-rich layers. Abe *et al.*[66] made a successful FIM study of aluminum based alloys such as Al–Ag, Al–Au, Al–Cu, Al–Zn, Al–Mg, Al–Sc, etc., where GP zones and various metastable phases can be seen, and also a preliminary atom-probe analysis of GP zones in Al–4 wt% Cu and Al–9.1 at.% Mg alloys. GP zones form bright line-like image structures. Mori *et al.*[67] obtained excellent field ion images of GP[1] zones in the {024} surfaces of an Al–Cu alloy where Cu atoms in GP zones are fully resolved. GP zones are believed to be only one to a few atomic layers in thickness. In atom-probe analysis, ion trajectories become very critical in finding the correct composition of the zones. So far, atom-probe data are inconclusive and comparisons with either theories or conclusions drawn from experiments using other techniques are not definitive, or even relevant. Field ion images, on the other hand, provide new insights for the structure models of GP zones even if some unresolved questions still persist.

One of the questions field ion microscope studies try to address is whether GP[1] zones in Al–Cu alloys are one {001} atomic layer of copper atoms in thickness or multi-layered. The classical model[68] of PG zones is that they are coherent {001} copper layers of only one atomic layer in thickness. However, Aurvray *et al.*[69] have recently proposed from an X-ray study that GP zones, though consisting mainly of single layer copper {001} platelets, may contain also an appreciable number of pure copper platelets of multiple atomic layers in thickness. Abe *et al.* report an FIM study which confirms this conclusion. On the other hand, the atomically resolved images of GP[1] zones at the {024} planes of Mori *et al.* seem to contradict such a conclusion. They observed four images of GP[1] zones at the {024} surfaces of an Al–1.7% Cu, aged with a procedure where GP[1] zones can be formed. Cu atoms in GP[1] zones are resolved into isolated image spots at the {024} planes, as shown in Fig. 5.9. By observing the position and shape formed by the spots corresponding to Cu atoms during successive field evaporations, it is concluded that a GP[1] zone consists of a single layer of Cu atoms on the {002} planes. On the (024) plane of an fcc crystal, the spacing between atoms in a single (200) atomic plane is $\sqrt{5}$ times the lattice constant, or 9.06 Å for Al. FIM images of atomic separations of Cu atoms in GP[1] zones on this surface are consistent with this distance. Cu atoms in GP[1] zones are slightly field evaporation resistant. Thus atoms sit in more protruding positions. Hono *et al.*,[70] however, further provide evidence from a field ion microscope study of images of {022} planes that GP[1] zones may have multi-layer structures. The pertinent questions one must answer in further studies are: (1) whether the number of GP[1] zones observed in Mori *et al.*'s experiment is much too small to be statistically reliable, and multi-layer structured zones perhaps have simply escaped

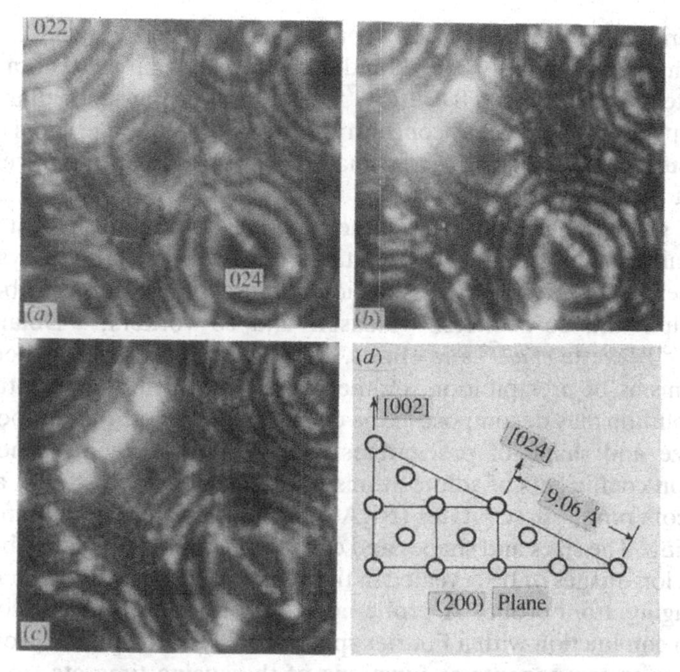

Fig. 5.9 Ne field ion images of a GP[1] zone emerging from a (024) plane of an Al–4 wt% Cu alloy. This alloy was grown by Bridgeman method. After a homogenizing treatment at 793 K for 1.7×10^6 s, single crystal samples with a longer axis parallel to [001] were prepared. They were quenched into ice water and then aged at 403 K for 6×10^4 s under the tensile stress of 73.5 MPa along the [001]. Between *a* and *b*, 11 (024) layers were field evaporated, and between *b* and *c*, 16 more layers were evaporated. The distance between two Cu atoms should be 9.06 Å as is shown in *d*. (Courtesy of M. Wada.)

their observation because of the poor statistics of the small sample; (2) whether the images of the so-called multi-layer structure GP zones represent insufficiently resolved images of GP[2] zones; and (3) whether the alternating image structures observed in layer by layer field evaporation are the artifacts of the field evaporation behavior of alloy layers as discussed by Wada *et al.*[71]

GP[2] zones, or two Cu {001} layers separated by a few aluminum layers, have also been observed in the field ion microscope in an aged Al–4 wt% Cu alloy.[71] A GP[2] zone formed on the (200) plane and observed on the (022) surface of the [001] oriented tip can be observed as two rows of bright image spots if an odd number of the (200) atomic layers are present between the Cu layers. The imaging condition at the (022) surface is more complicated if an even number of the (022) layers are present between the two Cu layers. The separations between the two Cu layers are estimated to be between 11 and 14 Å, which is larger than a TEM estimate of the spacing of 8 to 10 Å.[72] A GP[2] zone perpendicular to the [001] tip axis is observed as concentric rings of bright image spots on the (002) surface of the tip. It is estimated that 3 (002) atomic layers are present between two Cu layers.

The growth mechanisms, whether by nucleation and growth or by spinodal decomposition, of precipitates etc. in nickel base alloys have also been actively studied with field ion microscope and atom-probe FIM by Ralph and co-workers,[73] Haasen and co-workers,[74] Delargy & Smith,[75] Blavette *et al.*[76] and others. Their main concerns have been the mechanisms of precipitation, or the routes by which a supersaturated solid solution may decompose to the equilibrium phases, the composition and size and shape of precipitates as functions of ageing time, and diffusion coefficients of solute atoms in the matrix etc. In Ni–Al alloys, small coherent γ' precipitates (Ni_3Al with $L1_2$ structure) can be formed by ageing. The sizes and shapes and constituent atoms can either be seen in field ion images or be revealed in time-gated field desorption images of an imaging atom-probe. A probe-hole type atom-probe has also been used in conjunction with a Fourier spectral analysis to find the growth of the Fourier components as functions of the ageing time etc. for comparing with theories.[77] There have been some arguments of whether the data presented by Hill & Ralph truly represent the early stage of decomposition, or rather already the stage of coarsening, since the shortest ageing time in that study (5 h at 625°C) is already long enough for discrete γ'–Ni_3Al precipitates to be seen in the images of the alloy.[78]

Field ion images showing three-dimensional compositional modulations in Fe–24.8 at.% Be alloy have been obtained by Miller *et al.*[79] This alloy spinordally decomposes within a low temperature metastable miscibility gap to produce crystallographically aligned ⟨100⟩ composition modulations. When the alloy is annealed at 350°C for 2 to 15 h, images similar to the one shown in Fig. 5.10 are obtained. The axis of the specimen is parallel to ⟨100⟩. The three orthogonal sets of composition modulations are clearly evident in the vertical, horizontal and circular bands in the micrograph shown. The brightly imaging regions are the isolated cubicles of iron rich particles while the darkly imaging ones are the beryllium rich phase. From the micrograph it is clear that the modulations are not very regular, or parallel on a fine scale. In fact, some of the bands intersect with one another, giving rise to a distribution of interphase spacing. This distribution as well as its annealing time dependence, or the coarsening behavior, can be studied by properly accounting for the variation in magnification of field ion images with the help of computer image simulations.

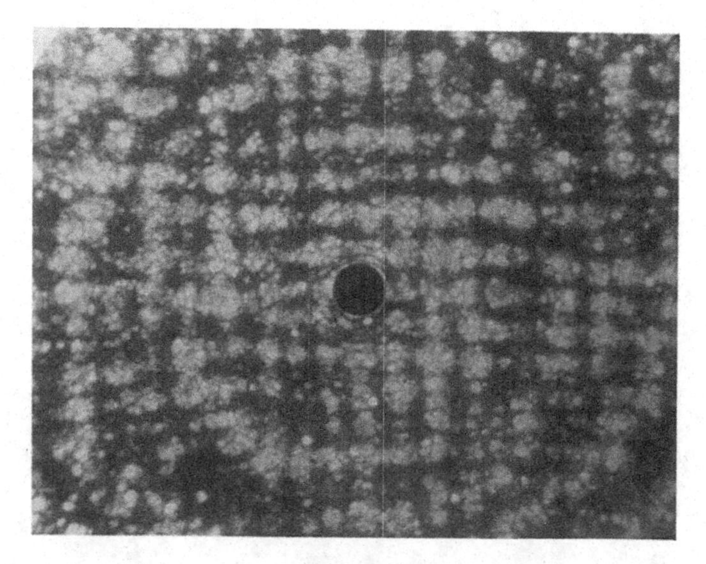

Fig. 5.10 Neon field ion image of Fe–25% Be aged for 2 hours at 350°C showing decomposition of an iron rich phase along three {001} directions, thus the decomposed phase forms small cubicles, obtained by Miller *et al.*[79] (Courtesy of S. S. Brenner.)

5.3 Ordered alloys and amorphous materials

5.3.1 Ordered alloys

Ordered alloys and amorphous materials represent two extreme ends of materials as far as the atomic structures are concerned. In ordered alloys, atoms of alloy components are arranged in periodic orders in a well defined lattice, or the atomic structure exhibits both short range order and long range order. In amorphous materials such as metallic glasses, atoms of alloy components are nearly randomly distributed, and there is no well defined crystal lattice either. While there may still be short range order, there is no long range order. Field ion images of random solid solutions, which still possess well defined crystal lattices, already appear very irregular, and distributions of alloy components in the crystals are difficult to extract from field ion images. Figure 5.11(*a*) shows a helium field ion image of a disordered equiatomic PtCo alloy. In an attempt to find a method for distinguishing alloy components directly from field ion images, Müller resorted to ordered alloys.[80] He found that FI images of ordered PtCo were perfectly regular, similar to those of pure metals. Unfortunately, for the ordered PtCo, which had an $L1_0$ fct structure with $c/a \sim 0.97$, so that an image pattern similar to that of an fcc metal could be expected, the image pattern was completely different from the usual

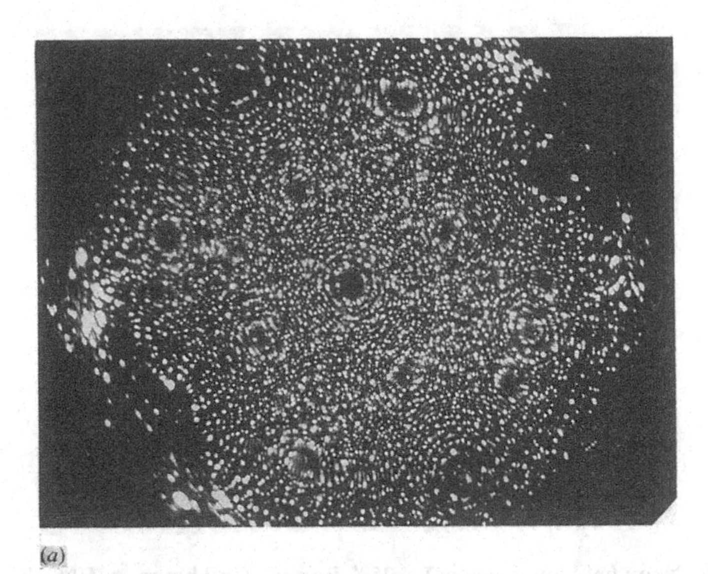

(*a*)

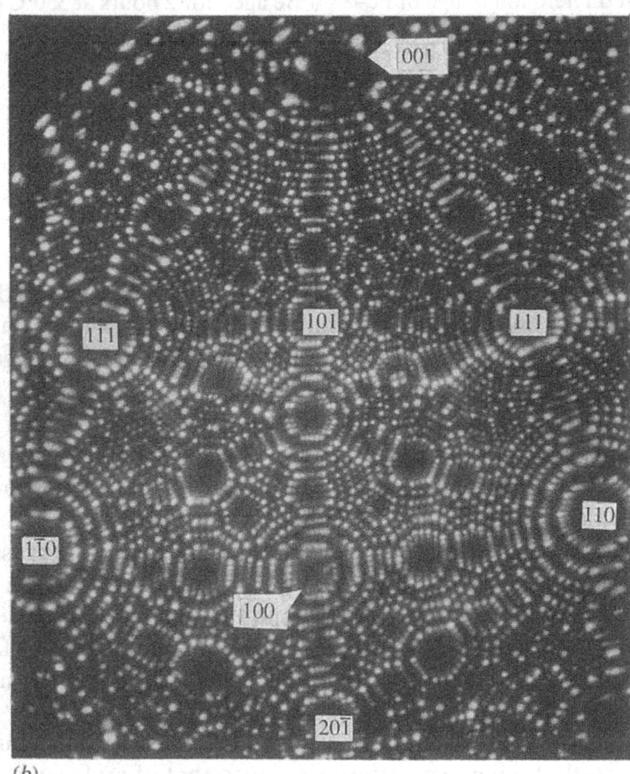

(*b*)

image pattern of fcc metals such as Pt, Ir or Rh, and a correct identification of crystallographic orientation of the tip was most difficult and was therefore incorrectly done. In 1966 Southworth & Ralph[81] correctly identified a large facet developed in ordered PtCo to be the (002) plane, which was a fundamental plane having a stacking sequence of alternating pure Pt and pure Co fcc (001) layers. By comparing the number of rings in the field ion image around the (002) pole and the radius of curvature around the pole, obtained from an electron microscope profile image of the tip, they concluded that the number of rings was too small by a factor of about two from that expected from a complete imaging of the ring structure of the lattice. They therefore suggested that Co layers were not imaged around the (002) pole of PtCo because of the preferential field evaporation of the (001) Co layers to the step edges of the pure Pt layers. In other words, the field evaporated structure of this alloy tip surface around the (002) was double-stepped. Tsong & Müller[82] approached the problem from a slightly different direction. They first tried to identify all the major crystallographic planes in the field ion images from their angular relations. Figure 5.11(*b*) shows a FIM image of a very well ordered PtCo surface where crystallographic planes are identified. Once all the crystallographic planes were properly identified, the atomically resolved image structures of some of these crystal planes were compared with the atomic structures of the $L1_0$ crystal. It was found that the image structures agreed with the atomic structures only if it was assumed that one of the alloy species was completely missing from the image. In other words, in the field ion microscope, only one of the sublattices, either the Pt sublattice or the Co sublattice, was imaged. It was first assumed that only Pt atoms were imaged, which was subsequently confirmed with a study of the asymmetric ordered Pt_3Co alloy having a $L1_2$ structure.[83] This is clear if one compares the image structure of the {111} with the atomic structure as shown in Fig. 5.12. If preferential field evaporation of Co to the (002) layer edge was the only reason for the missing Co images as proposed by Southworth & Ralph in their first study, then the image pattern of PtCo would still be essentially the same as an fcc metal, and Co atoms in {111} and some other high index planes of these alloys should

Fig. 5.11 (*a*) Helium field ion image of a disordered equiatomic PtCo alloy. It has an fcc structure. Although ring structures can be seen on the (001), {110} and {121} planes, there is no long range order of the FIM image. However, short range order can be seen from well ordered atomic rows and small ordered domains of about 10 Å sizes in the image.

(*b*) Helium field ion image of a well ordered PtCo alloy. Although it has an fct structure of $c/a \simeq 0.97$, the image appears completely different from those of fcc metals. In ordered PtCo, only the Pt sublattice is imaged. The very dim top (001) layer is a Co layer, but decorated with brightly imaged misplaced Pt atoms.

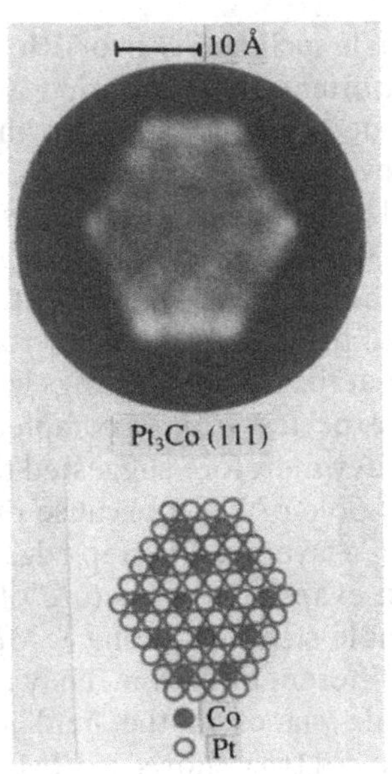

Fig. 5.12 Helium field ion image of a Pt_3Co (111) plane where Co atoms are not imaged. Co sites appear in the image as 'vacancies'.

still be imaged. Tsong & Müller interpreted their observation of missing Co atoms as arising from the lack of field ionization above this alloy component. This interpretation has since been confirmed with an atom-probe atomic layer by atomic layer compositional analysis.[84] Yamamoto *et al.* later[85] also concluded from a study of the relation between step sizes and facet sizes in the field evaporation end form of Ni_4Mo that the non-imaging of Mo atoms in Ni_4Mo was also due to a selective field ionization effect. Further FIM and atom-probe studies of ordered alloys[86] and single atomic layers of alloys artificially created on pure metal surfaces[87] indicate that both processes can occur, depending on alloy systems and crystal planes. In Ni_4Mo and Ni_4W, if the Ni layer is the top layer, nickel atoms at some of the fundamental planes can sometimes be very faintly seen, as shown in Fig. 5.13.

Regardless of whether the non-imaging of a species is due to preferential field evaporation or to preferential field ionization, the distinguishability of alloy components in ordered alloys makes much easier the identification of lattice defects and of all types of domains, such as orientational and translational domains, and the discernment of order–disorder phase boundaries in ordered alloys, as well as facilitating the study of clustering and order–disorder phase transformation, etc.[88] In most cases, image interpretations become self-obvious. For example in PtCo, which has the $L1_0$ structure, a Co layer can be distinguished from a

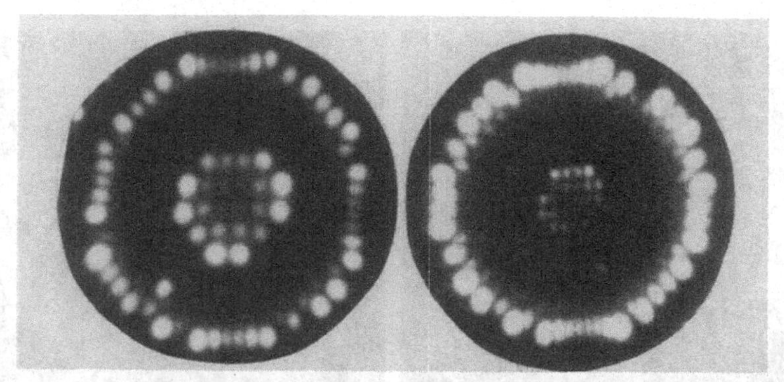

Fig. 5.13 Helium field ion images of a fundamental plane of Ni_4Mo. The very bright top layer in the left-hand side image is a Mo layer. When this bright Mo layer is field evaporated, only two to three very dim Ni layers can be seen even though there should be 4 Ni layers sandwiched between two successive Mo layers.

Pt layer in fundamental planes since Co layers are invisible. Thus when an antiphase domain passes across a fundamental plane, half-ring image structures 180° out of phase can be seen at the boundary. Even the (001) and (100) planes can now be distinguished since (001) is a fundamental plane whereas (100) is a superlattice plane, or a plane containing 50–50 of Pt and Co atoms for every layer. Since Co atoms are not imaged, all the (100) layers will show a c(2 × 2) structure. Thus 90° rotational domains, where the (100) and (001) may meet, can easily be recognized. Figure 5.14 shows a PtCo tip with a slice of 90° rotational domain. Note the (100) and (001) facets of the slice match, respectively, the (001) and (100) of the matrix, and the (101) of both the domain and the matrix coincide. Once the image interpretation is solved, only geometrical relations are left to be found for translational and rotational domains. Order–disorder phase boundaries can also be easily recognized since the ordered phase shows ordered images whereas the disordered phase shows almost random images. As ordered alloys are the only materials where both the structural and chemical identities of atoms can be recognized from field ion images alone, a considerable amount of FIM studies have been reported. At least 15 ordered alloys have been either imaged or studied in some detail. These include CoPt, $CoPt_3$, Ni_4Mo, Ni_4W, PtFe, Pt_3Fe, Au_4V, CuAu, Cu_3Au, Ni_3Al, Ni_3Mn, Fi_3Fe, Ni_3Ti, Fe_3Si, Fe_2TiSi, NiBe, etc.

Field ion images of Ni_4Mo are particularly well resolved and appear particularly beautiful because only one fifth of atoms, Mo atoms, are imaged, or only the Mo sublattice is imaged.[89] In addition, the structure of this alloy is a little more complicated than either the $L1_0$ or $L1_2$, and therefore it is richer in geometrical relations and domain structures. Ordered Ni_4Mo and Ni_4W have a body centered tetragonal structure of

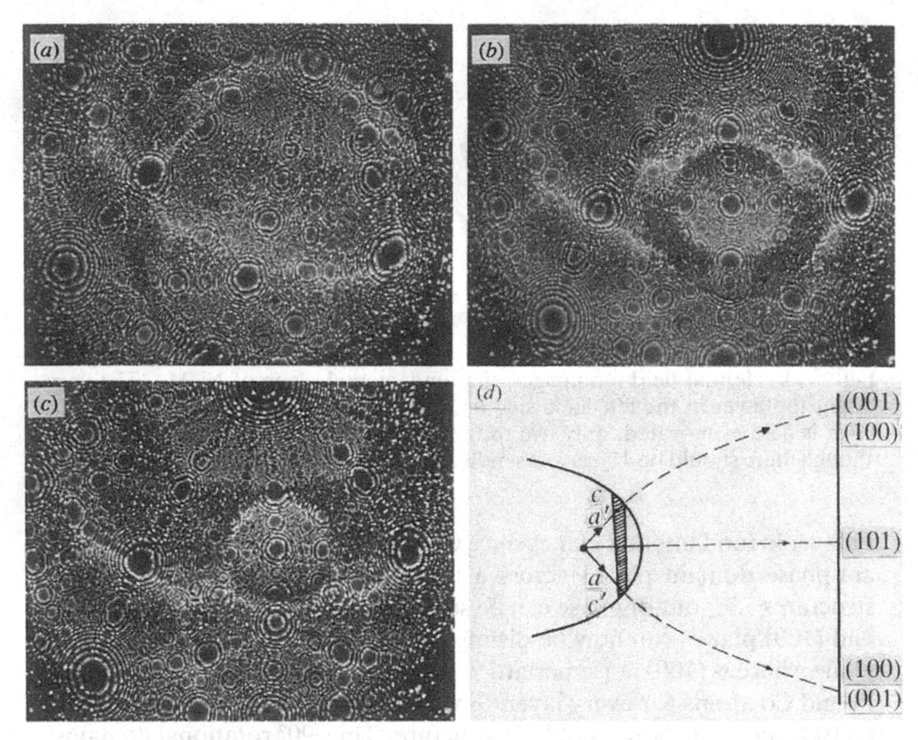

Fig. 5.14 Slice of 90° rotational domain in a PtCo tip. The geometrical relationship is shown in (*d*). Note the complete matching of the (101) plane of the two domains.

Ordered Ni_4Mo and Ni_4W have a body centered tetragonal structure of the $D1_a$ type (β phase), and in the disordered phase (α phase) the structure is face centered cubic. There are thirty possible ways for the β phase to nucleate in the α matrix, however only six of them can occur.[90,91] It is found that before ordered nuclei meet each other, six kinds of domains are distributed randomly. After they impinge they coalesce. At this stage a high density of small, plate-like domains is often observed. Many different kinds of domain boundaries can be seen. On further annealing, two domains may form an antiphase boundary (APB), or a perpendicular twin boundary (PTB); both of these boundaries show well matched images across the domain boundaries. An antiperpendicular twin boundary (APTB) does not show good matching of the images of the two domains. Identification of geometrical relations of these domain boundaries is greatly facilitated by the intrinsic dark zones[92] appearing in field ion images of field evaporation end forms of this ordered alloy.

In ordered alloys there is the question of whether disorder–order transition occurs by homogeneous ordering reaction or by nucleation and growth of the ordered phase. In PtCo and Pt_3Co, Tsong & Müller[82,83] made a preliminary study and found that if the annealing temperature was below ~550°C, or much lower than the transition temperature, which was about 650°C, then these alloys appeared to order via a homogeneous ordering process, and field ion images gradually showed better and better ordered images with the annealing time. No clear order–disorder phase boundaries and domains of appreciable sizes could be observed. On the other hand, if the annealing was done near the transition temperature, good-sized ordered domains, greater than 100 Å, were nucleated in the disordered matrix and sharp phase boundaries could be observed. These ordered domains grew in size with the expense of the disordered phase. When these ordered nuclei eventually met, well matched 90° rotational and antiphase domain boundaries could be observed. Similar studies of disorder–order transformations have been reported by Yamamoto *et al.*[85,91] for Ni_4Mo and Taunt & Ralph for Ni_3Fe.[93] The as-quenched Ni_4Mo, from 1100°C, already shows some short range order, and also exhibits two image characteristics: (1) low index planes are fairly well developed although high index planes are not, (2) bright-spot clusters of 10 to 30 Å sizes are already randomly distributed over the entire surface. Computer simulation FI images show that principal, or low index, superlattice planes develop gradually with an increasing long range order parameter, while principal fundamental planes develop well even in the disordered state. When the alloy is further annealed at 700°C, a large number of small ordered domains nucleate homogeneously at the early stage, and eventually coalesce. In general, at low temperatures ordering occurs by nucleation and growth of randomly distributed, very small ordered nuclei. Thus many small domains are formed in the ordered state. At 800°C annealing, a small number of large ordered domains are nucleated and grow with the expense of the disordered phase. FIM images clearly show sharp phase boundaries. The ordering mechanism seems to depend on the heating rate at which a sample is brought up to the annealing temperature. Figure 5.15 shows some of the results obtained by Yamamoto *et al.*

It is quite obvious from the above discussion that if the annealing temperature is very close to the order–disorder transition temperature, then the transformation is clearly by nucleation and growth of fairly large domains. On the other hand, if the annealing temperature is much lower than the transition temperature, the transformation may go though a nearly homogeneous, statistical nucleation process, or, in simpler terms, small ordered clusters may be formed initially by random statistical fluctuations of atoms and gradually grow in size and coalesce into larger domains. An X-ray study shows that when $CoPt_3$ is in a partially ordered state the atoms on wrong sublattices, or misplaced atoms, are not

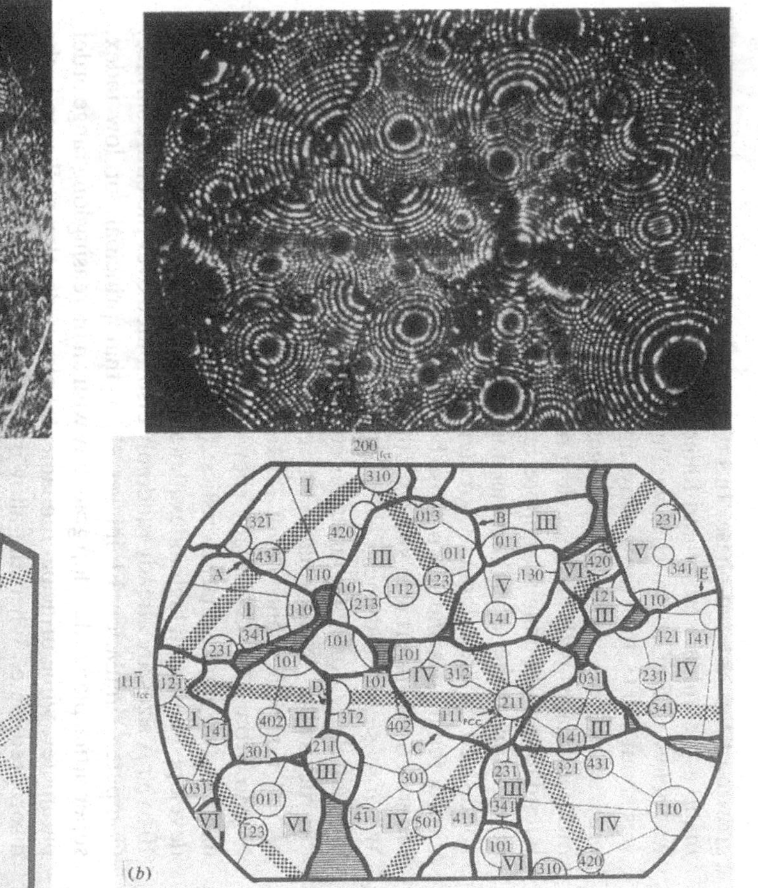
(b)

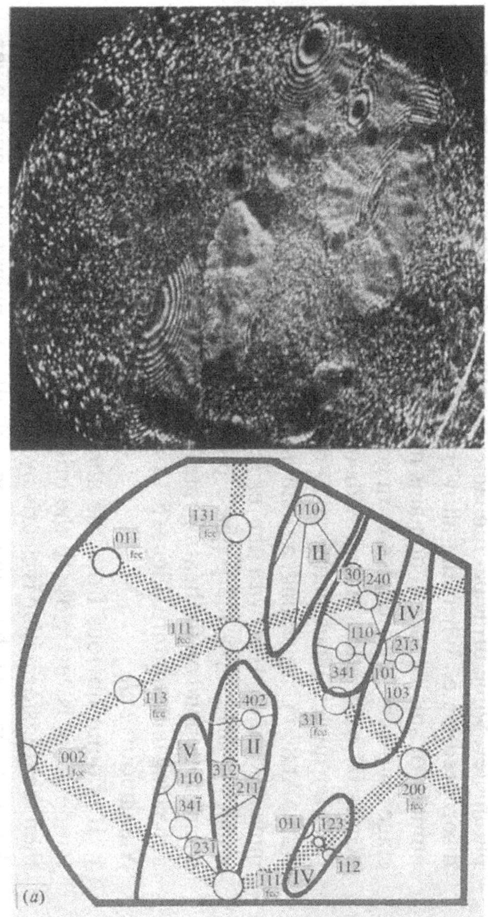
(a)

randomly arranged. The diffuse scattering is concentrated near superstructure peaks, even after the effects due to atomic displacements have been removed. Thus misplaced atoms tend to group together into small, ordered regions antiphase to the ordered matrix.[94] Since misplaced Pt atoms in this alloy can be seen directly in superlattice layers as bright spots (thus will not be confused with vacancies produced by field induced chemical etching as in misplaced Co atoms), it is possible to construct a map of the distribution of misplaced Pt atoms within a superlattice layer in a partially ordered alloy and compare these data with those of X-ray diffraction. This experiment has been reported by Berg *et al.*,[95] as shown in Fig. 5.16. Good agreement is obtained between the degree of long range order (S) determined with FIM study by direct counting of misplaced atoms and that using X-ray diffraction. Oscillations in the long-range order of amplitude in the 0.05 to 0.1 range, over 100 Å, can be detected. Quantitative measures from FIM images of the misplaced Pt atoms appearing as rods and two-dimensional regions on $\{110\}$ planes for $S = 0.8$ indicates that atoms on wrong sublattices are not randomly arranged, but are mostly in the form of small platelets on the $\{100\}$ planes. This is in agreement with computer simulations based on the Warren local order parameters which are determined from diffuse scattering of X-rays. The difficulty with FIM technique in this type of study is that only alloys with a fairly large degree of long range order can be studied, and the experiment is slow and tedious, although the data are more direct. Further detailed studies of ordered alloys on the subjects of temperature dependent ordering mechanisms and the kinetics of ordering appear very attractive and may be worth pursuing.

5.3.2 Amorphous materials

With materials of this type FIM finds its limitations. Several attempts have been made to use field ion microscopy to study amorphous materials such as metallic glasses and amorphous silicon or hydrogenated amorphous silicon thin films deposited on metal tip surfaces.[96–98,100–102] Since there is no well defined crystal lattice, the structure of an amorphous material is usually described by the pair distribution function of the

Fig. 5.15 (*a*) Field ion image and a domain map of a Ni_4Mo alloy which has been annealed for 5 min at 800°C with a slow heating rate to the final annealing temperature. Long range ordered domains are formed within the disordered matrix which, however, already exhibits a small degree of short range order. (Courtesy of M. Yamamoto.)

(*b*) Field ion image and domain map of a Ni_4Mo which have been annealed for 30 min at 800°C. Dotted bands are the $\langle 110 \rangle_{FCC}$ dark bands and shaded areas represent the α-phase. Poles are indexed on the basis of the bct lattice (β). Poles with an FCC subscript are indexed on the basis of the fcc lattice (α).

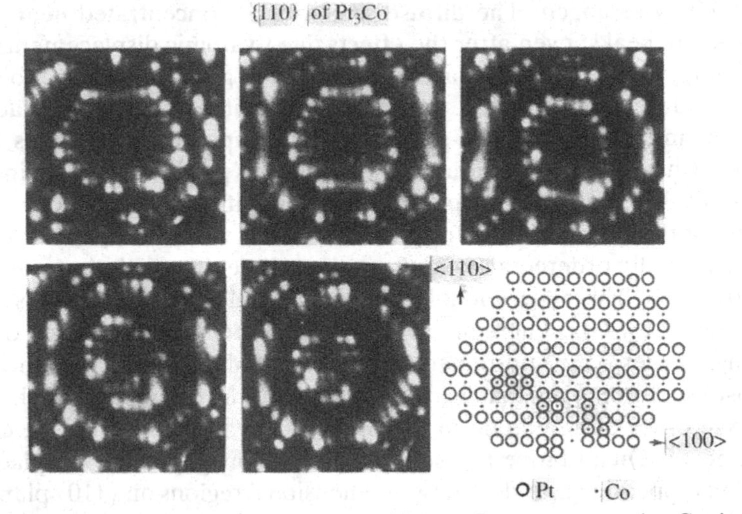

Fig. 5.16 Mapping of misplaced Pt atoms, which are Pt atoms occupying Co sites, in a {110} plane of partially ordered Pt_3Co alloy by slow field evaporation of a surface layer. From the work of Berg *et al.*[95]

atomic positions, which can be derived with diffraction techniques, and the higher order distribution functions, which are difficult to derive by any technique.[99] Since FIM can give atomic images of solid surfaces, it is natural that one can expect to be able to derive both the pair distribution function and the higher order distribution functions directly by mapping the atomic positions in field ion images, even if field ion images of amorphous materials do appear very irregular. Unfortunately, in FIM, some of the alloy components are not imaged. In addition the image magnification is very non-uniform, especially on field evaporated surfaces of amorphous materials where the surface is by no means smooth. Because of the random bond and chemical distributions in the alloy, the field evaporated surface must contain a random distribution of small atomic protrusions due to selective field evaporation. Therefore local magnification can vary by a factor of a few from the average magnification expected from the average radius of the tip. A comparison of a pair distribution function derived from real field ion images of $Ni_{40}Pd_{40}P_{20}$ with one derived from computer simulation FI images shows that they do not agree with each other. Structures in the pair distribution function are washed out in FIM data because of the non-uniform image magnification.

In $Fe_{80}B_{20}$, Jacobaeus *et al.*[96] claim to be able to derive a triplet correlation of atomic positions from field ion images. Specifically, they find that the distribution of bond angles between neighbor atoms exhibits a fairly distinctive peak at 60°, a broad peak between 90 and 125°, and a shoulder around 150°. These observations are said to be in good agreement with predictions of a model for metallic glass based on dense random packing of hard spheres. In this FIM study they find the imaged species to be Fe, in contradiction to the studies of Piller & Haasen[101] and Bhatti *et al.*[97] These latter authors conclude from atom-probe analyses that a very high percentage of bright image spots in boron-containing Fe and Fe–Ni metallic glasses are either single boron atoms or small clusters of boron atoms. The conclusion of Jacobaeus *et al.* is therefore criticized by Nordentoft[102] and by others. It is generally concluded that the accuracy of deriving even a pair distribution function from field ion images is too limited, and the possibility of deriving a higher order distribution function from FIM images is completely out of the question at the present moment.

It is, however, possible to study short-range compositional fluctuations in metallic glasses resulting from a decomposition of the amorphous alloy during an ageing process below the crystallization temperature. Piller & Haasen find in the as-quenched state, and after isothermal and isochronous annealing below 370°C of $Fe_{40}Ni_{40}B_{20}$, precipitates of stoichiometric $(Fe\ or\ Ni)_3B$. Their radii, originally in the range around 15 Å, coarsen to about 40 Å after annealing to 350°C for 300 minutes. The number density of precipitates does not change very much, but maintains a value of about $(2 \pm 1) \times 10^{18}$ cm^{-3}. These clusters are embedded in amorphous regions whose average boron content deviates as much as ± 3 at.% from the normal boron concentration of 20.1 at.%.

5.4 Radiation damages and recovery

One of the important applications of field ion microscopy is the study of radiation damages produced by the bombardment of energetic particles, and how these damages are recovered during the annealing process. In the 1960s and 1970s, a considerable amount of work in field ion microscopy was devoted to this subject. Field ion microscopy was the only microscopy capable of providing us with a lattice image of the primary state of radiation damage, i.e. the three-dimensional distribution of point defects produced by one energetic primary knock-on particle. Atomically resolved structures of defects produced by particle bombardments, such as vacancies, interstitials, Frenkel pairs, which were composed of a vacancy and a self-interstitial, depleted zones and voids, etc., could be observed only with the field ion microscope. In early experiments, Müller[103] used a ^{210}Po α-source for bombarding the tip with α-particles of 5.4 MeV. The range in tungsten of these particles was about 5000 Å. It was found that two damages on the emitter surface, one at the incident side and one at the exit side, could be observed. Sinha & Müller[104] reported one of the earliest comprehensive studies of this subject. An ion beam source was installed into a field ion microscope with which a field evaporated tip could be bombarded with various ion species of kinetic energy of a few keV to about 20 keV from one side of the tip. Since then a very large number of studies have been reported, using either an *in-situ* ion source or an ion beam from a particle accelerator. In earlier studies, attentions were mainly focused on defect structures and image interpretations of radiation induced defects such as vacancies, self-interstitials, and voids and depleted zones, etc., in tungsten. Later studies by Seidman and co-workers and others focus on range profiles, detailed atomic arrangements within depleted zones, size distributions of defects, and different stages of recovery of various defects in various metals and dilute alloys. Energetic particles of all kinds, including neutrons, have been used for the irradiation. Here only some of the later works will be very briefly described. For details of these and other FIM studies of radiation damages, we will refer to some of the reviews on this subject in the literature.

In studying defects created by particle bombardments and their spatial and size distributions, the particle dose has to be very high, in the range of 10^{11} to over 10^{15} particles per cm^2 for ions and over 10^{19} per cm^2 for neutrons, because of the small volume of the tip. It is also important to recognize the different temperature ranges at which some of the defects may already be able to migrate and can thus be annealed out. Thus the irradiation temperature of the tip has to be low enough where defects to be studied are stable. For mapping out the detailed atomic arrangements within a defect such as a depleted zone, it is necessary to field evaporate surface atoms slowly one by one and to take many photographs at different stages of the field evaporation, sometimes over ten thousand photographs, from which atomic arrangements can be reconstructed with the aid of a computer, provided that no accidental field evaporation of some of the atoms around a defect occurs between two recorded images.

It is found that the atomic arrangement, or a vacancy network, in a depleted zone in a refractory metal or a dilute alloy of a refractory metal, created by bombardment of an ion can be reconstructed on an atomic scale from which the shape and size of the zone, the radial distribution function of the vacancies, and the fraction of monovacancies and vacancy clusters can be calculated. For example, Wei & Seidman[108] studied structures of depleted zones in tungsten produced by the bombardment of 30 keV ions of different masses, W^+, Mo^+ and Cr^+. They find the average diameters of the depleted zones created by these ions to be 18, 25 and 42 Å, respectively. The fractions of isolated monovacancies are, respectively, 0.13, 0.19 and 0.28, and the fractions of vacancies with more than six nearest neighbor vacancies (or vacancy clusters) are, respect-

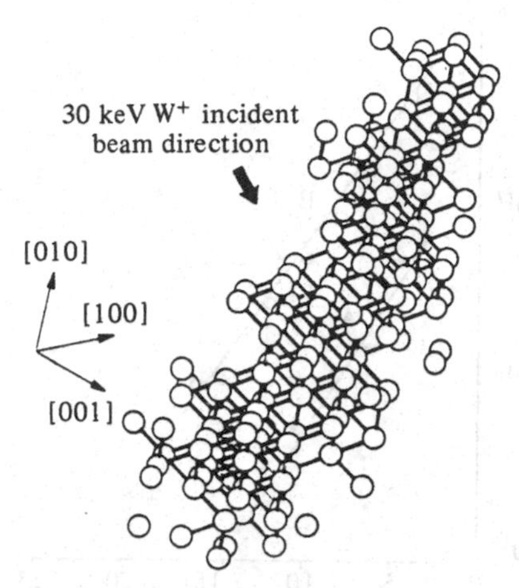

Fig. 5.17 Distribution of vacancies in a depleted zone in a tungsten tip produced by the direct impact of a 30 keV W^+ ion. From the work of Wei & Seidman.[108]

ively, 0.83, 0.70 and 0.54. It is found that as the mass of these 30 keV ions is increased, the spatial extent of the depleted zones (DZs) increases, the vacancy concentration with the DZs decreases, the fraction of isolated monovacancies increases, and subcascades are formed within the DZs. In general, the morphology of DZs in tungsten is independent of the radiation temperature, and they are usually extended in the ⟨110⟩ directions, or the directions of the closely packed atomic rows in bcc lattice, indicating the occurrence of a channeling effect. An example is shown in Fig. 5.17.

Size distributions of microvoids in neutron irradiated Fe–0.34 at.% Cu have been studied as shown in Fig. 5.18.[109] The neutron dose is about 3×10^{19} cm^{-2} and the neutron energy is greater than 0.1 MeV. It has been known that excessive amounts of residual copper in ferritic steels enhance the embrittlement of ferritic steels when they are irradiated by neutrons. The usual analytical techniques have been unable to reveal the cause of this enhanced embrittlement. By means of field ion microscopy it is possible to demonstrate that the embrittlement is caused by the formation of extremely small voids, of the order of 10 Å, in neutron irradiated Fe–0.3 at.% Cu alloy. Similar microvoids are believed to form in the ferritic steel, causing the embrittlement. It is also suggested that the copper is instrumental in stabilizing the microvoids by diffusion to the microvoid surface. However, an atom-probe study of void surface segregation in a Mo–1 at.% Ti–0.24 at.% C alloy by Wagner & Seidman[110] did not observe segregation of Ti to the void surfaces.

Recovery of radiation damages of a material by isochronal annealing

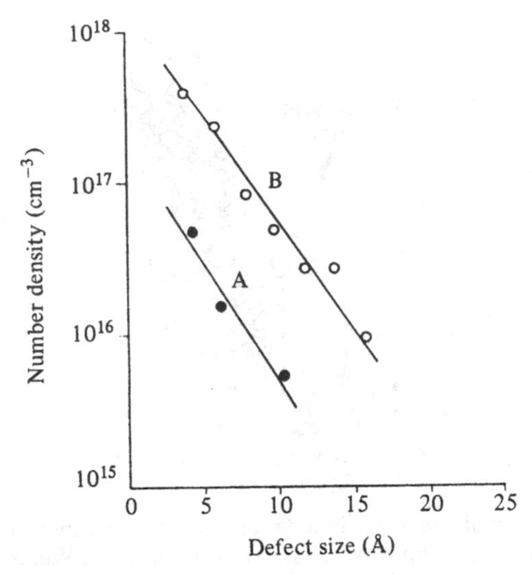

Fig. 5.18 Size distribution of microvoids in neutron irradiated Fe–0.34 at% Cu. Neutron dose is 3×10^{19} cm^{-2}, and the energy is greater than 0.1 MeV. Line A is for unirradiated samples and line B is for irradiated samples. After Brenner *et al.*[109]

occurs at distinctive temperatures, or the recovery exhibits distinctive stages, because of the different onset temperatures where diffusion and recombination of different types of defects can occur. FIM is particularly suited for analyzing the recovery spectra of irradiated metals and alloys because radiation damages of all types can be seen directly. FIM studies place emphasis on the question of whether or not the motion of a particular defect type is responsible for each of the distinctive stages of recovery and spectrum, and on the role played by the impurities in the recovery process. Discussion of these studies is beyond the scope of this book and the reader is referred to the reviews.[106–110]

Range profiles of implanted species can obviously be found with an atom-probe FIM where the chemistry of samples at different depths can be analyzed with true single atomic layer depth resolution, as has been discussed earlier in studies of surface segregations. As far as this author is aware, however, studies of range profiles always calibrate the depth by the cumulative number of ions collected, as in most other metallurgical applications. Wagner & Seidman[110] studied the range profile of 300 eV He^+ ions along the [110] direction of tungsten at 80 K. Since the maximum energy that can be transferred to a W atom is only about 25 eV, smaller than the 47 eV bulk displacement energy of tungsten at 80 K, no formation of self-interstitials can be expected. Only He can be imbedded into tungsten in interstitial positions. It was found that these He ions can reach a depth of about 140 Å, with a maximum concentration at about 25 Å. The mean range is 47 Å and the variance is 35 Å. These values compare favorably with theoretical values of Haggmark,[112] which are 40.5 and 22.4 Å, respectively. The number of He detected decreases dramatically when the tip temperature is kept at 110 K, indicating that interstitial He atoms are starting to diffuse at between 80 and 110 K. Panitz[113] studied range profiles of 80 eV D^+ ions in tungsten with an imaging atom-probe, and found excessive numbers of deuterium atoms at the surface. This may be due to diffusion of interstitial D atoms at the tip temperature. Igata *et al.*[114] studied the depth profile of radiation damages and vacancy clusters, produced in tungsten by the bombardment of 400 keV Ar^+ and 200 keV Ni^+ ions. Depth profiles of vacancy clusters of different sizes are derived, which are found to be in good agreement with a theory by Lindhard *et al.*[115]

5.5 Formation of compound layers

5.5.1 Oxidation of metals and alloys

Oxidation of metals and alloys is a subject of great technological interest[116] which has been studied by a large number of techniques, including surface-sensitive analytical techniques such as AES, LEED, APS, SIMS, XPS, and UPS, etc. Field ion microscopy has also been applied to study the formation of oxide layers and the initial stages of oxide formation on metal and alloy surfaces. Fortes & Ralph[117] reported the first detailed FIM study of oxidation of iridium. Very well ordered helium ion images of iridium oxide layers of a few hundred Å in thickness can be obtained by heating field evaporated Ir tips in an oven between 400 and 950°C in one atmospheric pressure of air or oxygen for from a few to over 24 h. However, below 500°C the growth rate is very slow. The polycrystalline iridium oxide films are assumed to have the stoichiometry of IrO_2, which has a rutile type structure (tetragonal P lattice of $a = 4.5$ Å and $c/a = 0.7$). Oxygen atoms were found to be not imaged. FIM images of oxide–metal interfaces are consistent with this crystal structure of the oxide. Interface relations between IrO_2 and Ir crystals are investigated and compared with field ion images. For an example, the (111) plane of Ir should coincide with the (100) and (111) surfaces of IrO_2, and indeed a field ion micrograph shows a good match between an Ir (111) and an IrO_2 (100). It is concluded that oxide layers are formed by nucleation process from probably {102} regions of the tip. The stoichiometry of the oxide layers has later been analyzed in an atom-probe to be $IrO_{2.13\pm0.15}$,[118] in agreement with the original assumption of IrO_2 based on an image analysis.

The first atom-probe analysis of oxide layers in the initial stage of oxidation was reported by Ng *et al.*[119] They employed a flight-time focused atom-probe to derive composition depth profiles in oxidation of Fe, Co and Ni. These tips were first field evaporated, and then heated in 10^{-3} Torr to 1000 μTorr of oxygen or air. These composition depth profiles all show metal rich surfaces, followed by oxide layers of non-uniform oxygen content. No sharp interface is observed.

An FIM study of oxidation of ruthenium, similar to Fortes & Ralph's study but done at a much lower oxygen pressure, is reported by Cranstoun & Pyke.[120] They again found good correlation between the major poles of the oxide films and the substrate. Between 600 and 450 K, the thickness of oxide film formed is nearly constant, at about 20 to 35 layers, regardless of the oxidation period, indicating that the oxide film formed behaves like a protective film. This protective behavior by the oxide layer is attributed to its close structural fit to the metal and the lack of high angle grain boundaries. Above 950 K, protection is less effective and oxide thickening occurs, probably owing to the less precise nucleation on the metal and a more highly defective oxide layer. The composition of the protective tarnish films were analyzed in an atom-probe.[121] The oxide film is found to consist essentially of RuO_2, but with a narrow oxygen deficient region at the interface with the substrate. This region consists of a few atomic layers and can accommodate the structural misfit between metal and oxide.

None of the above studies focuses on the very early stages of oxide layer formation. Kellogg[122] reported such a study for Rh using an imaging atom-probe. A Rh tip is first field evaporated and then heated to a temperature between 350 and 550 K for 15 min in 1 Torr of oxygen. The oxide layer formed is then analyzed in the form of cumulative plots, i.e. plotting the cumulated number of oxygen detected against the cumulative number of rhodium detected. The slopes of the plots represent the O/Rh ratios at different depths, or the composition depth profile. It is found that at the surface the oxygen content is consistent with Rh_2O_3 for oxides formed at these temperatures. The oxygen content decreases smoothly into the bulk. The total number of O-atoms increases with the heating temperature, and is linear in a $\ln N_0$ vs. $1/T$ plot, or an Arrhenius plot. The activation energy in the formation of an oxide layer, which can be derived from the slope of the Arrhenius plot, is found to be 4.0 to 5.0 kcal/mole, or only about 0.18 eV. Although this activation energy agrees with the dissociation energy of O_2 on Rh surface into O atoms, it is thought to be associated rather with the dissolution of oxygen into the bulk, the surface rearrangement process, or the migration of oxygen atoms to nucleation sites. In this experiment, data are collected from the entire emitter surface. Since initial nucleation of the oxide layer is likely to be very sensitive to crystallographic regions, atomistic interpretation of the data is obscure at best. The smooth composition depth profiles obtained pose the question of why the stoichiometry of the oxide layer should change so smoothly into the bulk. This may in fact occur merely owing to an average effect of the data, which are collected from all the different crystallographic regions of the emitter surface.

Oxidation of an alloy has also been investigated in connection with surface segregation of the alloy as has already been mentioned in Chapter 4. Tsong *et al.*[123] find that when Ni–5 at.% Cu is annealed to 823 K for 5 to 10 minutes in N_2 of pressure in the range between 10^{-4} and 10^{-6} Torr, the copper enrichment at the top surface layer is greatly enhanced for both the (111) and the (001) surfaces. On the other hand, when the annealing is done in oxygen of pressure of the same range, the situation is completely reversed. Not only are no copper atoms detected on the near surface layers; few are detected down to depth of 200 Å. FIM images show the formation of polycrystalline oxide layers at the alloy surface and only ions of nickel, oxygen and nickel oxide can be detected. This experiment was extended by Kamiya *et al.*,[124] who studied surface segregation of both Ni solute and Cu solute Ni–Cu alloys in the presence of CO and O_2. Their conclusions were: both CO and O_2 suppress Cu segregation in Cu solute alloys, CO shows no influence on Ni segregation in Ni solute alloys and O_2 enhances Ni segregation in Ni solute alloys; also, Ni oxides are formed on the (111) surface but on the (100) surface less oxides are formed. These observations are in agreement with a theoretical calculation,[125] which also concluded that Cu segregation in Ni–Cu alloys are suppressed by chemisorption of oxygen.

5.5.2 Formation of compounds on solid surfaces

Formation of metal carbonyls and subcarbonyls has been investigated with a pulsed field desorption mass spectrometer.[126] In this instrument, the tip is kept at a high positive d.c. voltage and negative high voltage pulses (at a maximum amplitude of 20 kV, a repetition rate of up to 100 kHz and a half width of some 100 ns) are applied to a counter-electrode of the tip to field desorb the surface species. The system is filled with $\sim 10^{-6}$ Torr of CO, and during the time when there is no high negative voltage of the pulses at the counter-electrode gas species can be adsorbed on the surface, perhaps greatly enhanced by the d.c. holding field even if this field is kept at a low value. When a high voltage pulse is applied, all the surface compounds are assumed to be desorbed. Thus by changing the pulse repetition rate, or the time between two pulses where adsorption and surface reaction can occur, and then measuring the intensities of desorbed species, reaction kinetics can be studied and analyzed on the basis of some proposed models. They found that $Ru(CO)_x^{n+}$ species of up to $x = 4$ can be detected. These ions are assumed to be formed from their corresponding neutrals formed on the tip surface. Field variation measurements showed that $Ru(CO)_4$ is not formed under the influence of the d.c. field. Thus the $Ru(CO)_4$ is the product of surface reaction in the time period between two HV pulses. By changing the duration of this time period and the tip temperature, it was concluded that formation of $Ru(CO)_2$ is associated with an activation process.

Chuah *et al.*[127] studied the decomposition reaction of methanol over a Ru field emitter surface between 300 and 580 K. In addition to gaseous species of all kinds, such as CH_3OH^+, $CH_3OH_2^+$, COH_x^+ ($x = 0$ to 3) and CH_3^+, ruthenium oxide ions RuO_y^{n+} ($y = 1$ to 3, $n = 1$ and 2) were observed. Below 460 K a high intensity of CH_3^+ is observed, which corresponds to the oxygen content of the detected RuO_y^{n+}. Thus these oxide ions are presumably formed from CH_3O_{ad}. Below 460 K, the intensities of CH_3^+, COH_2^+ and RuO_y^{n+} drop but those of COH^+ and CO^+ increase and reach maxima just at the thermal desorption temperature of CO on Ru. It is therefore concluded that the steady decomposition of methanol below 460 K is prevented by the adsorbed CO. Studies of this type are intimately related to surface reactions and heterogeneous catalysis. The artifacts of the applied field in these studies are identical to those in high voltage pulse atom-probe studies of gas–surface interactions. The difference is that in atom-probe studies, emitter surfaces are first characterized by field ion images, and data are usually collected from a specific region or plane of the emitter surface by aiming of the probe-hole. If the effect of the applied field and field dissociation of compound ions can be properly evaluated in all these studies, the data obtained and conclusions drawn will be most valuable. Atom-probes and related instruments are very powerful microanalytical tools for studying compound formation on material surfaces because of their sensitivity and spatial resolution in three-dimensional analysis. Studies have just begun, and we can expect further contributions in this area of research.

5.5.3 Electrodeposition

Field ion microscopy has been used to study the structures of electrodeposited layers by Rendulic & Müller[128] in connection with an attempt to image biological molecules by embedding them into the electrodeposited metal layers. As the deposited layers have to be able to withstand the field stress of $\sim 10^{11}$ dynes/cm^2 during the FIM imaging, special care has to be taken to obtain a clean surface to assure a strong adherence at the interface. Iridium is found to be least corrosive. When a field evaporated clean Ir tip is exposed to atmosphere for 15 min, less than three contaminant layers are formed. Dipping a cleaned Ir tip to a platinum plating solution of sulfato-dinitrito-platinous acid without any voltage creates cavities of 20 to 80 Å diameters and 2 to 5 atomic layers in depth. When electrodeposition is done with the solution with a concentration of 5 g Pt/liter, good deposits can be obtained with an a.c. voltage of 1.2 volts at 500 μA for 30 to 60 s with the bath kept at 50°C. The current density is presumably about 10 to 0.1 A/cm^2, depending on the position of the tip surface. The thickness of the coating so obtained is of the order of 500 Å, and the dominant orientation of the small crystallites is having the [111] normal to the tip surface independent of the direction of the base surface. Figure 5.19 shows an example of such a surface where many lattice defects, particularly vacancies and vacancy clusters, can be seen. Using the electrodeposition techniques, small biomolecules can be embedded into metal layers, but the images do not show many details of the atomic structures; perhaps fragmentation of the embedded molecules cannot be avoided.

5.5.4 Interfaces of quantum well heterostructures

Interfaces are of critical importance in determining the electronic and optical properties of quantum well heterostructures. It is necessary to

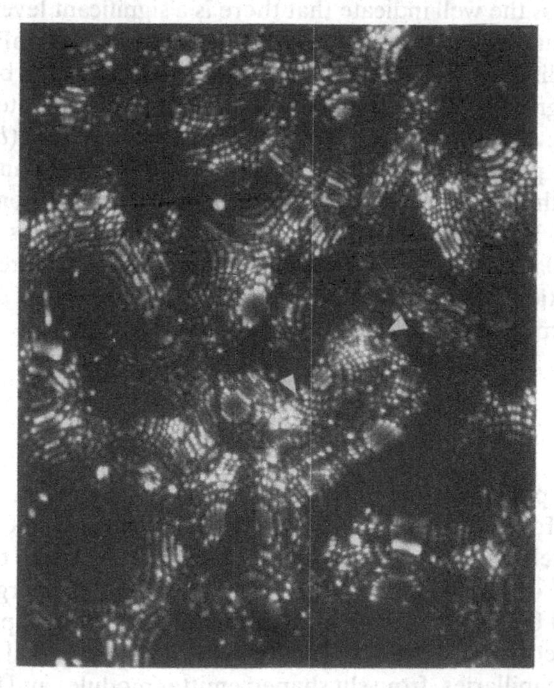

Fig. 5.19 A helium field ion image of a tungsten tip electroplated with a thick layer of platinum. Many lattice defects such as grain boundaries, twin boundaries, vacancies and vacancy clusters can be seen. From the work of Rendulic & Müller.[128]

find out how these properties are correlated to their microstructures so that the fabrication methods can be chosen to reliably produce the desired properties. Unfortunately it is very difficult to distinguish between chemical abruptness and morphological roughness of quantum well heterostructures with sufficient spatial resolution using conventional techniques. This problem has recently been studied with a position sensitive atom-probe (POSAP)[129] for GaInAs/InP samples by Liddle *et al.*[130] Tips are prepared by first selectively etching the InP substrate away from the quantum well stack, and then using a chemical polish to produce a tip of ~1000 Å radius out of the layer.

Two sets of quantum wells were studied. Both samples were 30-period stacks of GaInAs wells with InP barriers, grown at 660°C by organometallic vapor phase epitaxy (OMVPE) on an InP substrate. The first set was grown with a quartz wool baffle in the reactor, in an attempt to make the gas flow more uniform over the area of the wafer. The second set was grown in an unmodified reactor. TEM images of the first set show a relatively poor morphology, with variations in the thickness of the GaInAs layers from 30 to 70 Å. The POSAP composition maps of Ga and As also indicate poor quality with rough interfaces. The composition profiles across the well indicate that there is a significant level of P in the well and that the interfaces between the GaInAs and InP layers are chemically diffuse. The second set of wells appears much better in the TEM micrograph with uniformly thick layers and smooth interfaces. The POSAP Ga and As maps, shown in Fig. 5.20(*a*) and (*b*), and the composition profile, Fig. 5.20(*c*), all show better defined and smoother interface with the Ga and As concentrations closer to the nominal values of the wells. It appears that quantum well heterostructures with widths less than 100 Å can be studied with a spatial resolution better than 10 Å with a position sensitive atom-probe, and possibly with an ordinary imaging atom-probe also.

5.6 Liquid metal ion sources

5.6.1 Basics of liquid metal ion sources

A practical application coming out of field ion emission is the liquid metal ion source. Ion sources of a wide variety of chemical elements, most of them low melting point metals, can be produced by using either liquid metals[131,132] or liquid alloys.[133] The idea of extracting charged droplets out of liquid by application of an electrostatic field is perhaps older than field ion microscopy. But the development of liquid metal ion sources from liquid capillaries, from slit shaped emitter modules and from wetted field emission tips, etc., as well as the understanding of the mechanisms of ion formation in terms of field evaporation and field ionization theories,

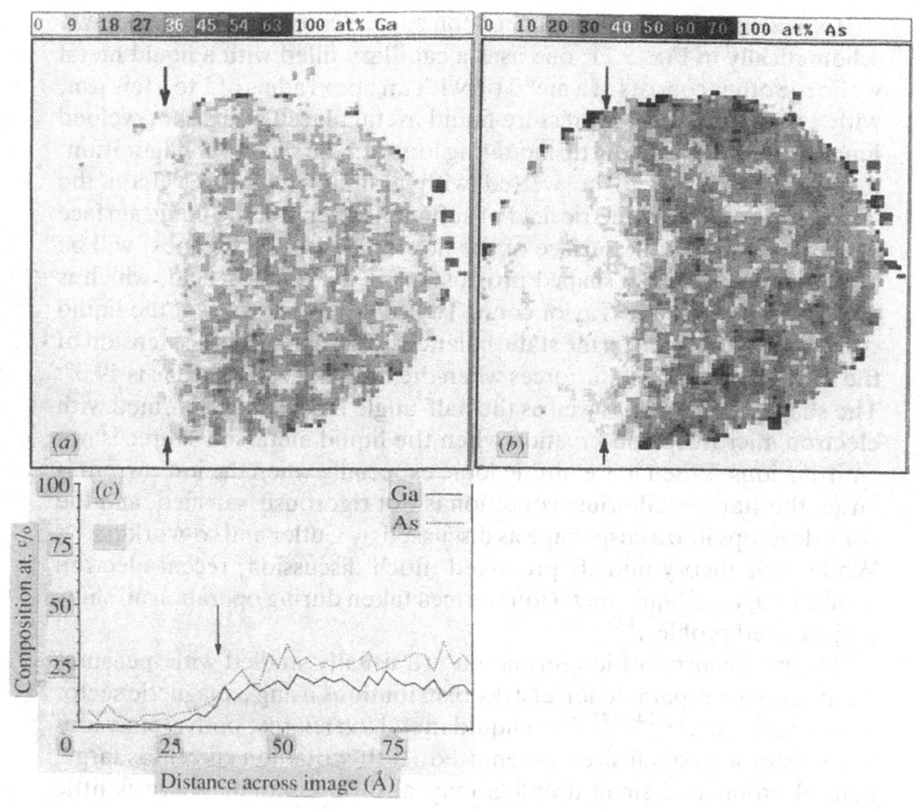

Fig. 5.20 (*a*) and (*b*) are POSAP composition maps of Ga and As at an interface of a GaInAs/InP quantum-well heterostructure. (*c*) is a composition profile showing Ga and As concentrations across the interface. The approximate position of the interface is shown by arrows in these figures. Data are from Liddle *et al.*[130] (Courtesy of G. D. W. Smith.)

must have come from knowledge gained from field ion emission and microscopy. In the 1960s the main aim for developing liquid metal ion sources was the production of electrostatic thrusters for spacecraft propulsions,[134] and perhaps also the chemical analysis of liquids.[135] Nowadays, focused ion beams from liquid metal ion sources can be used for many purposes. Some of them are:

1. scanning ion microscopy,[136,137]
2. X-ray and optical mask repair,[138]
3. secondary ion mass spectroscopy with sub-μm spatial resolution for chemical analysis of sub-μm structures,[139]
4. ion beam lithography, and
5. maskless implantation doping of semiconductors, etc.[140]

Two commonly used liquid metal ion source configurations are shown schematically in Fig. 5.21; one uses a capillary filled with a liquid metal while the other consists of a metal tip with an apex radius of 1 to a few μm, with a drop of low vapor pressure liquid metal placed at the spot-welded junction of the tip and the tip mounting loop. In the second configuration, the tip surface has to be wetted with a liquid film first. Upon the application of a high electric field of sufficient strength, the liquid surface in the capillary, or the surface of the liquid film on the tip apex, will be drawn into a conically shaped protrusion by the applied field, which is generally referred as a Taylor cone. Taylor[141] has shown that the liquid cone can be stabilized by the static balance between the surface tension of the liquid and electrostatic forces when the half-angle of the cone is 49.3°. The shape of the cone as well as the half-angle have been confirmed with electron microscope observation when the liquid metal ion source is not emitting ions. When it is emitting ions, especially when the ion current is large, the static equilibrium condition is not rigorously satisfied, and the cone develops into a cusp shape as discussed by Cutler and co-workers.[142] While their theory initially provoked much discussion, recent electron profile images of liquid metal ion sources taken during operation all show cusp-shaped profiles.[143]

The mechanisms of ion formation are usually studied with measurements of mass separated ion energy distributions using a magnetic sector mass spectrometer.[144,145] For a liquid metal ion source, both atomic ions and cluster ions of all sizes are emitted. If the total ion current is large, neutral atoms and small droplets may also be emitted. There is little question that most of the atomic ions in a liquid metal ion source at low

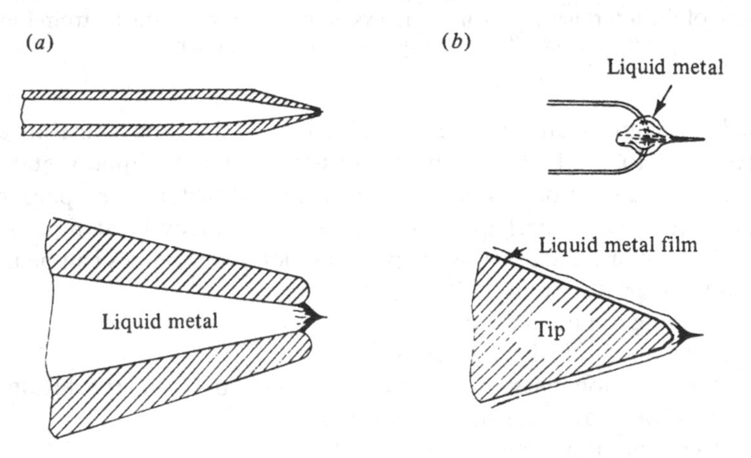

Fig. 5.21 Two commonly used configurations of liquid metal ion sources.

emission current regime are produced by field evaporation. Thus the critical ion energy deficits in the ion energy distributions for atomic ions agree with those in low temperature field evaporation to within about 1 to 3 eV, or to within, presumably, the activation energy in high temperature field evaporation of the liquid metal ion sources. The critical ion energy deficits for atomic ions are also independent of the total ion current. On the other hand, the FWHM of the energy distributions of atomic ions spread continuously with the ion current. At the small current limit, the FWHM for all atomic ions all approach the same limit of less than 5 eV; presumably this value is only limited by the resolution of the ion kinetic energy analyzers, and the true limit is probably less than 2 eV, as found in pulsed-laser ToF atom-probe energy analysis of low temperature field evaporated ions as discussed in Chapter 2. For multiply charged ions it is possible that post field ionization plays an important role, since in high temperature field evaporation ions formed in initial field evaporation are dominantly singly charged.

The critical ion energy deficits of cluster ions increase with ion current, indicating that as the ion current increases these ions are formed further away from the liquid surface. In addition, the FWHM also increases with current. These observations suggest that cluster ions are most likely produced by field ionization of neutral atomic clusters. Even a fraction of atomic ions may in fact be produced by field ionization of neutral atoms when the total ion current is large. Under such conditions the liquid surface is continuously bombarded by electrons released in field ionization and post field ionization. The emitting part of the liquid surface can be heated to a much higher temperature than the liquid by the bombardment of these electrons and thus vaporization can occur. The vapor pressure near the emitter surface must be greatly enhanced by the field gradient induced attractive polarization force. Therefore formation of atomic clusters by condensation of vaporized atoms can occur near the liquid emitter surface. Some of these clusters and atoms can then be field ionized or ionized by electron impacts. Ion energy distribution data showing these features are summarized in Fig. 5.22.

The great advantage of liquid metal ion sources as compared with other types of ion sources is the very high brightness. Typically, an axial current intensity well in excess of 20 μA/sr can be realized with a total ion current in the 1 to over 100 μA range. As ions are emitted from a liquid metal emitter of radius less than ~100 Å, the virtual source size is very small, also of less than ~100 Å, and the ion beam can be focused to a spot diameter of less than 0.1 to a few μm with a current density of the order of A/cm^2 at 20 keV. Thus the ion source can be used for scanning ion microscope and microlithography with a spatial resolution of better than 0.1 μm. Although ions of all kinds, such as Li, B, Na, Al, Si, K, Fe, Ga,

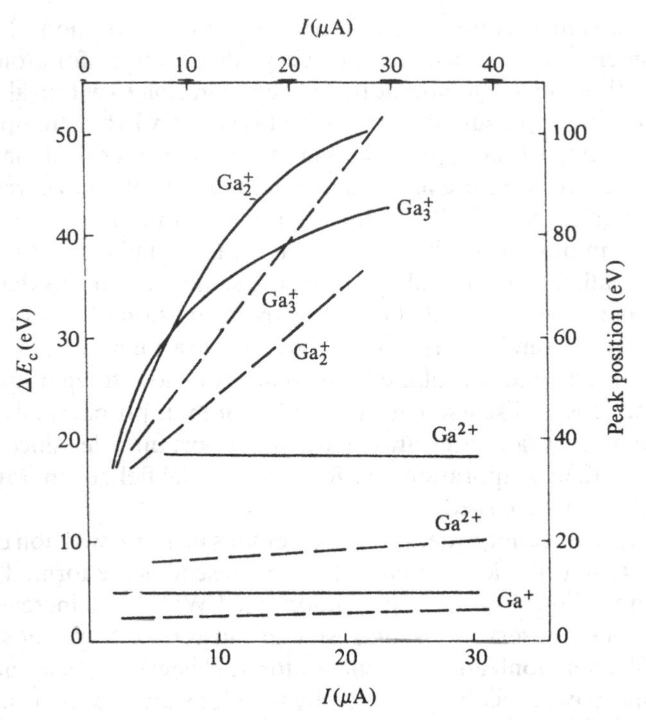

Fig. 5.22 Critical energy deficits of various ion species in a liquid Ga ion source as functions of the ion currents (solid lines and left and lower scales), from Swanson.[144] Dashed lines are the peak positions of the ion energy distributions (upper and right scales), from Culbertson *et al.*[145]

Ag, In, Sn, Cs, Hg, Au, Bi, etc., have been produced, the most commonly used liquid metal ion sources use Ga and In because of the very low melting points of these metals, 29.9 and 156.2°C, respectively. Cs ions, on the other hand, are especially suited for propulsion purposes because of their larger mass, which is more efficient in momentum transfer in propulsion.[146] One of the difficulties of liquid metal ion source is the long term stability and high frequency noises in the ion current, which presumably can be improved by using a better controlled shape, surface morphology and temperature of the tip, and also the electrode geometry. Presumably, in slit emitter modules, ions are emitted from a large number of cones, and the total ion current is therefore less subject to the instability of an individual cone.

In the 1960s emission of charged droplets was of primary interest because of the possible use in spacecraft propulsion, although for efficiency ions should be the preferred emitted species for propulsion purposes. There is a renewed interest in this phenomenon because of possible applications in ultra-fine powder metallurgy and focused beam deposition, both broad area and localized area, of uniform thin films. D'Cruz *et al.*[147] report that plenty of charged droplets can be formed in a liquid gold ion source if the emission current is of the order of several tens of μA. The beam divergence has a FWHM of ~2° for the charged droplets as compared with ~45° for ions. The upper limit of the virtual source size of charged droplets is 8 μm. At an emission current of 150 μA typical deposition fluxes are $5 \times 10^5\ \mu m^3\ s^{-1}\ sr^{-1}$ with a total deposition rate of $\sim 3 \times 10^3\ \mu m^3/s$ and an average droplet diameter of ~1 μm. The average droplet size decreases with smaller current and angle. Gold films with a fine textured morphology suitable for sub-micron patterning can be formed on a silicon wafer surface at an emission current smaller than 135 μA.

5.6.2 Scanning ion microscope

A focused ion beam from a liquid metal ion source can be used to image solid surfaces with good spatial resolution, ~200 to 700 Å. A schematic of a scanning ion microscope (SIM) co-developed by University of Chicago Ion Microscope Research Group and GMHE/Hughes Research Laboratories is shown in Fig. 5.23. This system uses a 40 keV scanning ion microprobe equipped with a liquid gallium ion source. It is capable of yielding mass-resolved elemental distribution images (maps), or maps of the distribution of chemical species, of the surface of a material with a lateral resolution just given. At the same time, the SIM can provide topographic and voltage contrast images, with comparable spatial resolution, by utilizing the copious ion-induced secondary electron (ISE) signal. In other words, this type of instrument is capable of providing both the topography as well as the chemical distribution of the surface of a material with a spatial resolution of ~500 Å. In addition, the ion beam can be used to etch away surface layers slowly, thus revealing the chemical distribution inside a material with the same lateral resolution and perhaps a depth resolution of about 50 to 100 Å.

The system consists basically of two design-integrated and yet decoupled systems, the primary and secondary ion optical columns. In the primary column, ions are extracted from a liquid gallium ion source of high brightness ($\sim 10^6\ A\ cm^{-2}\ sr^{-1}$) through a beam defining aperture that collimates the ions to within a small angular acceptance half angle of 0.16 to 0.8 mrad to reduce spherical and chromatic optical aberrations. To reduce chromatic aberration, a total ion current of only a few μA is extracted from the ion source of which only 2 to 50 pA passes through the beam-defining aperture to ultimately bombard the sample. The ion spot size, or probe size, is determined primarily by the chromatic aberration;

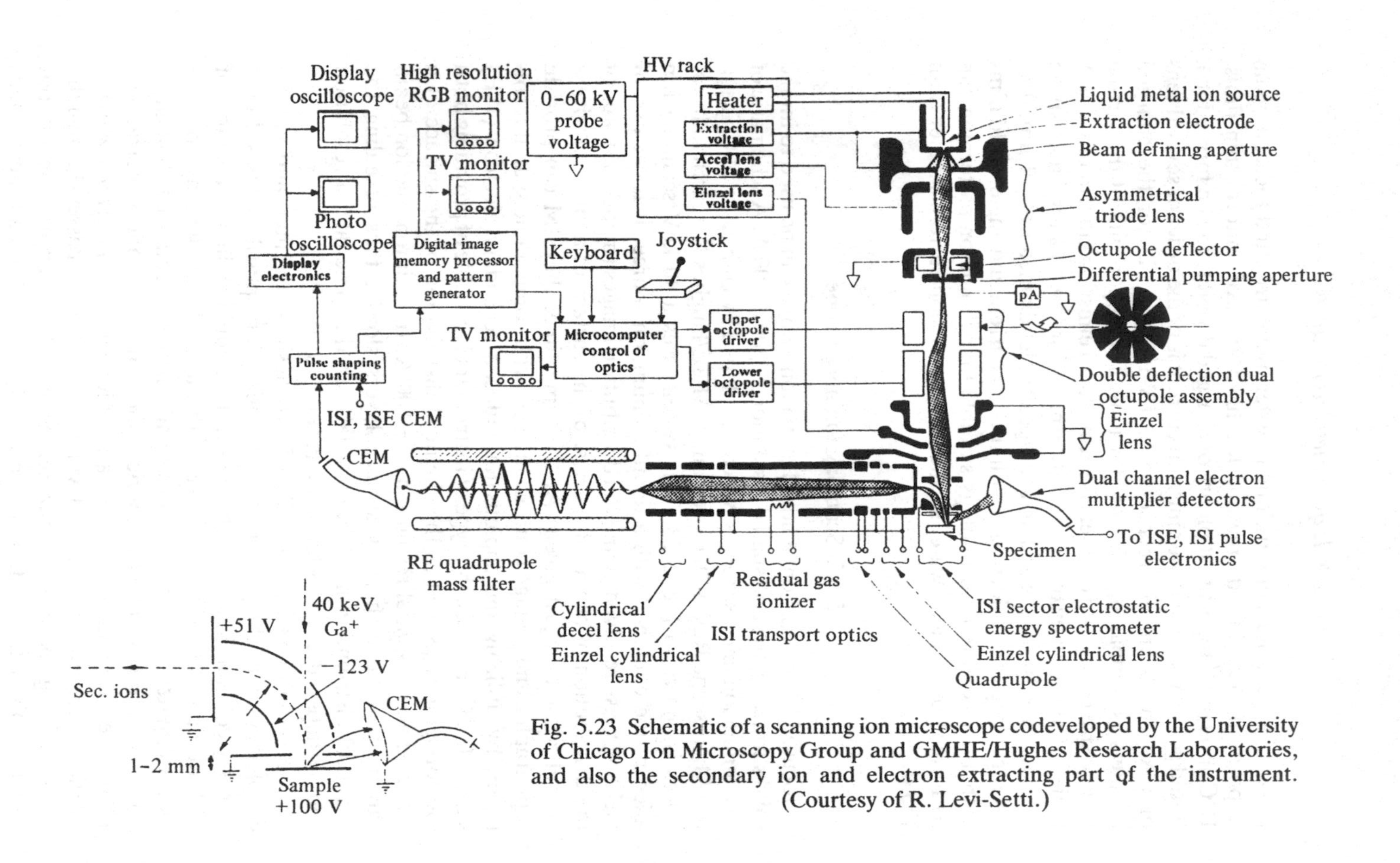

Fig. 5.23 Schematic of a scanning ion microscope codeveloped by the University of Chicago Ion Microscopy Group and GMHE/Hughes Research Laboratories, and also the secondary ion and electron extracting part of the instrument. (Courtesy of R. Levi-Setti.)

the probe current decreases with the square of the probe diameter. Theoretically the probe could be made smaller than 100 Å, but the beam current would have to be less than 1 pA, which is of limited practical use.

The primary ion beam is incident normal to the sample surface. For secondary ion mass spectrometry (SIMS) the secondary ions are also extracted normal to this surface. These ions traverse an 8 mm radius cylindrical energy analyzer positioned 1–2 mm above the sample. The energy analyzer is usually operated with a 10 eV ion energy acceptance window. Beyond the exit slit of the analyzer are transport optics which carry and shape the secondary ion beam for optimum insertion into an RF quadrupole mass filter. For imaging the topography of the sample, secondary electrons induced by primary ion beam or non-mass resolved secondary ions can be collected with a channeltron electron multiplier. The ion and electron extraction region is also shown in Fig. 5.23. To study voltage contrast in ion induced secondary electron images, the sample can be biased to different voltages.

Scanning ion microscopes have been used to image a wide variety of samples, which include structures of circuits in microelectronic chips, fossil diatoms, microcrystals and biological objects such as the eyes of *Drosophila melanogaster*, human erythrocytes and other microstructured materials, etc. In Fig. 5.24 are shown two mass separated elemental images, one of $^{27}Al^+$ ions and one of $^{48}Ti^+$ ions, of a test circuit obtained by Levi-Setti and co-workers. Whereas scanning ion microscopy cannot achieve atomic resolution and the damage of samples by ion bombardment can never be avoided, it has the advantage of being able to analyze the three-dimensional distribution of chemical species in samples by

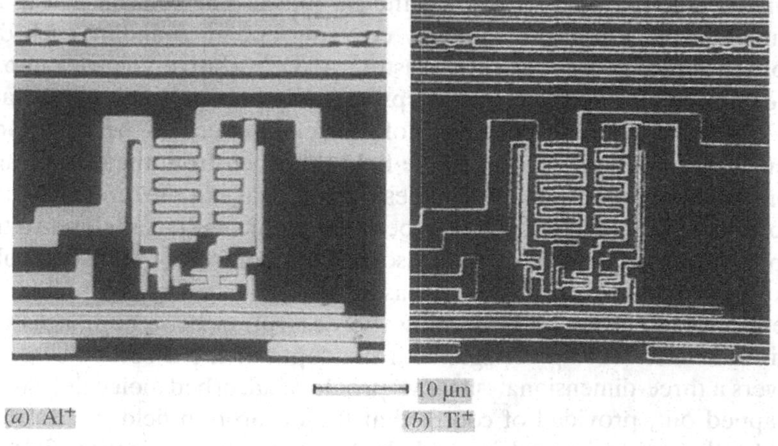

Fig. 5.24 SIM elemental images of a test circuit; 40 μm full scale. (*a*) is an image of $^{27}Al^+$ ions and (*b*) is an image of $^{48}Ti^+$ ions. (Courtesy of R. Levi-Setti.)

gradual ion etching of surface layers, which is similar to SIMS but with a sub-micron spatial resolution. This instrument should be particularly useful in practical structural and chemical analysis of microstructures of materials of micron sizes. So far SIM uses ions from a liquid metal ion source and these ions are usually gallium ions. It should also be possible to use a field ionization source from a single atom tip for Ar or Ne ions; these ions are what are generally used in SIMS. The field ion current from a low temperature single atom tip should be of the order of 10 pA, just sufficient for SIM. The ion energy spread is very small compared with liquid metal ion sources, and the virtual ion source size is of the order of only 1 to 3 Å. Another advantage is that inert gas atoms will not condense on the sample surface.

5.7 Imaging biomolecules by field desorption tomography

A point projection microscopy, sometimes referred to as field desorption tomography, has been developed by Panitz to image biomolecules adsorbed on metal emitter surfaces.[148] The point projection microscope is identical to a field ion microscope or an imaging atom-probe where a Chevron double channel plate is used for displaying field desorbed ions. The basic idea is as follows. A smooth surface of a tip of large radius is first prepared by field evaporation or by thermal annealing in UHV. Molecules are then deposited onto the field emitter surface by deposition from aqueous solution.[149] After the tip is placed in UHV it is cooled to below 30 K. An infrared laser pulse briefly raises the tip temperature to about 300 K to thermally desorb contaminant species which have adsorbed on the cold tip surface. A precise quantity of benzene is then condensed onto the tip apex from the gas phase. The benzene acts as an immobile blanket gas, completely covering and surrounding each deposited molecule or molecular cluster. A d.c. positive voltage ramp is then applied to the tip in small steps. At a tip voltage when the surface field reaches the desorption field of benzene molecules, benzene molecules within a finite depth will be field desorbed from the surface and form an almost contrastless field desorption image at the screen. As the tip voltage is gradually increased, benzene molecular layers closer to the tip surface will be desorbed and adsorbed biomolecules will be gradually exposed in the form of dark areas in the field desorption images of benzene ions. Thus by correlating the desorption benzene images obtained at different tip voltages with the depth of the desorbed benzene layers a three-dimensional 'shadow' profile of adsorbed molecules can be mapped out, provided of course that the desorption field of adsorbed molecules to be imaged is much higher than the desorption field of benzene molecules. Figure 5.25 shows a computer reconstructed image of

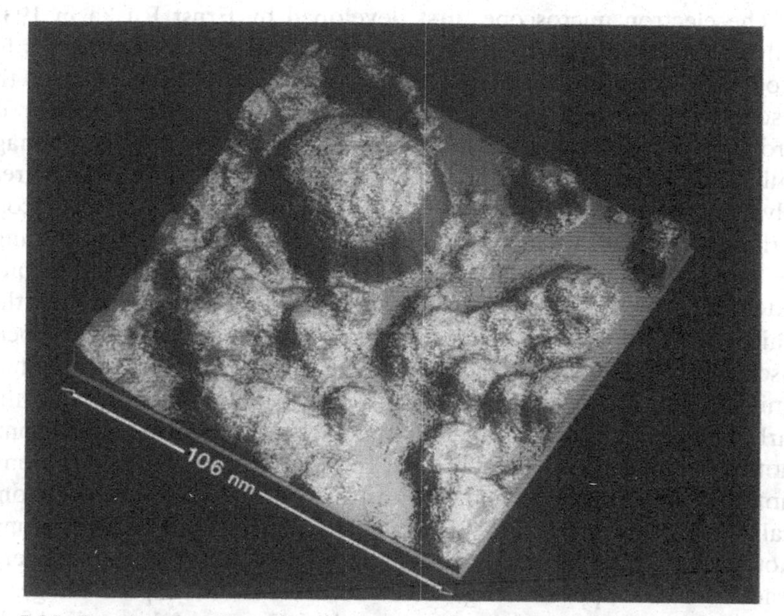

Fig. 5.25 Quasi-three-dimensional, point projection image of ferritin on tungsten at 30 K before applying a critical desorption field of ferritin. The image is a digital composite of seven images taken at different field strengths, thus at different depths of the covered methane layer. (Courtesy of J. A. Pantiz.)

ferritin monolayer on tungsten. The image resolution is estimated to be about 20 to 30 Å, and in principle macromolecules with a desorption field greater than 0.4 V/Å, can be imaged with benzene as the blanket gas.

5.8 Comparisons with other atomic resolution microscopies

5.8.1 Electron microscopies

Several other microscopies have recently also achieved atomic resolution, and some of them promise great versatilities and wide applications. These microscopies include the scanning electron microscope with a field emission tip,[150] the scanning tunneling microscope and the atomic force microscope,[151] and the high resolution transmission electron microscope,[152] etc. Each microscopy has its own advantages and shortcomings and therefore also has its best suited areas of application. Although the areas of application will change and expand with time, the basic principles of these microscopies will not change drastically and therefore one should be able to make some general comments about the similarities and differences of these microscopies and where each of them will be most useful. We consider here only their atomic resolution aspects.

The electron microscope, first developed by Ernst Ruska in 1931, utilizes the wave nature of electrons and electron focusing lenses for producing a greatly magnified image of a sample. To improve the resolution, electrons of larger kinetic energy, and hence shorter de Broglie wavelength, should be used. Unfortunately radiation damage will become a problem and image contrast will also deteriorate. A great advantage of electron microscopy as compared with field ion microscopy is that the magnification can be changed continuously over a wide range for examining samples of different sizes. In FIM, image magnification is determined by the tip radius and there is a very narrow range within which the FIM can be conveniently operated, as has already been discussed in Chapter 3. In the scanning electron microscope with a field emission gun, single atoms of high-Z elements deposited on very thin carbon films a few atomic layers in thickness can be imaged and atomic motion in real time can be observed. In the field ion microscope, atomic motion of single atoms in real time can also be observed as in directional walk experiments. In these FIM experiments, however, the atomic motion is performed mostly under the influence of a potential energy gradient of the applied image field. Of course, time lapsed images of atomic motion can be obtained without the influence of the applied field. In SEM observations of atomic motion, the effect of electron bombardments and excitations cannot be excluded despite some initial conclusions that the heating effect of electrons is not important. Thermal motion of atoms on a carbon film a few atomic layers thick should also be quite different from that in a macroscopic sample since the phonon spectrum for the atomic motion perpendicular to the carbon film should be quite different from that of a macroscopic sample. One of the difficulties of this SEM is that low-Z atoms cannot be seen, either because of the lack of image contrast with the background or because of the small scattering power of these atoms.

In high resolution transmission electron microscopy, atomic structures are always derived from a lattice image of a very thin film. In the investigation of surface atomic reconstructions, cross-sectional views of the atomic structure of the knife edge, or the 'surface', of a very thin film can be observed. Examples, one from a study of the (1×2) reconstruction of the Au (110) by Marks,[152] and one from a study of the (1×5) atomic reconstruction of the Au (001) surface by Takayanagi *et al.*,[153] are given in Fig. 5.26. Field ion microscopy, on the other hand, gives the top atomic view of a very small surface which contains no information on the vertical distance of these atoms with respect to the underlying layers. Thus TEM and FIM provide complementary information on surface atomic structures: one gives the cross-sectional views while the other gives the top views. In the (1×5) reconstruction of an fcc (001) surface,

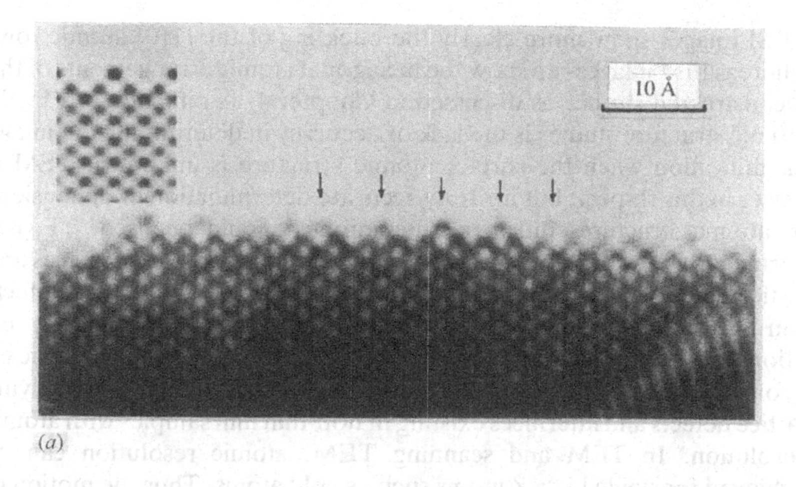

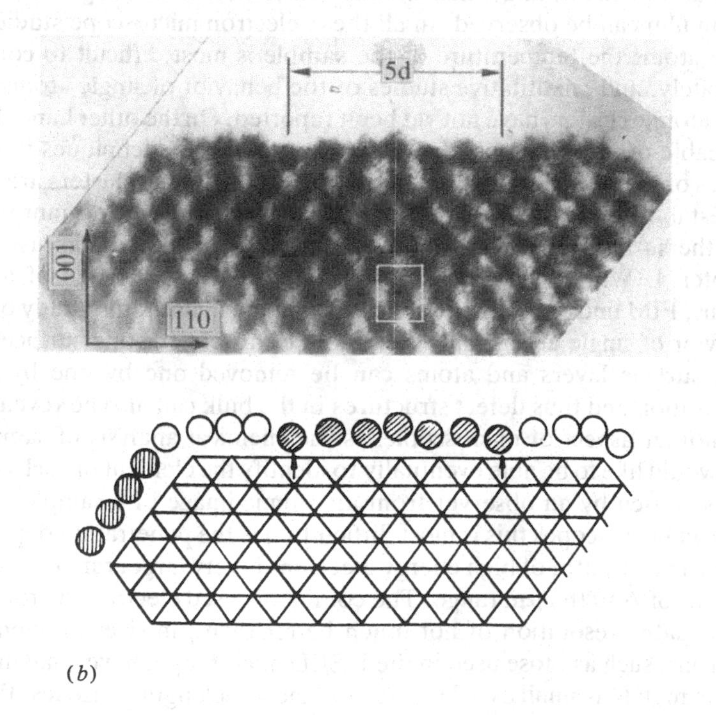

Fig. 5.26 (*a*) High resolution TEM image of a (1 × 2) reconstructed Au (110) surface. The inset is a computer simulated image. (Courtesy of L. D. Mark.) (*b*) High resolution TEM profile image of a (1 × 5) reconstructed Au (001) surface and an illustration of atomic row positions in the image. (Courtesy of K. Takayanagi.)

TEM images show more clearly the buckling of the [110] atomic rows whereas FIM images can show the hexagonal atomic arrangements of the reconstructed surface as discussed in Chapter 4. A difficulty of FIM in atomic structure studies is the lack of accuracy in determining the image magnification when the surface atomic structure is unknown. TEM is better in this respect, but for truly accurate determination of dimensions of atomic structures diffraction techniques should be used. TEM is particularly powerful for studying atomic structures at the interfaces and lattice defects in very thin films.[154] In FIM, interfaces and lattice defects can also be studied by field evaporation techniques. Image interpretations are much more complicated because of the much less symmetric tip geometry. With TEM, on the other hand, one has difficulty in studying lattice defects and interfaces existing in non-thin film samples with atomic resolution. In TEM and scanning TEM, atomic resolution can be achieved for single high-Z atoms such as gold atoms. Thus the motion of these atoms in forming small atomic clusters near the edge of a thin carbon film can be observed. In all these electron microscope studies of single atoms the temperature of the sample is most difficult to control accurately, and quantitative studies of the behavior of single atoms and small atomic clusters have not yet been reported. On the other hand, FIM is capable of imaging most non-gaseous atoms, and techniques in FIM studies of the behavior of single atoms and small atomic clusters are very well established; most of the data derived in these studies are quantitatve, with the surface temperature finely controlled, as already discussed in Chapter 4. While electron microscopes have a wider range of applications, FIM finds its unique applications in the quantitative study of the behavior of single atoms and small atomic clusters on solid surfaces. In FIM, surface layers and atoms can be removed one by one by field evaporation and thus defect structures in the bulk can also be revealed.

Another aspect of microscopies is the chemical analysis of samples. One would like to be able eventually to identify the element of each of the atoms chosen by an observer from an atomic image of a sample. With electron microscopes this is most difficult since the penetration depth, or the mean free path, of high energy electrons is very large, ranging from a few tens of Å to the μm range. The commonly used electron microprobe has a spatial resolution of not much better than μm size. Low energy electrons, such as those used in the LEED microscope, have a mean free path in metals as small as ~3 to 5 Å, and the wavelength is also less than 2 Å. If one can overcome the diffculty of focusing a low energy electron beam to a beam size of the order of ~Å, one may be able to localize electrons to a volume as small as a few atoms in size. But chemical analysis with low energy electron excitation is inherently difficult. In contrast, in field ion microscopy, there is the atom-probe field ion microscope with which atom by atom chemical analysis is possible. This technique is, however, destructive, and is also confined to samples with a tip geometry. One of the great limitations of both electron microscopy and field ion microscopy is that these instruments can be operated only in vacuum. In contrast, scanning tunneling microscopes and atomic force microscopes can be operated in either vacuum or atmosphere, or even in liquid.

5.8.2 Scanning tunneling microscopy

Scanning tunneling microscopy (STM) has been enjoying an immense popularity since its success in imaging the (7 × 7) reconstructed silicon (111) surface by Binnig & Rhorer and their co-workers and by many others. Figure 5.27 gives an example of the image.[155] STM is now a fully fledged atomic resolution microscopy.[156] The development of the scanning tunneling microscope can be traced to field emission microscope and in fact an early instrument of its kind was already developed by R. D. Young in the early 1970s which he called a topographfinder.[157] A resolution of better than a few hundred Å was achieved. Young did his Ph.D. research under Müller, and had worked with a field emission microscope for many years when he developed the idea of finding the topography of a sample by scanning a field emission tip above the sample with an XYZ-piezo controlled scanner, identical to that used in modern STM. Owing to the lack of mechanical stability of his instrument and the rather high tip voltage applied between the sample and the tip, he did not achieve atomic resolution even though vacuum tunneling at a distance of perhaps several Å was observed. Field ion microscopy is also an outgrowth of field emission microscopy, both of them having been developed

Fig. 5.27 Three-dimensional view of the reconstruction of the clean Si (111) (7 × 7) surface across an atomic step, showing the characteristic 12 adatoms per unit cell and a corrugation of 2 Å. The size of the image is about 290 × 170 Å. (Courtesy of S. Chiang.)

by Müller, and the FIM was the first microscope to achieve atomic resolution.

There are many similarities, and of course also differences, between these two atomic resolution microscopies, i.e. scanning tunneling and field ion. Both STM and FIM use electron tunneling for forming an image of a solid surface. In FIM an image is formed by radial projection of image gas ions, which are produced by field ionization of image gas atoms above the surface. The initial states in field ionization are atomic states of image gas atoms and the final states are electronic states of the tip, or the sample material. In STM an image is formed by electron tunneling between the probing tip and the sample, and the x and y scanning and z motion of either the probing tip or the sample. STM uses a tip for probing the electronic charge density of states of the sample. As both microscopes use an electron tunneling effect for forming an image, resonance tunneling features through standing wave states can be observed. In STM these features manifest themselves in the dI/dV vs. V curve,[158] whereas in FIM they appear in energy distributions of field ions and field desorbed ions of gases.[159]

In STM samples can be of any shape and the microscope can be operated in different environments such as in vacuum, in atmosphere or in liquid. The scanning range can also be varied widely, as with the electron microscope. Thus there is no limitation on the size of the sample in STM, and the area of a flat surface which can be studied can also be very large. In FIM, with the tip geometry of the sample, the size of a flat surface plane available for experiments rarely exceeds 80 to 100 Å. The size of a surface in a FIM experiment can be controlled and the size limitation can sometimes be beneficially exploited and can be used as an advantage where plane size effects of surface phenomena can be studied. In some surface phenomena, size effect is not important. Thus in the (1×2) surface reconstruction of the Pt and Ir (110), a surface so small as to contain only several atoms makes the same reconstruction as a much larger one.[160] On the other hand, an FIM study of the (1×5) reconstruction of the Ir (001) surface finds that the buckling of the [110] atomic rows in the reconstructed surface always forms a symmetric pattern,[161] as is also seen in the high resolution TEM image of Fig. 5.26. Thus the detailed buckling of these atomic rows in the reconstruction, as well as the domains formed, have to depend on the plane size. It is also now well recognized that even the most carefully prepared macroscopic surface, which shows a perfect LEED pattern, may contain islands of good atomic structures surrounded by areas of high irregularities. Thus studies performed on small surfaces of field ion emitter tips are also quite relevant to surfaces of macroscopic samples.

An important consideration of any microscopy is the ultimate resolution achievable. As both STM and FIM rely on tunneling of electrons between a probing atom – an image gas atom in FIM and an atom of the scanning tip in STM – and an atom of the sample, the *ultimate resolution* should be less than the average of the two atomic diameters, or the lateral extent of the matrix element of the overlapping wavefunctions of the probing and imaged atoms. Discussion of the resolution of STM assumes a tip radius of a certain size;[162] such resolution does not really represent the ultimate resolution of STM. In FIM, the resolution can be slightly deteriorated by the lateral motion of the ensuing image gas ions. On the other hand FIM uses He atoms as the image gas which have a very small diameter compared to ordinary tip atoms. In both microscopes the best lateral resolution should be of the order of 2 Å. The ideal vertical resolution of both microscopes, again, should be comparable. It should depend on the noise to signal ratio of these instruments; when the noises are very low a small change in the signal due to a small change in the corrugation height of the surface can be detected. However, in FIM the image intensity of a flat surface is not uniform because of the non-uniform field distribution. Thus image intensity variation due to surface corrugation is overshadowed by the effect of the non-uniform field. A limitation of FIM is that atoms on a large flat surface cannot be resolved. Atomic arrangements can be revealed only if the plane sizes are small. No such limitation exists for STM, and atomic arrangement of a large flat surface containing lattice steps can be resolved for the entire surface, as seen in Fig. 5.27.

In FIM, the sample tips and adsorbed molecules have to be able to withstand the high image field in order that stable images can be obtained. As STM does not have this limitation, adsorbed organic and biomolecules such as benzene and Cu-phthalocyanine molecules can be imaged.[163] Figure 5.28(*a*) and (*b*) show two images of isolated Cu-phthalocyanine molecules adsorbed on a Cu (001) surface, one at submonolayer coverage and one at one monolayer coverage. These images exhibit atomic scale features which agree well with Hückel molecular orbital calculations. In STM the stability of the probing tip is essential for having a reproducible and stable image of a sample. To achieve atomic resolution, there should be a protruding atom on the tip and this atom, irrespective of whether it is an adatom, a kink atom, or an atom of a small atomic cluster on the tip, should be stable during the scanning. At room temperature an adatom is stable on rough surfaces such as the W (111), but not on smooth surfaces such as the fcc (111), where surface diffusion of single adatoms can already start below 100 K. Small atomic clusters such as W_5 and W_6 on the W (110) surface become mobile only at around 400 K. Tips with such or similar clusters can be considered for STM scanning.

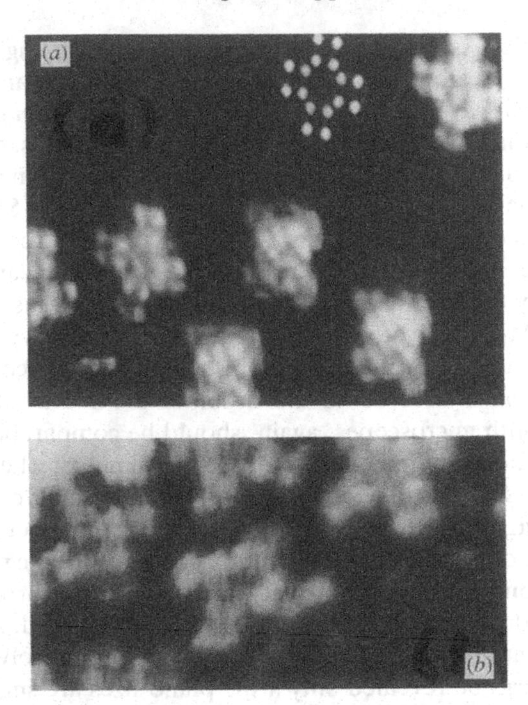

Fig. 5.28 STM images, (*a*) at submonolayer coverage and (*b*) at one monolayer coverage, of Cu–phthalocyanine molecules adsorbed on a Cu (001) surface. A gray scale representation of the highest occupied molecular orbital, evaluated 2 Å above the molecular plane, has been embedded in the image at (*a*). (Courtesy of S. Chinag.)

There are of course many other similarities and differences, and some of them are listed in Table 5.1 without further explanations. In general, STM is very versatile and flexible. Especially with the development of the atomic force microscope (AFM), materials of poor electrical conductivity can also be imaged. There is the potential of many important applications. A critically important factor in STM and AFM is the characterization of the probing tip, which can of course be done with the FIM. FIM, with its ability to field evaporate surface atoms and surface layers one by one, and the capability of single atom chemical analysis with the atom-probe FIM (APFIM), also finds many applications, especially in chemical analysis of materials on a sub-nanometer scale. It should be possible to develop an STM–FIM–APFIM system where the sample to be scanned in STM is itself an FIM tip so that the sample can either be thermally treated or be field evaporated to reach into the bulk or to reach to an interface inside the sample. After the emitter surface is scanned for its atomic structure, it can be mass analyzed in the atom-probe for one atomic layer,

Table 5.1 *Characteristics of FIM and STM*

Items	Field ion microscope	Scanning tunneling microscope
Image formation	Tunneling of atomic electrons into sample which is a tip	Tunneling of electrons between the probing tip and sample
Sample	Tip shape only	Any shape
Resolution		
lateral	~2.5 Å, *r*-dependent	~2.0 Å
vertical	~0.2 Å (?), difficult to calibrate	<0.1Å
Imaged area	$\sim 1.5\pi^2(\text{radius})^2$	Any (in principle)
Plane size	$<(80\ \text{Å})^2$	Any (in principle)
Operation conditions	Vacuum; low temperature	Vacuum, air or liquid; variable (in principle)
Image distortion	Non-linear; difficult to correct	Mostly linear; easier to correct
Stability		
Thermal	Not a problem, tip temp. adjustable	Careful design required
Mechanical	Not a problem	Vibration isolation required
Sample treatment	Field desorption, annealing, sputtering	Annealing, sputtering, cleaving a single crystal
Distance calibration		
Vertical	?	Z-piezo for relative distance
	Resonance peaks in field ionization distance to x_c	Resonance tunneling peaks for absolute distance calibration
Horizontal	Magnification by radial projection	X- and Y-piezo controlled motion
Chemical analysis		
By images	A species invisible	By changing bias voltage, a species can be made visible
By device	Time-of-flight atom-probe field ion microscope	None
Atomic site spectroscopy	Maybe	Yes
Biomolecule imaging	Maybe	Yes

and then scanned again in the STM mode. For the STM probing tip, one may use one of the sharp protrusions made on a flat surface of a solid, which can be produced by natural lithography[164] or a similar method. Since the sharp protrusions can be made with a very large density, the chance of having the sample tip scanned by a protrusion can be made to be appreciable. A critical factor in this type of instrument will be the sharpness and uniformity of these protrusions one can consistently fabricate.

It appears that there is no limit as to what one can do with atomic resolution microscopy as long as our imaginations are not deterred by technical difficulties, which can always be gradually overcome by our ingenuities and perseverance. Finally this author believes that the ultimate goal in scientific studies is to understand physical phenomena in terms of microscopic theories. In addition, the sizes of many practical devices such as electronic ones are becoming smaller and smaller. Understanding materials of nanometer and sub-micron sizes is therefore becoming more and more important. In this sense, field ion microscopy as well as other atomic resolution microscopies have a lot to contribute both to basic science and to technology in the days ahead.

References

1 M. K. Miller & G. D. W. Smith, *Atom-Probe Microanalysis: Principles and Applications to Materials Science*, to be published by *Mat. Res. Soc.* (1989); see also the references listed in the last reference of Chapter 1.
2 E. W. Müller, *Z. Physik*, **156**, 399 (1959).
3 C. A. Speicher, W. T. Pimbley, M. J. Attardo, J. M. Galligan & S. S. Brenner, *Phys. Lett.*, **23**, 194 (1966).
4 J. F. Mulson & E. W. Müller, *J. Chem. Phys.*, **38**, 2615 (1963).
5 W. DuBroff & E. S. Machlin, *Act. Met.*, **23**, 919 (1968).
6 A. S. Berger, D. N. Seidman & R. W. Balluffi, *Act. Met.*, **21**, 123 (1973).
7 D. Paris, P. Lesbats & J. Levy, *Scripta Met.*, **9**, 1373 (1975).
8 E. Gold & E. S. Machlin, *Phil. Mag.*, **18**, 453 (1968); E. S. Machlin, *Phil. Mag.*, **18**, 465 (1968).
9 C. G. Chen & R. W. Balluffi, *Act. Met.*, **23**, 911, 931 (1975).
10 H-O. Andren, U. Rolander & G. Wahlberg, 35th Intern. Field Emission Symp., Oak Ridge, TN, 1988. *J. de Physique Coll.*, **49**, C6–323 (1989).
11 M. Drechsler, G.Pankow & R. Vanselow, *Z. Physik. Chem. N. F.*, **4**, 249(1955); M. Drechsler, *Z. Physik. Chem.*, **6**, 272 (1956).
12 E. W. Müller, *Act. Met.*, **6**, 620 (1958).
13 S. Ranganathan, *J. Appl. Phys.*, **37**, 4346 (1966).
14 D. W. Pashley, *Rep. Prog. Phys.*, **28**, 291 (1965).
15 R. C. Sanwald, S. Ranganathan & J. J. Hren, *Appl. Phys. Lett.*, **3**, 393 (1966).
16 See for example the book by Bowkett & Smith, listed in ref. 13(c) of Chapter 1.
17 See for examples J. Friedel, *Dislocations*, Addison-Wesley, Reading, MA,

1964; J. Weertman & J. R. Weertman, *Elementary Dislocation Theory*, Macmillan, London, 1964.
18 O. Nishikawa & E. W. Müller, *J. Appl. Phys.*, **38**, 3159 (1967).
19 E. O. Hall, *Twinning*, Butterworth, London, 1954.
20 K. D. Rendulic & E. W. Müller, *J. Appl. Phys.*, **37**, 2593 (1966).
21 B. Loberg, H. Norden & D. A. Smith, *Phil. Mag.*, **24**, 89 (1971).
22 S. B. McLane & E. W. Müller, *9th Field Emission Symp.*, *Univ. Notre Dame*, 1962.
23 M. K. Kronberg & F. H. Wilson, *AIME Trans.*, **185**, 501 (1949).
24 D. G. Brandon, B. Ralph, S. Ranganathan & M. Wald, *Act. Met.*, **12**, 813 (1964).
25 S. Ranganathan, *Acta Cryst.*, **21**, 197 (1966); *J. Appl. Phys.*, **37**, 4346 (1966).
26 See the books of Bowkett & Smith and of Wagner for discussion of field ion microscope analysis of grain boundary structures (reference 13 of Chapter 1).
27 Y. Ishida, ed., *Grain Boundary Structures and Related Phenomena*, *Jpn Inst. Metals Int. Symp.*, Vol. 4 (1986).
28 P. L. Bolin, R. J. Bayuzick & B. N. Ranganathan, *Phil. Mag.*, **32**, 891 (1975).
29 G. A. Chadwick & D. A. Smith, *Grain Boundary Structures and Properties*, Academic Press, London, 1973; D. McLean, *Grain Boundaries in Metals*, Oxford University Press, London, 1957; W. C. Johnson & J. Blakely, *Interface Segregation*, *Am. Soc. Met.*, Metals Park, OH, 1979.
30 M. Forte & B. Ralph, *Act. Met.*, **15**, 707 (1967).
31 P. R. Howell, D. E. Fleet, A. Hildon & B. Ralph, *J. Microscopy*, **107**, 155 (1976).
32 T. T. Tsong, unpublished data of 1967; it is discussed and two micrographs are presented on pp. 256–7 of the book by Müller and Tsong (ref. 13(a) of Chapter 1).
33 P. J. Turner & J. M. Parazian, *Metal Sci. J.*, **1**, 82 (1973).
34 S. S. Brenner & M. K. Miller, *J. Metals*, **35**, 54 (1983).
35 B. Ralph, S. A. Hill, M. J. Southon, M. P. Thomas & A. R. Waugh, *Ultramicroscopy*, **8**, 361 (1982).
36 L. Karlsson & H. Norden, *Acta Met.*, **36**, 13, 35 (1988); L. Karlson, *Acta Met.*, **36**, 25 (1988); H. Norden & H. O. Andren, *Surface and Interface Anal.*, **12**, 179 (1988).
37 S. R. Goodman, S. S. Brenner & J. R. Low, *Met. Trans.*, **4**, 2363, 2370 (1973).
38 M. K. Miller & G. D. W. Smith, *Metal Sci.*, **11**, 249 (1977); M. K. Miller, P. A. Beaven, R. J. Lewis & G. D. W. Smith, *Surface Sci.*, **70**, 470 (1978).
39 P. R. Williams, M. K. Miller, P. A. Beaven & G. D. W. Smith, *Proc. Conf. Phase Transform.*, Inst. Metal. Conf. Ser. 3, **2**, 98 (1979); P. R. Williams, M. K. Miller & G. D. W. Smith, *Proc. Int. Conf. Solid–Solid Phase Transform. Metal Soc.*, AIME, Warrendale, PA (1981).
40 S. S. Brenner, M. K. Miller & W. A. Soffa, *Scripta Met.*, **16**, 83 (1982); L. Chang, G. D. W. Smith & G. B. Olson, *J. de Physique, Coll.*, **47**, C2–265 (1986); F. Zhu, L. v. Alvensleben & P. Haasen, *Scripta Met.*, **18**, 337 (1984); F. Zhu, P. Haasen & R. Wagner, *Acta Met.*, **34**, 457 (1986).
41 J. A. Horton & M. K. Miller, *Acta Met.*, **35**, 133 (1987).
42 C. T. Liu, C. L. White & J. A. Horton, *Acta Met.*, **33**, 213 (1985).
43 S. S. Brenner & S. D. Walck, Proc. 27th Intern. Field Emission Symp., Tokyo, 1980, 328.

44 A. R. Waugh & M. J. Southon, *Surface Sci.*, **68**, 79 (1977); **89**, 718 (1979).
45 T. T. Tsong, S. B. McLane, M. Ahmad & C. S. Wu, *J. Appl. Phys.*, **53**, 4182 (1982).
46 M. K. Miller, P. A. Beaven, C. J. Miller & G. D. W. Smith, *Proc. 26th Intern. Field Emission Symp.*, Berlin, 1979; K. E. Biehl, R. Wagner & J. Piller, in *Proc. 27th Intern. Field Emission Symp.*, Tokyo, 1980.
47 S. A. Hill & B. Ralph, *Proc. 26th Intern. Field Emission Symp.*, Berlin, 1979.
48 Workshop on statistical analysis of atom-probe data, organized by M. K. Miller & G. D. W. Smith at *35th Intern. Field Emission Symp.*, Oak Ridge, TN, 1988.
49 D. M. Schwartz & B. Ralph, *Phil. Mag.*, **19**, 1069 (1969); *Met. Trans.*, **1**, 1063 (1970).
50 R. Kampmann, R. Wagner, F. Frisius & A. Bunemann, *Rept of GKSS*, Gesthact, Germany, 1983.
51 G. D. W. Smith, 1988 Boston MRS meeting.
52 A. Youle & B. Ralph, *Met. Sci. J.*, **6**, 149 (1972).
53 H. Wendt & R. Wagner, *Acta Met.*, **30**, 1561 (1982).
54 M. Wada, Y. Yuchi, R. Uemori, M. Tanino & T. Mori, *Acta Metal.*, **36**, 333 (1968).
55 R. Wagner & S. S. Brenner, *Acta Met.*, **26**, 197 (1978).
56 G. P. Huffman & H. H. Podgurski, *Acta. Met.*, **23**, 1367 (1975).
57 S. S. Brenner, P. P. Camus, M. K. Miller & W. A. Soffa, *Acta Met.*, **32**, 1217 (1984).
58 F. Zhu, L. v. Alvensleben & P. Haasen, *Scripta Met.*, **18**, 337 (1984); F. Zhu, P. Haasen & R. Wagner, *Acta Met.*, **34**, 457 (1986).
59 M. G. Hetherington, A. Cerezo, J. P. Jakubovics and G. D. W. Smith, *J. de Physique, Coll.*, **C9**, 429 (1984).
60 M. K. Miller & G. D. W. Smith, *Met. Sci.*, **11**, 249 (1977).
61 B. J. Regan, P. J. Turner & M. J. Southon, *J. Phys. E*, **9**, 187 (1976).
62 H. O. Andren & H. Norden, *Proc. 27th Intern. Field Emission Symp.*, Tokyo, 1980, 250.
63 R. Uemori & M. Tanino, *J. de Physique, Coll.*, **C6**, 399 (1987).
64 See for example M. Wada, R. Uemori & O. Nishikawa, *Surface Sci.*, **134**, 17 (1983).
65 E. D. Boyes, A. R. Waugh, P. J. Turner, P. F. Mills & M. J. Southon, *Proc. 24th Intern. Field Emission Symp.*, Oxford University, 1977, 26.
66 T. Abe, K. Miyazaki & K. I. Hirano, *Proc. 27th Intern. Field Emission Symp.*, Tokyo, 1980, 81; also *Acta. Met.*, **30**, 357 (1982).
67 T. Mori, M. Wada, H. Kita, R. Uemori, S. Horie, A. Sato & O. Nishikawa, *Jpn J. Appl. Phys.*, **22**, L203 (983).
68 V. Gerold, *Z. Metallkd.*, **45**, 599 (1954).
69 X. Aurvray, P. Georgopoulos & J. B. Cohen, *Act. Met.*, **29**, 106 (1981).
70 K. Hono, T. Sato & K. Hirano, *Phil. Mag.*, **53**, 495 (1986).
71 M. Wada, H. Kita, T. Mori & O. Nishikawa, *J. de Physique, Coll.*, **C9**, 251 (1984).
72 A. V. Philips, *Act. Met.*, **23**, 751 (1975).
73 R. G. Faulkner & B. Ralph, *Acta Met.*, **20**, 703 (1972); S. A. Hill & B. Ralph, *Acta Met.*, **30**, 2219 (1982).
74 H. Wendt & P. Haasen, *Acta Met.*, **31**, 1649 (1983); Z. G. Liu & R. Wagner, *J. de Physique, Coll.*, **C6**, 361 (1987).
75 K. M. Delargy & G. D. W. Smith, *Metal Trans.*, **14A**, 1771 (1983).

76 D. Blavette, A. Bostel & J. M. Sarrau, *Metal Trans.*, **16A**, 1703 (1985); S. Chambreland, D. Blavette & M. Bouet, *J. de Physique, Coll.*, **C6**, 361 (1987).
77 J. W. Cahn, *Trans. AIME*, **242**, 166 (1968); *Acta Met.*, **14**, 1685 (1966).
78 R. Wagner, *Field Ion Microscopy in Materials Science*, Springer, Berlin, 1982, 86.
79 M. K. Miller, S. S. Brenner, M. G. Burke & W. A. Soffa, *Scri. Metall.*, **18**, 111 (1984).
80 E. W. Müller, *8th Field Emission Symp.*, Williamstown, Mass. (1961); *Bull. Am. Phys. Soc.*, **117**, 27 (1962).
81 H. N. Southworth & B. Ralph, *Phil. Mag.*, **14**, 383 (1966).
82 T. T. Tsong & E. W. Müller, *Appl. Phys. Lett.*, **9**, 7 (1966).
83 T. T. Tsong & E. W. Müller, *J. Appl. Phys.*, **38**, 3531 (1967).
84 T. T. Tsong, S. V. Krishnaswamy, S. B. McLane & E. W. Müller, *Appl. Phys. Lett.*, **23**, 1 (1973).
85 M. Yamamoto, M. Futamoto, S. Nenno & S. Nakamura, *Jpn. J. Appl. Phys.*, **13**, 1461 (1974).
86 For reviews of earlier FIM works on ordered alloys, image interpretations, domain structures, and mechanisms of order–disorder phase transformations, see B. G. LeFervre, *Surface Sci.*, **23**, 144 (1970) and articles by B. G. LeFervre and R. W. Newman, by T. T. Tsong, by H. N. Southworth and B. Ralph in *Applications of Field Ion Microscopy*, eds. R. F. Hochman, E. W. Müller & B. Ralph, Georgia Inst. Technology Press, 1969, also M. Yamamoto & D. N. Seidman, *Surface Sci.*, **118**, 535 (1982); T. F. Page & B. Ralph, *Proc. Roy. Soc. (London)*, **A339**, 223 (1974).
87 Q. J. Gao & T. T. Tsong, unpublished data (1987).
88 T. T. Tsong & E. W. Müller, *J. Appl. Phys.*, **38**, 545 (1967).
89 R. W. Newman & J. J. Hren, *Phil. Mag.*, **16**, 211 (1967).
90 E. Ruedl, P. Delavignette & S. Amelinckx, *Phys. Status Solidi*, **a28**, 305 (1968).
91 M. Yamamoto, S. Nenno, Y. Izumi & F. Shohno, *Proc. Intern. Symp. Appl. FIM to Metallurgy*, Lake-Yamanaka, Japan, 112, 1977; M. Yamamoto, S. Nenno, M. Futamoto & S. Nakamura, *J. Phys. Soc. Jpn*, **36**, 1330 (1974).
92 R. W. Newman & B. G. LeFervre, *Phil. Mag.*, **19**, 241 (1969).
93 R. J. Taunt & B. Ralph, *Phys. Status Solidi*, **24**, 207 (1974); *Surface Sci.*, **47**, 569 (1975).
94 H. Berg, Jr & J. B. Cohen, *Acta Met.*, **21**, 1575 (1975).
95 H. Berg, Jr, T. T. Tsong & J. B. Cohen, *Acta Met.*, **21**, 1589 (1973).
96 P. Jacobaeus, J. U. Madsen, F. Kragh & R. M. J. Cottrill, *Phil. Mag.*, **B41**, 11 (1980).
97 A. R. Bhatti, B. Cantor, D. S. Joag & G. D. W. Smith, *Phil. Mag.*, **B52**, L63 (1985).
98 O. T. Inal, L. Keller & F. G. Yost, *J. Mat. Sci.*, **15**, 1947 (1980).
99 P. A. Egelstaff, *An Introduction to the Liquid State*, Academic Press, London, 1967.
100 H. B. Elswijk, PhD thesis, University of Groningen (1988).
101 J. Piller & P. Haasen, *Acta Met.*, **30**, 1 (1982).
102 L. Nordentoft, *Phil. Mag.*, **B52**, L21 (1980).
103 E. W. Müller, *Adv. Electronics and Electron Phys.*, **13**, 89 (1960); also see pp. 278–86 of the 1969 book by Müller & Tsong.
104 M. K. Sinha & E. W. Müller, *J. Appl. Phys.*, **35**, 1256 (1964).

105 E. W. Müller, in *Vacancies and Interstitials in Metals*, eds A. Seeger, D. Schumacher, W. Schilling & J. Diehl, North-Holland, Amsterdam, 1970, 557; J. M. Galligan, *ibid.*, 575.
106 D. N. Seidman, *J. Phys. F (Metal Phys.)*, **3**, 393 (1973); *Surface Sci.*, **70**, 532 (1978).
107 L. A. Beaven, R. M. Scanlan & D. N. Seidman, *Acta Met.*, **19**, 1339 (1971).
108 C. Y. Wei & D. N. Seidman, *Appl. Phys. Lett.*, **34**, 622 (1979).
109 S. S. Brenner, R. Wagner & J. A. Spitznagel, *Met. Trans.*, **9A**, 1961 (1978).
110 A. Wagner & D. N. Seidman, *J. Nucl. Mat.*, **83**, 48 (1979).
111 See second of ref. 86.
112 L. Haggmark (1977). See second of ref. 106.
113 J. A. Panitz, *J. Vac. Sci. Technol.*, **14**, 502 (1977).
114 N. Igata, S. Sato, T. Sawai & N. Tanabe, *Proc. 27th Intern. Field Emission Symp.*, University of Tokyo, 1980.
115 J. Lindhard, M. Scharff & H. E. Schiott, *Mat. Fys. Medd. Dan. Vid. Selsk.*, **33**, 14 (1963).
116 For example see P. Kofstad, *High Temperature Oxidation of Metals*, Wiley, New York, 1966.
117 M. A. Fortes & B. Ralph, *Proc. Roy. Soc.*, **A307**, 431 (1968).
118 M. P. Thomas & B. Ralph, Proc. 28th Intern. Field Emission Symp., 1981, 204, 207, unpublished.
119 Y. S. Ng, S. B. McLane & T. T. Tsong, *J. Appl. Phys.*, **49**, 2517 (1978).
120 G. K. L. Cranstoun & D. R. Pyke, *Appl. Surface Sci.*, **2**, 359 (1979).
121 G. K. L. Cranstoun, D. R. Pyke & G. D. W. Smith, *Appl. Surface Sci.*, **2**, 375 (1979).
122 G. L. Kellogg, *Phys. Rev. Lett.*, **54**, 82 (1985).
123 T. T. Tsong, Y. S. Ng & S. B. McLane, *J. Appl. Phys.*, **51**, 6189 (1980).
124 I. Kamiya, T. Hashizume, A. Sakai, T. Sakurai & H. W. Pickering, *J. de Physique Coll.*, **C7**, 195 (1986).
125 D. Tomanek, S. Mukherjee, V. Kumar & K. H. Bennemann, *Surface Sci.*, **114**, 11 (1982).
126 D. B. Liang, G. Abend, J. H. Block & N. Kruse, *Surface Sci.*, **126**, 392 (1983); N. Kruse, G. Abend, J. H. Block, E. Gillet & M. Gellet, *J. de Physique Coll.*, **C7**, 87 (1986).
127 G. K. Chuah, N. Kruse, G. Abend & J. H. Block, *J. de Physique, Coll.*, **C7**, 435 (1986).
128 K. D. Rendulic & E. W. Müller, *J. Appl. Phys.*, **38**, 550 (1967); E. W. Müller & K. D. Rendulic, *Science*, **156**, 961 (1967).
129 A. Cerezo, T. J. Godfrey & G. D. W. Smith, *Rev. Sci. Instrum.*, **59**, 862 (1988).
130 J. A. Liddle, A. G. Norman, A. Cerezo & C. R. M. Grovenor, *Appl. Phys. Lett.*, to be published; *J. de Physique, Coll.*, **49**, C6–509 (1989).
131 R. Clampitt, K. L. Aitken & D. K. Jeffries, *J. Vac. Sci. Technol.*, **12**, 1208 (1975).
132 J. F. Mahoney, A. Y. Yahiku, H. L. Daley, R. D. Moore & J. Perel, *J. Appl. Phys.*, **40**, 5101 (1969).
133 K. Gamo, T. Ukegawa, Y. Inomoto, K. K. Ka & S. Namba, *Jpn J. Appl. Phys.*, **19**, L595 (1980); T. Ishitani, K. Uemura & H. Tamura, *Nucl. Instr. Meth. Phys. Res.*, **218**, 363 (1983).
134 V. E. Krohn, *Prog. Astronautics Rocketry*, **5**, 73 (1961); *J. Appl. Phys.*, **45**, 1144 (1974).
135 C. A. Evans, Jr & C. D. Hendricks, *Rev. Sci. Instrum.*, **43**, 1527 (1972).

136 R. Levi-Setti, *Adv. Electron. Electron Phys., Suppl.*, **13A**, 261 (1980); R. Levi-Setti, P. H. La Marche, K. Lam, T. H. Shields & Y. L. Wang, *Nucl. Instr. Meth. Phys. Res.*, **218**, 368 (1983); R. Levi-Setti, J. M. Chabala, P. Hallegot & Y. L. Wang, in *Microelectronics Engineering*, to be published by Elsevier, North-Holland, Amsterdam (1989).
137 R. L. Saliger, J. W. Ward, V. Wang & R. L. Kubane, *Appl. Phys. Lett.*, **34**, 310 (1979).
138 A. Wagner, *Nucl. Instr. Meth. Phys. Res.*, **218**, 355 (1983).
139 G. R. Ringo & V. E. Krohn, *Nucl. Instr. Meth.*, **149**, 735 (1978).
140 V. Wang, J. W. Ward & R. L. Saliger, *J. Vac. Sci. Technol.*, **19**, 1158 (1981).
141 G. I. Taylor, *Proc. Roy. Soc. (London)*, **A280**, 383 (1964).
142 N. Sujatha, P. H. Cutler, E. Kazes, J. P. Rogers & N. M. Miskovsky, *Appl. Phys.*, **A32**, 55 (1983); P. H. Cutler, M. Chung, T. E. Feutchwang & E. Kazes, *J. de Physique, Coll.*, **47**, C2-87 (1986).
143 G. B. Assayag & P. Sudraud, *J. de Physique, Coll.*, **C9**, 223 (1984).
144 L. W. Swansom, *Nucl. Instr. Meth. Phys. Res.*, **218**, 347 (1983).
145 R. J. Culbertson, G. H. Robertson & T. Sakurai, *J. Vac. Technol.*, **16**, 1868 (1979).
146 J. Mitterauer, *IEEE Trans. Plasma Sci.*, **PS-15**, 593 (1987).
147 C. D'Cruz, K. Pourreza & A. Wagner, *J. Appl. Phys.*, **58**, 2724 (1985).
148 J. A. Panitz, *Ultramicroscopy*, **7**, 241 (1982).
149 J. A. Panitz & I. Giaever, *Ultramicroscopy*, **6**, 3 (1981).
150 A. V. Crewe, *Chem. Scrip.*, **14**, 17 (1978).
151 For review of scanning tunneling microscopy, see P. K. Hansma & J. Tersoff, *J. Appl. Phys.*, **61**, R1 (1987).
152 For review of atomic resolution electron microscopy, see D. J. Smith, in *Chemistry and Physics of Solid Surfaces VI*, eds R. Vanselow and R. Howe, Springer, Berlin, 1986.
153 K. Takayanagi, Y. Tanishiro, K. Kobayashi, K. Akiyama & K. Yagi, *Jpn J. Appl. Phys.*, **26**, L957 (1987); K. Takayanagi, Y. Tanishiro & K. Morooka, *J. de Physique, Coll.*, **C6**, 525 (1987).
154 R. T. Tung, J. M. Gibson & J. M. Poate, *Phys. Rev. Lett.*, **50**, 429 (1983).
155 S. Chiang & R. J. Wilson, *Anal. Chem.*, **59**, 1267A (1887).
156 G. Binnig & H. Rohrer, *Rev. Mod. Phys.*, **59**, 615 (1987); *IBM J. Res. Dev.*, **30**, 355 (1985).
157 R. D. Young, *Physics Today*, **24**, 42 (1971); R. D. Young, J. Ward & F. Scire, *Phys. Rev. Lett.*, **27**, 922 (1971); *Rev. Sci. Instrum.*, **43**, 999 (1972).
158 G. Binnig, K. H. Frank, K. H. Fuchs, H. Garcia, B. Reihl, H. Rohrer, F. Salvan & A. R. Williams, *Phys. Rev. Lett.*, **55**, 991 (1985); R. S. Becker, J. A. Golovchenko & B. S. Swartzentruber, *Phys. Rev. Lett.*, **55**, 987 (1985).
159 See Chapter 2 for works on FIM.
160 G. L. Kellogg, *Phys. Rev. Lett.*, **55**, 2168 (1985); Q. J. Gao & T. T. Tsong, *Phys. Rev. Lett.*, **57**, 452 (986).
161 Q. J. Gao & T. T. Tsong, *Phys. Rev.*, **B36**, 2547 (1987).
162 J. Tersoff & D. R. Hamann, *Phys. Rev. Lett.*, **50**, 25 (1983).
163 H. Ohtani, R. J. Wilson, S. Chiang & C. M. Mate, *Phys. Rev. Lett.*, **60**, 2398 (1988); P. H. Lippel, R. J. Wilson, M. D. Miller, Ch. Woll & S. Chiang, *Phys. Rev. Lett.*, **62**, 171 (1989).
164 H. W. Deckman & J. H. Dunsmuir, *Appl. Phys. Lett.*, **41**, 377 (1982).

Index